G000166724

WATER RESOURCE PLANNING, DEVELOPMENT AND MANAGEMENT

WATER QUALITY

INDICATORS, HUMAN IMPACT AND ENVIRONMENTAL HEALTH

WATER RESOURCE PLANNING, DEVELOPMENT AND MANAGEMENT

Additional books in this series can be found on Nova's website under the Series tab.

Additional E-books in this series can be found on Nova's website under the E-book tab.

ENVIRONMENTAL HEALTH - PHYSICAL, CHEMICAL AND BIOLOGICAL FACTORS

Additional books in this series can be found on Nova's website under the Series tab.

Additional E-books in this series can be found on Nova's website under the E-book tab.

WATER RESOURCE PLANNING, DEVELOPMENT AND MANAGEMENT

WATER QUALITY

INDICATORS, HUMAN IMPACT AND ENVIRONMENTAL HEALTH

YOU-GAN WANG
EDITOR

New York

Library of Congress Cataloging-in-Publication Data

Water quality : indicators, human impact, and environmental health / [edited by] You-Gan Wang, Centre for Applications in Natural Resource Mathematics (CARM), School of Mathematics and Physics, the University of Queensland, Queensland, Australia).
pages cm
Includes bibliographical references and index.
ISBN:978-1-62417-111-6 (hardcover)
1. Water quality. 2. Water quality--Environmental aspects. I. Wang, You-Gan, editor of compilation.
TD370.W3953 2013
333.91'16--dc23

2012038765

Published by Nova Science Publishers, Inc. † New York

CONTENTS

PREFACE

Water quality is fundamental for our health and affects the environment we share with other animals including marine, freshwater and terrestrial species. Water quality is often managed based on indicators for levels of bacteria and other chemical/physical contents.

To assist in better management and monitoring of water quality, we are proud to present his book collecting 11 chapters providing the state of the art in assessment of water quality, understanding how water quality is affected, and improving water quality for irrigation, drinking and recreation activities.

Assessment of water quality is challenging due to spatial and temporal variability and its multivariate nature arising from a number of parameters representing physical and biological measures. Novel statistical approaches to establish representative annual indices are proposed and illustrated by Raincan *et al.* (Ch. 1). These issues are also dealt with by Sparks *et al.* (Ch. 2) who focused on designing new monitoring programs.

The chemical process affecting water quality is investigated using a sediment-associated process in lakes by Chao and Jia (Ch. 3). This is followed by a number of studies on *Groundwater quality*. This is studied via ion chemistry by Kumar and Avtar (Ch. 4) in the context of drinking consumption in Japan. Hamzaoui et al. (Ch. 5) also investigate underground water quality based on a number of physical and chemical factors mainly for irrigation purposes which will indirectly impact our health. Gibrilla et al. (Ch. 6) consider a stable isotopes approach and show us how to establish quality indices for drinking and agricultural purposes.

Water monitoring and management issues are well presented by Nare and Odiyo (Ch. 7). Their chapter presented a very interesting study in South Africa, where the community takes the major responsibility in monitoring and management.

Recycled or reclaimed water is studied by Qiu, Saint and Barton (Ch. 8) who provide a comprehensive review on removal of triclosan. Water quality can also be assessed by evaluating the physiological condition of fish, which provides a practically effective approach for water monitoring (Moiseenko, Ch. 9). Water quality indices were also found correlated with heavy metal accumulation in human bodies (Moiseenko, Ch. 10).

Finally, Wei and Wolfe (Ch. 11) present state-of-the-art technology in applying chitosan for improving water quality by absorbing heavy metals.

I sincerely believe that these 11 book chapters will be helpful for our researchers, managers and other decision makers in understanding the current research topics and important issues in making effective use of and managing our water resources.

You-Gan Wang, D.Phil. (Oxon)
Professor of Applied Statistics
Director, Centre for Applications in Natural Resource Mathematics (CARM)
School of Mathematics and Physics
The University of Queensland
Queensland 4072, Australia

LIST OF CONTRIBUTORS

You-Gan Wang, Water Quality Indices from Unbalanced Spatio-temporal Monitoring Designs. E-mail: you-gan.wang@uq.edu.au.

Ross Sparks, Estimates of Likelihood and Risk Associated with Sydney Drinking Water Supply from Reservoirs, Local Dams and Feed Rivers. E-mail: Ross.Sparks@csiro.au.

Xiaobo Chao, Three-dimensional Numerical Modeling of Water Quality and Sediment-Associated Processes in Natural Lakes. E-mail: chao@ncche.olemiss.edu.

Pankaj Kumar, Integrating Major Ion Chemistry with Statistical Analysis for Geochemical Assessment of Groundwater Quality in Coastal Aquifer of Saijo plain, Ehime prefecture, Japan. E-mail: pankajenvsci@gmail.com.

F. Hamzaoui, Suitability of Groundwater of Zeuss-Koutine Aquifer (Southern of Tunisia) for Domestic and Agricultural Use. E-mail: fadoua_fst@yahoo.fr.

Abass Gibrilla, Application of Water Quality Indices (wqi) and Stable Isotopes ($\delta^{18}O$ and $\delta^{2}H$) for Groundwater Quality Assessment of the Densu River Basin of Ghana. E-mail: gibrilla2abass@yahoo.co.uk.

L. Nare, Evaluation of Community Water Quality Monitoring and Management Practices, and Conceptualization of a Community Empowerment Model: A Case Study of Luvuvhu Catchment, South Africa. E-mail: leratonare@yahoo.com.

Chris Saint, The Fate and Persistence of the Antimicrobial Compound Triclosan and its Influence on Water Quality. E-mail: Christopher.Saint@ unisa.edu.au.

T. I. Moiseenko, Water Quality Assessment Methods: the comparative analysis. E-mail: moiseenko.ti@gmail.com.

T. I. Moiseenko, Water Quality Impacts on Human Population Health in Mining-and-metallurgical Industry Regions, Russia. E-mail: moiseenko.ti@gmail.com.

Xinchao Wei, Chitosan Biopolymer for Water Quality Improvement: Application and Mechanisms. E-mail: weix@sunyit.edu.

In: Water Quality
Editor: You-Gan Wang

ISBN: 978-1-62417-111-6

Chapter 1

WATER QUALITY INDICES FROM UNBALANCED SPATIO-TEMPORAL MONITORING DESIGNS

Sarah M. Raican[1], You-Gan Wang[1,2,*] and Bronwyn Harch[1,3]

[1]CSIRO Mathematics, Informatics and Statistics, Australia
[2]Centre for Applications in Natural Resource Mathematics (CARM), School of Mathematics and Physics, The University of Queensland, Australia
[3]CSIRO Sustainable Agriculture Flagship, Australia

ABSTRACT

This chapter investigates a variety of water quality assessment tools for reservoirs with balanced/unbalanced monitoring designs and focuses on providing informative water quality assessments to ensure decision-makers are able to make risk-informed management decisions about reservoir health.

In particular, two water quality assessment methods are described: non-compliance (probability of the number of times the indicator exceeds the recommended guideline) and amplitude (degree of departure from the guideline). Strengths and weaknesses of current and alternative water quality methods will be discussed. The proposed methodology is particularly applicable to unbalanced designs with/without missing values and reflects the general conditions and is not swayed too heavily by the occasional extreme value (very high or very low quality).

To investigate the issues in greater detail, we use as a case study, a reservoir within South-East Queensland (SEQ), Australia. The purpose here is to obtain an annual score that reflected the overall water quality, temporally, spatially and across water quality indicators for each reservoir.

* Corresponding author.

1. INTRODUCTION

Water quality monitoring is becoming increasingly important in order to meet public health and safety regulations, for the protection and sustainability of our natural resources, as well as to meet local and state government guidelines and regulations. Appropriate assessment tools are required to summarize the annual health of the reservoir, accommodating seasonal effects, health/environmental guidelines and the often unbalanced nature of the monitoring design.

Water quality assessment tools that summarize measurements taken as part of a monitoring program are a simple and concise way of providing information about the health of the resource. There are a number of general issues that need to be considered when undertaking water quality assessments including:

The question/reason for the water quality assessment;
Desired timescale for reporting water quality;
Seasonal differences;
Sampling frequency/timescale within the period of interest;
Balanced/unbalanced designs;
The choice of water quality indicators (variables which reflect the health of the reservoir);
Selection of sites;
Data quality; and
Impact of missing values.

However, evaluating overall water quality status from a large number of variables and samples is often difficult [Kannel et. al, 2007]. The type of water quality assessment technique used depends on the objective of the monitoring program, the specific questions being asked and the implemented monitoring/sampling design. Often, the design is required to fit within financial budgetary constraints or relies on using existing data and monitoring designs.

Extra consideration therefore needs to be given to the impact on water quality assessment when dealing with unbalanced designs (i.e. increased sampling during certain periods/months and/or an inappropriate sampling frequency that does not adequately capture the temporal variation) and/or missing values.

The type of water quality assessment technique used depends on, the objective of the monitoring program, the specific questions being asked and the implemented monitoring/sampling design. Often, the design is required to fit within financial budgetary constraints or relies on using existing data and monitoring designs. Extra consideration therefore needs to be given to the impact on water quality assessment when dealing with unbalanced designs (i.e. increased sampling during certain periods/months and/or an inappropriate sampling frequency that does not adequately capture the temporal variation) and/or missing values.

1.1. Current Approaches to Water Quality Assessment

A water quality index (WQI) is one assessment technique that enables a complex system of water quality measurements to be summarized by a single number. This simple tool aims to inform managers and decision makers about the overall water quality status, improves general understanding of the water quality issues and is therefore likely to result in investment to effectively manage our water resources [Cude, 2001]. These indices can be used to assess changes and trends in overall water quality [Cude, 2001]. Sarkar and Abbasi [2006] identify four main steps in the development of a water quality index:

1. Selection of an optimum set of parameters (indicators)
2. Transformation of parameters onto a common scale
3. Assignment of parameter weights
4. Aggregation of scores to form a final, single score.

A water quality index has been developed within an Ecosystem Health Monitoring Program (EHMP) in South-East Queensland (SEQ). The EHMP [2007] was created as part of the South-East Queensland Regional Water Quality Management Strategy [SEQRWQMS; Smith and Storey, 2001] to monitor and report on the ecosystem health of freshwater systems. A report card score (grade) is given to a number of reporting areas and is based on a corresponding water quality index for each area. Indicators of ecosystem health were chosen that responded to known disturbance gradients and a set of ecosystem health guidelines developed using a set of minimally disturbed reference sites [Smith and Storey, 2001]. The report card is calculated each year using samples taken during spring and the following autumn and standardises for both spatial variability and measurement scale. The standardisation of parameters occurs by calculating the ratio of the distance between the measurement and the guideline to the distance between the worst-case scenario (WCS) and the guideline [EHMP, 2007].

The WCS is calculated to be a theoretical limit or the $10^{th}/90^{th}$ percentile of data for all sites and assessment periods for a given stream type. Calculation of the guideline and WCS values for each stream group with similar physical conditions accommodates spatial variability. A final water quality index is calculated by taking a series of arithmetic means across various groupings.

Calculating the final score in this way means that each indicator, index and season are given equal weighting in the final score regardless of how many indicators are within each index. This approach is appropriate if few missing values occur, the sampling design is balanced and there is likely to be minimal differences in the WCS levels between seasons or data is collected from only one season. If this is not the case, then alternative scoring procedures may be needed.

Instead of considering just one aspect of water quality, the Canadian Council of Ministers of the Environment (CCME) develop a WQI taking account of: scope, frequency and amplitude [CCME, 2001], with each factor ranging between 0 and 100. The use of more than one aspect allows a more representative view of the water quality. Scope measures the percentage of water quality variables for which the guideline is not met in at least 1 sample during the period of interest.

Frequency measures the percentage of measurements in which the guideline was exceeded and amplitude measures the departure from the guideline for measurements that did not meet the guideline. The amplitude is computed by firstly calculating the number of times the measurements do not meet the guideline (excursion). The normalized sum of excursions is then calculated by dividing the excursion by the total number of observations. Finally an amplitude value is computed by using an asymptotic function that scales the normalized sum of excursions to range between 0 and 100.

The final water quality index is then the Euclidean distance of the three scores scaled by a constant so the final score also ranges between 0 and 100. The scope, frequency and amplitude all rely on information about the objective or guideline value. All indicators/indicator values are combined in the calculation of the scope, frequency and amplitude so there is no need to use averaging, weighted sum or the like to combine the scores across indicators or water quality categories.

Many indices are however developed for each indicator (or sub-index) and therefore require aggregation to form a single final score. Sarkar and Abbasi [2006] note that many different variants of aggregating scores for indicators (or sub-indices) are used including weighted sums, weighted geometric means, weighted products, an unweighted harmonic mean square formula and modified additive aggregation functions. Aggregation across multiple indicators and/or index categories and the combining of multiple water quality aspects (i.e. scope, frequency and amplitude etc.) requires the indices/scores to be on a common scale.

Scores are often normalised or scaled using a linear/non-linear function. Sargaonkar and Deshpande [2003] develop non-linear relationships for each indicator to transform the indicator values to a common scale ranging between 0 and 16. Each curve is constructed using predetermined indicator limits for 5 water quality categories (excellent, acceptable, slightly polluted, polluted and heavily polluted). More commonly the scores are scaled to be on a range of 0 to 1 or 0 to 100. EHMP [2007] transform their scores on a 0 to 1 scale while the CCME [2001] transform the scores onto a 0 to 100 scale. Swamee and Tyagi [2000] look at different aggregation procedures and find that as result of the type of aggregation some indices may not reflect an overall poor water quality when a subindex shows poor water quality. Other aggregation indices may show the aggregated index to be unacceptable even though the water quality is in fact acceptable, while others fail to give a complete picture of the water quality. The choice of aggregation should depend on the reason for the summary measure, which is related to the objectives of the monitoring program and the specific questions being answered. Water quality indices can however be calculated without the need for computation and aggregation of sub-indices. Swamee and Tyagi [2000] list a number of approaches taken by various researchers including: factor analysis, principal component analysis, fuzzy clustering analysis, multivariate ranking procedure and uniformity indexing method. Said et al. [2004] also develop a water quality index not requiring aggregation techniques, sub-indices or standardization procedures. Instead the index is a simple numerical equation that is a function of the indicators with appropriate weights determined by level of significance and effect of indicator on the water quality. Such approaches avoid the need for and difficulties associated with aggregating across sub-indices but cannot reflect short-term changes and localised changes may not be immediately reflected in the score [Said et al., 2004].

1.2. Limitations and Strengths

Much thought has been given to index calculation, aggregation across sub-indices and transformation of indices to a common scale, however there seems to be little research on the development of indices for unbalanced designs or designs with missing values. There seems to be even less research on standardising scores through time. This is particularly important when dealing with multiple samples taken across the year in an annual water quality assessment.

Many water quality variables have seasonal cycles with peaks occurring at particular times of the year. Although an observed value may not be extreme during peak times, if observed during a period when conditions are normally quite low and stable, it could indicate potential problems with the water quality. When dealing with multiple samples in a given period we therefore believe that seasonality should be taken into account and the water quality index standardised through time.

Although standardised in both space and scale, it appears that the EHMP freshwater score has not been standardised through time. If significant changes in water quality could be expected between the spring and autumn samples then perhaps the score should also be standardised across seasons.

According to Swamee and Tyagi [2000] this perhaps reduces the possibility that a poor sub-index will be reflected in a poor overall water quality index. However aggregating in this way enables the user to easily trace the reason for a poor/excellent water quality score through the various stages of aggregation. The strength of the CCME WQI is its ability to measure not one but three different aspects of water quality and little aggregation is required as the sub-indices are calculated across all water quality indicators therefore removing the need for aggregation.

However, no averaging makes it difficult to trace which water quality indicator/(s) has (have) the greatest impact on the final index, which may impact on a decision makers ability to specifically target management actions. The WQI calculations also allow regular water quality monitoring data to be used where multiple samples may be recorded during the period of interest (i.e. once per month for an annual index). However there appears to be little consideration of the impact when the design is unbalanced and missing values occur or when a particular season or period may be measured more frequently than another within the period of interest. Seasonality is another aspect which does not appear to be well addressed within this or many other articles. The need for seasonality depends on the reason for the water quality assessment and the question being answered.

1.3. New Approaches to Water Quality Assessment

This Chapter investigates a variety of water quality assessment tools for reservoirs with balanced/unbalanced monitoring designs and focuses on providing informative water quality assessments to ensure decision-makers are able to make risk-informed management decisions about reservoir health. In particular, two water quality assessment methods are described: non-compliance (probability of the number of times the indicator exceeds the recommended guideline) and amplitude (degree of departure from the guideline). Strengths and weaknesses of current and alternative water quality methods will be discussed and a case study will be provided from a local reservoir within South-East Queensland, Australia. The Queensland Bulk Water Supply Authority (trading as Seqwater) manages and supplies water within SEQ. In total there are 14 sites that Seqwater sample as part of their campaign/drought monitoring program on this particular reservoir. A large number of water quality variables are monitored with some monitored since 1997.

Water quality samples are usually taken from the reservoir once or twice per month; however this does vary and can occur up to four times per month (or more on some occasions). The reasons for the increased samples are varied but may include increased sampling efforts during an event or increased sampling effort in periods of known high variability. Samples may also be missing at some sites for some months therefore also contributing to the unbalanced design problems.

The purpose for developing a new approach to water quality assessment was to calculate an annual score that reflected the overall water quality, temporally, spatially and across water quality indicators for each reservoir. The proposed methodology is applicable to unbalanced designs with/without missing values and reflects the general conditions and is not swayed too heavily by the occasional extreme value (very high or very low quality).

2. Water Quality Assessment Methods

We investigate a number of water quality assessment methods for the purpose of quantifying the health of the reservoir and investigate temporal and spatial components of the report card scoring procedures. Our aim is to provide an annual water quality assessment tool for decision makers to enable them to make risk-informed management decisions. We believe a two-step process is necessary to achieve this aim:

1. Calculate the frequency of samples exceeding the guideline (compliance);
2. Calculate the distance of exceedance from the guideline (amplitude).

Compliance is a common water quality assessment tool and is usually assessed using known public health or environmental guidelines. Non-compliance is a measure of how often water samples exceed these guidelines. Compliance/non-compliance is an important management tool, however to make targeted management decisions we believe that a measure of non-compliance is not sufficient on its own. We believe it is equally important to understand the magnitude of the non-compliance, that is, the degree of departure from compliance (amplitude). Different management actions may be required depending on the non-compliance and amplitude scores. It is possible for two different years to have the same non-compliance but have very different amplitudes (see Figure 1). The combination of both non-compliance and amplitude measures therefore provides greater insight into the water quality and gives deeper insight for the decision-makers. A non-compliance measure indicates whether the guidelines are exceeded frequently, regularly or occasionally. The amplitude informs on the extremity of the values that exceed the guideline.

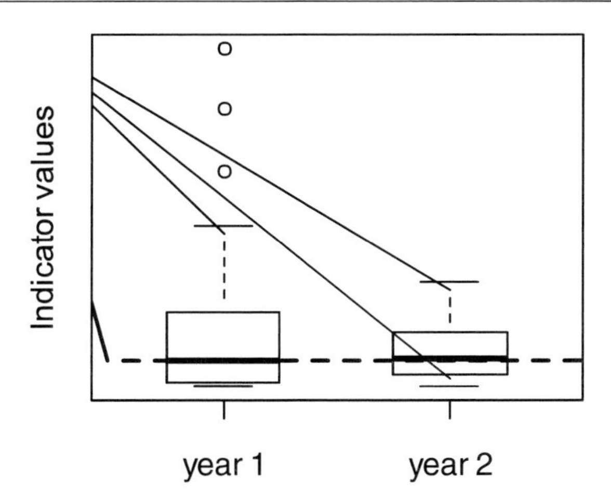

Figure 1. Boxplots showing an example distribution of values for two separate years. Both distributions have approximately the same non-compliance (a) 50% of samples > guideline but have very different amplitudes (b) quantile95(year 1) > quantile95(year 2). The median of the distribution is indicated by the solid black line and the guideline by the dashed line.

2.1. Non-Compliance Scores

We investigate four different methods for calculating non-compliance and detail the strengths/weaknesses of each method and their most appropriate use.

The first and most simple non-compliance method (NC1) can be calculated as:

$$NC1 = \frac{N_{above}}{N_{total}}, \quad (1)$$

where N_{total} is the number of samples taken for a particular water quality indicator during the period of interest (annual) and N_{above} is the number of these samples that exceed the recommended guideline. If there is more than one site being monitored, non-compliance is calculated separately for each site. Figure 2 shows the number of samples above and below the Queensland Water Quality guideline [Queensland, 2006] for Filtered Reactive Phosphorus.

This scoring procedure can be used with a small or large number of samples and could still be calculated even when only one sample is collected throughout the year or period of interest. Repeat samples or periods of increased monitoring (typically likely to occur during periods of increased variability) can bias the score towards the scores during the months/periods of increased monitoring. For example, if a particular indicator is sampled more during the summer period which is known to have increased variability and more extreme levels/concentrations than the bias may tend towards higher non-compliance scores. A score derived directly on this type of data is therefore likely to reflect the conditions during those months/periods that are highly monitored and not necessarily reflect the overall condition of the storage throughout the entire year. Bias could also result if there are missing samples throughout the year and the range of temporal conditions is not adequately captured.

Missing data from periods prone to be more non-compliant may be biased towards smaller non-compliance scores. This method is therefore probably more appropriate to continuous, very regular sampling that is evenly spaced throughout the year to capture the range of temporal conditions.

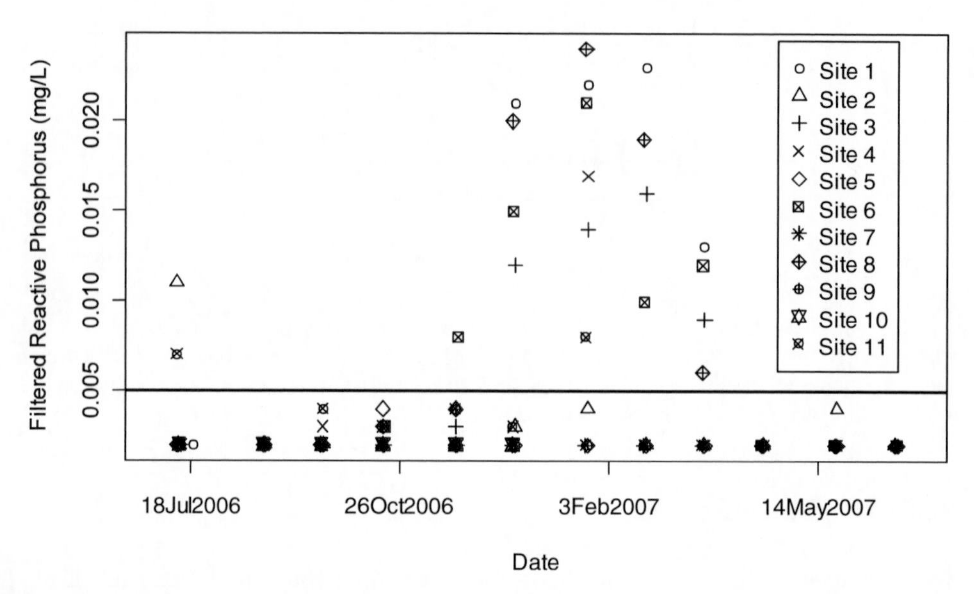

Figure 2. Data taken from 11 sites on a local reservoir for the 2006/2007 year for Filtered Reactive Phosphorus taken at the surface using a 3m integrated tube. A non-compliance score (NC1) is calculated separately for each site.

In order to remove/reduce bias resulting from over-sampling, under-sampling or missing values an average of the proportion of values above the guideline each month (or period) could be calculated (NC2):

$$NC2 = \overline{p_m}\,, \tag{2}$$

where p_m is the proportion of samples that exceed the guideline during month, m. Every month within the year has equal weighting and is therefore not biased by periods of increased monitoring (see Figure 3). When each month is sampled an equal number of times then NC2 is simply equal to NC1. When the monitoring/sampling design is unbalanced then the NC2 score appears to be a more appropriate method for calculating non-compliance. As each month is equally weighted in the final non-compliance score, the overall water quality throughout the year is much better represented. This method is suited to both unbalanced and balanced sampling designs and does not require samples to be taken at equal time intervals throughout the year. The weakness is that the year (or period of interest) must be divided into categories that represent the range of temporal conditions seen throughout the year and averaged across categories. Too many missing months may also still create bias in the non-compliance scores.

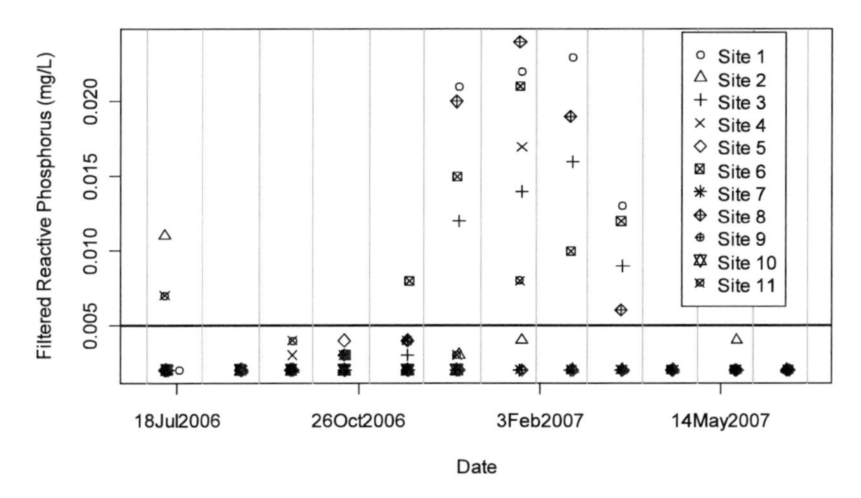

Figure 3. Calculation of NC2 occurs by calculating the proportion of values exceeding the guideline for each month (bounded by the gray lines) at each site. The months are then averaged within each site to derive a non-compliance score for each site.

The last two non-compliance scores (NC3 and NC4) aim to overcome difficulties caused by both unbalanced sampling designs and missing values. The first of these two approaches uses linear interpolation between sampling points. This method assumes that the values between samples can be approximated by a linear function of time. Instead of using the raw samples we then use the interpolated values to calculate the non-compliance score to be:

$$NC3 = \frac{L_{above}}{L_{total}} \tag{3}$$

where L_{total} is the total length of time (i.e. 365 days for an annual score) and L_{above} is the length of time the interpolated function is above the guideline (see Figure 4). This is simply a modified version of Equation 1 that uses interpolated values rather than the collected samples. This method represents the range of temporal variation without the need for categorisation into smaller sub-periods and does not require the sampling to be equally spaced throughout time. Multiple or repeat samples (samples taken at the same time) are simply averaged to create a single value for each sampling time. This method can be used with a large or small sample size throughout a given year however the more samples collected the greater the accuracy in interpolation. This method can be used for both balanced and unbalanced sampling designs. Bias may still result when a large proportion of the year is not adequately sampled, however smaller proportions of the year can be linearly interpolated using the next available adjacent samples. Bias may also result when samples are missing from key times within the period of interest (i.e. times of significant decrease/increase in concentrations/levels). For this reason we still recommend that sampling is done as regularly and consistently as possible.

In the case where we may have upper and lower guideline limits (e.g. % saturation of Dissolved Oxygen: 90 – 110% recommended range) we make the additional assumption that the direction from which the guideline is exceeded is not important (being too high or too low is of equal importance). Before calculating *NC*3 (Equation 3) we calculate

$V_i^* = |V_i - (u+l)/2|$ where u and l are the upper and lower guidelines respectively and redefine the guideline according to the transformed data $G = u - (u+l)/2$. Calculation of L_{above} and L_{total} is then based on the transformed values (V_i^*) and redefined guideline. A more elegant approach for calculating non-compliance is to use a smooth function of time rather than a linear interpolation between samples. We investigate fitted smooths using the family of Generalized Additive Models [GAMs; Wood, 2006] where the response distribution belongs to the family of exponential dispersion models (EDM). If samples are only taken at a single site than a smooth could be fit to just one site (provided there is sufficient data). Alternatively the following model could be fit:

$$g(u) = \beta_0 + \beta_j + s(t) \tag{4}$$

where $s(t)$ is a spline-based smooth to represent the trend throughout the year, β_j is a categorical variable representing the shift in mean at site j and $g(.)$ is a link function. It seems reasonable to initially assume that sites within the same lake may have similar temporal trends with a shift in site mean. In fitting the smooth (Equation 4), strength can be borrowed from other sites when sites are missing samples. This helps with the infilling and interpolation of values required for calculating the non-compliance score. Bias is reduced as missing values are replaced by interpolated or infilled values and equal weights are then applied throughout the year.

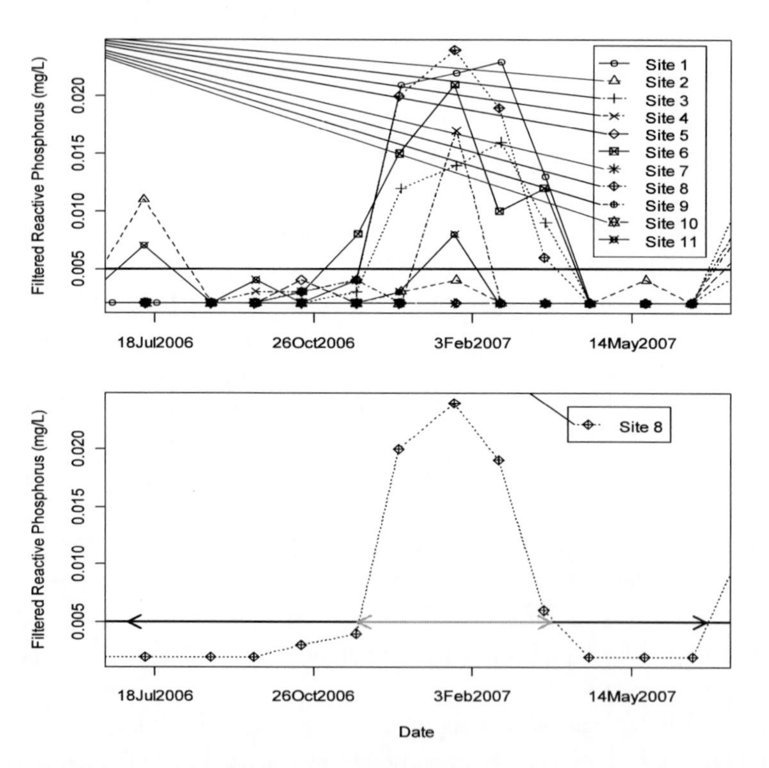

Figure 4. Linear interpolations for (a) all sites and (b) site 8. Non-compliance can be calculated as the proportion of time the interpolations are above the guideline (length of gray arrow) in reference to the total length of time (length of black and gray arrows).

Figure 5 however shows, there are some limitations with this. Some curves are overestimated and some are underestimated resulting in non-compliance scores for some sites greater than non-compliance based on the raw samples and some non-compliance scores less than expected based on raw samples. An alternative would be to fit a separate smooth for each site; however this may require a substantial amount of additional samples to enable the model to be appropriately fitted. This underestimation at sites and overestimation at others may be may be averaged out when the scores are combined across all sites. It is also possible to adjust the smoothness parameter, however this would require user interaction and for all 16 variables could be very time consuming. As we also intend to automate the generation of report card scores in the future we want as little user interaction as possible.

The non-compliance score (*NC4*) can be calculated as:

$$NC4 = \frac{S_{above}}{S_{total}} \quad (5)$$

where S_{above} is the length of time the smooth exceeds the recommended guidelines and S_{total} is the length of the period of interest (i.e. 365 days for annual scores).

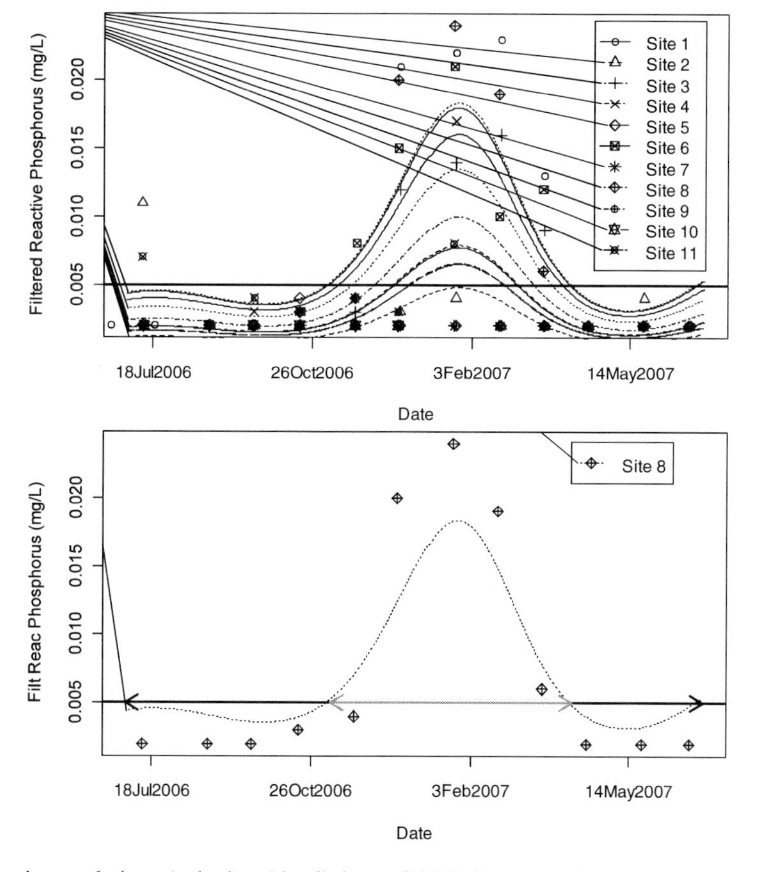

Figure 5. Spline interpolations (calculated by fitting a GAM) for (a) all sites and (b) site 8. Non-compliance can be calculated as the proportion of time the spline curves are above the guideline (length of gray arrow) in reference to the total length of time (length of black and gray arrows).

Table 1. Table summarises strengths, weaknesses and when the conditions under which the non-compliance methods are recommended for use

Non-Compliance Method	Strengths	Weaknesses	Recommended for use when …
NC1	Simple and easy to calculate. Can be calculated with small or large sample sizes.	Easily biased by repeat samples or periods of increased monitoring. May not reflect the overall conditions throughout the period of interest. Missing values or under-sampling may mean temporal conditions throughout period are not adequately captured.	The sampling is balanced, continuous, very regular and evenly spaced throughout the period of interest with few missing values.
NC2	Reduces bias due to over-sampling, under-sampling and a few missing values. Every month throughout the year has equal weighting. Accounts for unbalanced sampling designs.	The year (or period of interest) must be divided into discrete categories that represent the range of temporal conditions. Too many missing values can be problematic.	The sampling design is balanced or unbalanced, and unevenly spaced throughout the year with few missing values.
NC3	Represents range of temporal variation. No need to categorise into sub-periods. Multiple or repeat samples do not bias results. Biased is reduced when missing values are present. Can be used with small/large sample sizes. Appropriate for balanced/unbalanced designs.	Assumes values between samples are a linear function of time. Biased when a large proportion of the year is not sampled. Bias may result when samples are missing from key times within period of interest.	The sampling design is unbalanced, irregular and unevenly spaced throughout the year with missing values. Sampling at key times (highs/lows) throughout period will improve model
NC4	Represents range of temporal variations. Smooth functions can be used rather than linear interpolations between samples. In fitting the smooth, strength can be borrowed from other sites when sites are missing samples. No need to categorise into sub-periods. Multiple or repeat samples do not bias results. Bias is reduced when missing values are present. Appropriate for balanced/unbalanced designs.	Sites are assumed to have similar temporal trends or there has to be a large number of samples at each site to fit separate curves for each site. Complex and computationally intensive. Smooths will probably underestimate non-compliant extremes. Smoothness of curves determined by choice of a smoothness parameter.	The sampling design is unbalanced, irregular and unevenly spaced, there are missing values and a large number of samples at each site. Sampling at key times (highs/lows) throughout the period of interest will improve the model.

Unfortunately, this model requires a reasonable sample size to fit an adequate model and this method is more complex and computationally intensive. Difficulties may also be encountered when a few sites experience unusually high concentrations or levels in comparison to the other sites and there aren't sufficient samples to fit a separate smooth for each site. The fitted/interpolated values may be smoothed out by the values from other sites so these extreme values may not be adequately captured. This is simply the nature of the model and this does not invalidate the methodology, however making the procedure robust is desirable.

Where upper and lower guideline limits are given, the data is transformed (V_i^*) and S_{above} and S_{total} calculated using the transformed values (as explained for $NC3$) and adjusted guideline value.

2.2. Amplitude Scores

The second component of this water quality assessment involves calculating amplitude scores (magnitude of non-compliance). We investigate six alterative amplitude scoring methods. As the non-compliance scores by definition range from 0 to 1 where 0 indicates all values were within the recommended guidelines and 1 indicates 100% non-compliance then the amplitude scores will be developed on a similar scale of 0 to 1.

The first and simplest method for calculating amplitude ($D1$) is to calculate the observed distance of each value from the guideline:

$$d_i = V_i - G$$

or expressed as the number of guidelines from the recommended guideline:

$$g_i = \frac{V_i - G}{G}.$$

Both of these scores need further scaling so the amplitude ranges between 0 and 1 where 1 indicates most extreme distance from the guideline. To scale this we need information about the limits or maximum values for each water quality indicator being assessed (akin to worst case scenario (WCS) and guideline values used by the EHMP freshwater scoring). These limits, d_{WCS} and g_{WCS} can be calculated as the $10^{th}/90^{th}$ percentile of d_i and g_i respectively. The 10th percentile value is calculated if lower levels/concentrations indicate poor water quality. Alternatively, the 90th percentile is used if higher concentrations are indicative of poor water quality. For the occasional times when the $10/90^{th}$ percentile is exceeded the score is scaled back to the maximum score of 1.

The scaled amplitude scores therefore are:

$$D1 = median\left(\frac{d_i}{d_{WCS}}\right) = median\left(\frac{g_i}{g_{WCS}}\right) \tag{6}$$

The limits could be based on a single limit that is representative of the limit throughout the year or could be a seasonally changing limit. Given the strong presence of seasonality seen within many of the water quality indicators, an amplitude measure that calibrates extremes for each season would seem to be more appropriate. An extreme value during the summer period may be quite different to what would be classed as extreme during the winter period. Extremes are therefore relative to the natural seasonal cycles/peaks that occur throughout the year. One disadvantage of deriving a limit for each season/month however is that there must be sufficient samples during each season/month in order to calculate the limits. An alternative approach is to model the limit as a continuous function through time using a quantile regression technique on all available data from all sites. The occasional season/month with small or no samples will be interpolated with greater weight on neighbouring seasons. This method will be discussed in more detail later.

Like the non-compliance method *NC1* this amplitude measure suffers bias caused by missing values, unbalanced designs and increased monitoring during particular periods of time. If greater emphasis should be placed on those months with increased monitoring then this bias is not problematic. However, our aim is to assess the overall condition of the lake throughout the year and so alternative methods to reduce this bias would be more appropriate.

The second amplitude measure avoids the need to determine limits for each indicator. Instead of a limit we simply determine the proportion of values above the guideline (non-compliance) and the proportion of values above a second level which is highly undesirable to exceed without severe ramifications. The second amplitude measure is calculated to be:

$$D2 = \sqrt{p1 \times \frac{(p1 + p2)}{2}} \tag{7}$$

where $p1$ is the proportion of samples that are above the guideline and $p2$ is the proportion above a second level which indicators should not exceed (see Figure 6). A score of 0 would indicate all samples were below the recommended guideline ($p1$) while a score of 1 would indicate all values exceed the second level ($p2$). This categorical approach already contains a combination of non-compliance and amplitude in the one score. A second level could be chosen to be a higher warning level or guideline or could be subjectively chosen for each indicator using expert knowledge. For examples within this Chapter we have chosen the second level to be twice the recommended guideline. As with most categorical methods, there is some information loss, in this case we do not use the actual distances from the guidelines just their location in respect to the guideline and a second guideline. A significant change in scores may also result from a change in the value of the second guideline.

Another alternative to amplitude measures requiring calculation of indicator limits is the probability associated with the location of the median distance in a reference distribution. This reference distribution could be an empirical distribution based on historical data from a number of years and sites or could be a known theoretical distribution. The third amplitude score ($D3$) can be calculated as:

$$D3 = \Pr(dist_{median} \leq d_i), \tag{8}$$

where $d_i = V_i - G$ for $V_i > G$ (see Figure 7).

This method is appropriate for ascertaining the typical conditions for the water body. The reference distribution may need to be altered if we want to represent what is happening throughout the entire period of interest. Alternatively, if we want to assess how far the extreme values are from the guideline perhaps the 90th percentile distance should be used instead of the median distance.

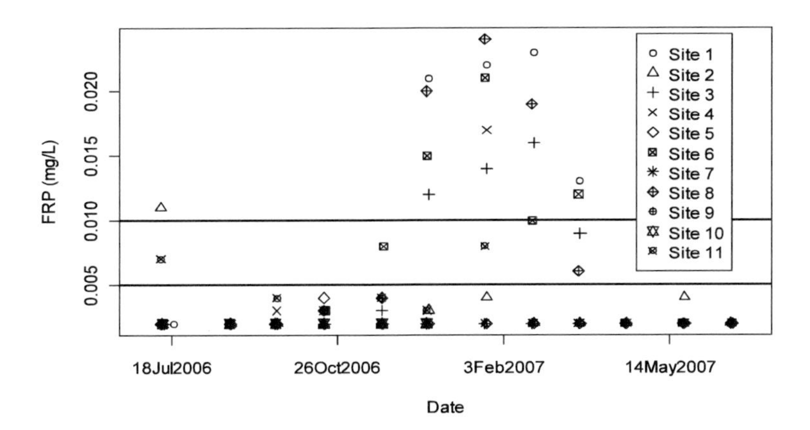

Figure 6. Calculation of an amplitude score using method 2. Two solid lines represent the first and second guidelines, G1 and G2 respectively.

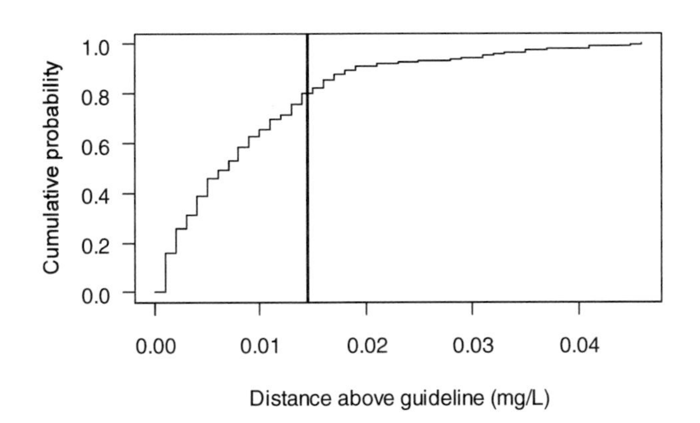

Figure 7. The empirical cumulative distribution (reference distribution) of distance values above the guideline based on all years across the 3 dams for FRP (bottom). The bold line indicates the median distance above the guideline (D3) for site 8 and shows the corresponding cumulative probability of D3 when compared to the reference distribution.

An alternative method uses the amplitude score from *D*4 to derive a corresponding probability from a beta distribution with mean 0.5. The beta distribution is more appropriate then *D*3 as we are able to incorporate the uncertainty associated with the original proportion (*D*3) in the calculation of the score. The added benefit is that the transformation also results in a proportion between 0 and 1. No scaling is necessary to create the index score and the beta distribution is very flexible in shape. The fixed points in the beta distribution are 0, 0.5 and 1 such that a proportion of *D*3=0 will map to *D*4=0, *D*3=0.5 will map to *D*4=0.5 and *D*3=1 will map to *D*4=1. The cumulative distribution function for the beta distribution is:

$$F(x;a,b)=\frac{B_x(a,b)}{B(a,b)}=I_x(a,b) \quad (9)$$

where $B(a,b)$ is the beta function, $B_x(a,b)$ is the incomplete beta function and $I_x(a,b)$ is the regularized incomplete beta function. The fourth amplitude score can be calculated as:

$$D4 = F(D3, a, a) \tag{10}$$

where a can be chosen as $(n+1)/2$ or other informative values to reflect the variability in n, where n is the number of values used to calculate $dist_{median}$ (see Figure 8).

Although both *D3* and *D4* do not require upper limits (or worst case scenarios), both are prone to bias as a result of increased monitoring during certain periods (unbalanced design) and missing values. Alternative methods that give equal weight throughout the entire period of interest and methods that are more robust against missing values and unbalanced designs would be more appropriate.

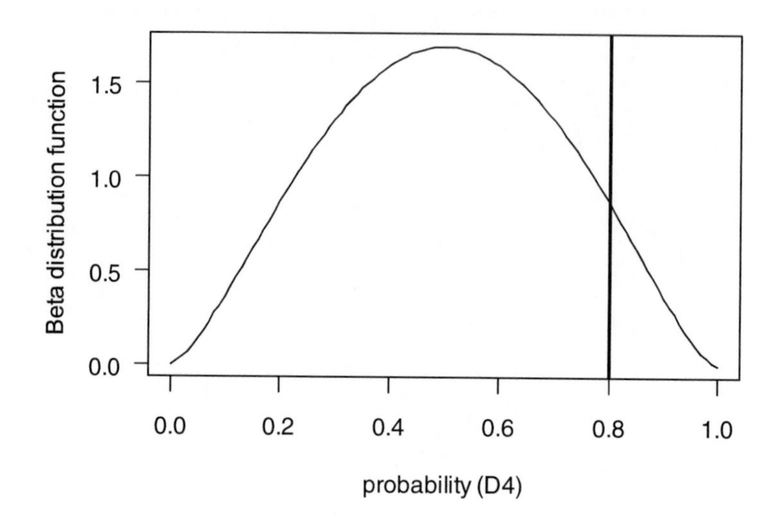

Figure 8. Probability from Beta distribution (mean = 0.5) used as an amplitude score D4. Bold line indicates D3 score at site 8 and used to calculate the corresponding p-value from a beta distribution.

Two methods for calculating non-compliance (*NC*3 and *NC*4) are developed to handle unbalanced designs and it is these approaches that are adapted to also provide our final two amplitude scores (*D*5 and *D*6). The unbalanced design could be a result of missing values or additional sampling. We again use the interpolated/infilled data instead of the collected samples. However limits will still be required for each indicator to scale the amplitude measurements to range between 0 and 1 inclusive.

For similar reasons as discussed previously, we would like the limits to reflect the seasonal changes throughout the year. We could simply use the $10^{th}/90^{th}$ percentile of values that exceed the guideline each month. As the amplitude scores use continuous functions it would be more appropriate to develop smooth functions for the limits. One such option involves using quantile regression techniques [Koenker and D'Orey, 1987; Koenker et. al., 1994].

Quantile regression techniques (parametric and non-parametric) have wide use in environmental modelling. Modelling of droughts and floods (rainfall extremes) requires knowledge of extreme quantiles to assist in the design of reservoirs, flood drains and run-offs

[Yu et al, 2003]. Morgan II et. al. [2006] uses the 50th quantile regression for modelling the benthic index of biotic integrity and fish index of biotic integrity as functions of other water quality parameters. As many water quality indicators do not have a Gaussian distribution and transformations may be inadequate a non-parametric technique would be more suitable. Little work has been done on non-parametric quantile regression techniques and software for such techniques is very limited. Pin and Maechler [2008] have developed a non-parametric quantile regression function for R [see http://www.r-project.org/]. This package computes constrained B-Splines (COBS) non-parametric regression quantiles using linear or quadratic splines [Bartels and Conn, 1980; Ng, 1996; Koenker and Ng, 1996; He and Ng, 1999; Koenker and Ng, 2005]. As this method is non-parametric no distributional assumptions on the errors are needed. The non-parametric quantile regression technique also allows us to perform analyses without having to categorise the data into seasons/months. By using this quantile regression technique we avoid the need for categorisation which often results in information loss while at the same time still accounting for seasonality. Amplitude scores $D5$ and $D6$ will therefore use the 95th non-parametric quantile regression based on historical data for indicator limits.

An equivalent amplitude score based on the linear interpolation technique developed for $NC3$ can be computed as:

$$D5 = \frac{\sum_{k} \sum_{f_k(t)>G} \{f_k(t) - G\}}{\int \max\{f_{95}(t) - G, 0\}\, dt}, \qquad (11)$$

where $f_k(t)$ is the linear function connecting the kth and $(k+1)$th sample points (V_i) sorted in time, $f_{95}(t)$ is the 95th non-parametric quantile regression and G is the guideline value (see Figure 9). Approximations to the sums/integrals are calculated by summing the distance of the fitted values from the guideline at a smaller timescale (e.g. days if creating an annual report).

If upper and lower guideline limits exist then the transformed data (V_i^*) is used in calculating $f_k(t)$ and $f_{95}(t)$ rather than the observed values (V_i). The guideline should also be adjusted to this transformed scale (see $NC3$ and $NC4$).

Although this method may be more computationally intensive then the previous methods for amplitude, there will be reduced bias when the design is unbalanced and there are missing values for some samples. The method can therefore be used for both balanced and unbalanced designs so long as there are enough samples to detect the general trend throughout the period of interest. Alternatively a smoothing approach (similar to that used for calculating $NC4$) could be used when non compliance score $NC4$ is used. Like the linear interpolation methodology, we simply calculate the ratio of areas underneath the limit and smooth based on samples for the year/period of interest. The sixth amplitude score can be calculated as:

$$D6 = \frac{\int \max\left(f(t) - G, 0\right) dt}{\int \max\left(f_{95}(t) - G, 0\right) dt}, \qquad (12)$$

where *f(t)* is the spline curve fitted to the observed samples as a function of time, *t*, in a GAM framework and $f_{95}(t)$ is the non-parametric 95th percentile regression and *G* is the recommended guideline. Approximations can be used to calculate the area underneath each of the curves (see amplitude method *D*5). The smoothness is controlled and set as a default parameter for all variables. This may seem too simple, however as the procedure is possibly to be automated we don't want to implement procedures that require intensive user input. Transformation of the data (see *D*5) will again be necessary when we have a range for the recommended guidelines. The functions *f(t)* and $f_{95}(t)$ are calculated on the transformed data (V_i^*) rather than observed values (V_i) as was done for amplitude method *D*5.

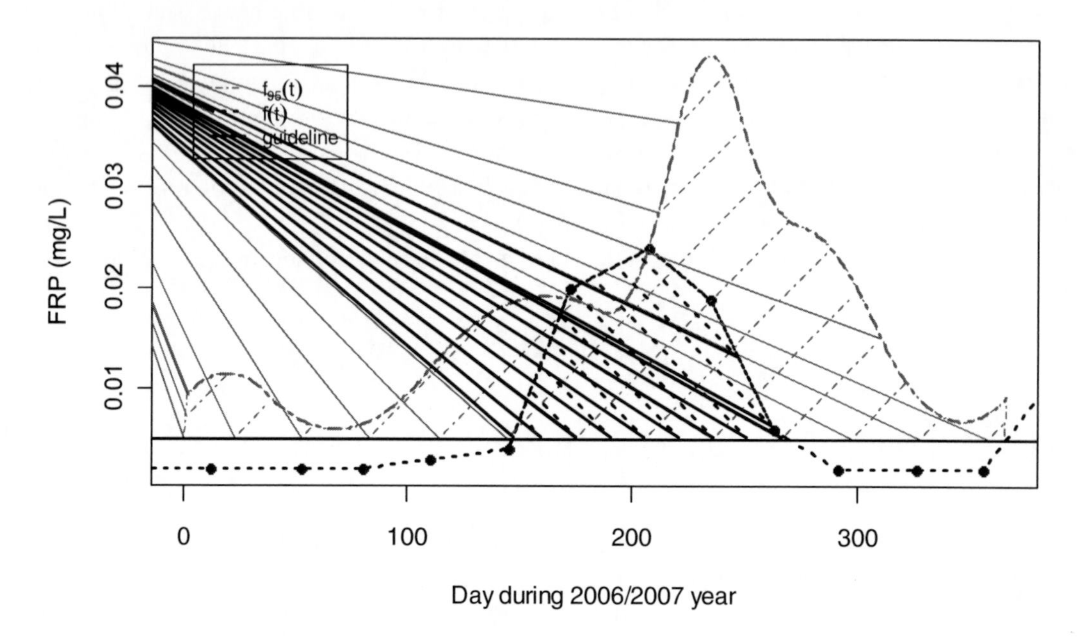

Figure 9. Amplitude measure calculation (D5) for FRP at site 8 on a local reservoir. The distance value is the ratio of the dark shaded are to the light shaded area.

This method uses the same GAM fit used to calculate non-compliance (*NC*4) and so suffers from the same under and over-estimation of the trend at sites. When averaging across sites it is possible that this over/under-estimation may be averaged out across sites resulting in a reasonable distance score representing the 'average' conditions on the lake. Providing a sufficient number of samples have been collected at each site, model fits could be improved by allowing separate curves to be fitted to each site.

Methods *D*5 and *D*6 are the preferred options for calculating amplitude however, due to the limitations in the GAM approach to fitting a smooth curve to each site, we recommend *D*5 as the best approach when dealing with unbalanced monitoring designs and missing values, so long as a sufficient number of samples have been collected to adequately capture the general temporal changes over the period of interest.

Table 2. Table summarises strengths, weaknesses and when the conditions under which the amplitude methods are recommended for use

Amplitude Method	Strengths	Weaknesses	Recommended for use when
D1	Simple Can be calculated with small or large sample designs.	Bias caused by unbalanced designs, missing values and increased monitoring during particular sub-periods. Need to determine limits for each indicator.	The sampling design is balanced, contains few missing values is regularly and evenly sampled throughout the period of interest.
D2	Avoids need to calculate indicator limits. Simple to calculate.	Categorical approach (above/below levels) results in some information loss. 2nd level may need to be chosen based on expert knowledge unless multiple guideline levels exist. A significant change in scores may result from a change in the second level. Missing values, repeat samples or sub-periods with more intensive monitoring may bias the results.	The sampling design is balanced, there are very few missing values and the sampling is regular and evenly spaced throughout the period of interest and multiple guidelines exist.
D3	Can be modified to use other statistics such as the 90th percentile instead of median distance. Relatively simple to calculate.	Requires sufficient historical data to generate empirical distribution if theoretical distribution of indicator values is unknown. Missing values, repeat samples or sub-periods with more intensive monitoring may bias the results. Appropriate for a balanced sampling design.	Assessing the 'typical' conditions for the reservoir or water body. It requires a regular and relatively even-spaced sampling throughout time to capture temporal variation and very few missing values.

Table 2. (Continued)

Amplitude Method	Strengths	Weaknesses	Recommended for use when
D4	Able to incorporate uncertainty associated with original proportion D3 into the score. Beta distribution is flexible in shape. Transformation of D3 occurs on a 0 to 1 scale so no scaling is necessary. Values of D3= 0, 0.5 and 1 are transformed to the same values for D4.	Requires sufficient historical data to generate empirical distribution if theoretical distribution of indicator values is unknown. Missing values, repeat samples or sub-periods with more intensive monitoring may bias the results. Sampling design must be regular and relatively even-spaced to capture temporal variation.	Assessing the 'typical' conditions for the reservoir or water body. It requires a regular and relatively even-spaced sampling throughout time to capture the temporal variation and very few missing values to reduce potential bias.
D5	Reduced bias as a result of missing values, unbalanced designs and sub-periods of intense monitoring. Limits change temporally to reflect seasonal changes in the definition of an 'extreme' event. Can be used with small/large sample sizes. Appropriate for balanced/unbalanced designs.	Sufficient historical data required for modelling indicator limits. Greater complexity and slightly longer computation time. Assumes values between samples are a linear function of time.	The sampling design is unbalanced, irregular and unevenly spaced throughout the year with missing values. Sufficient historical data is needed to generate indicator limits. Sampling at key times (highs/lows) throughout the period of interest will improve the model.
		Biased when a large proportion of the year is not sampled. Bias may result when samples are missing from key times within period of interest.	
D6	Reduced bias as a result of missing values, unbalanced designs and sub-periods of intense monitoring. Limits change temporally to reflect seasonal changes in the definition of an 'extreme' event. Smooth functions can be used rather than linear interpolations between samples. In fitting the smooth, strength can be borrowed from other sites when sites are missing samples. Multiple or repeat samples do not bias results. Appropriate for balanced/unbalanced designs.	Sufficient historical data required for modelling indicator limits. Most complex method and longer computation times. Sites are assumed to have similar temporal trends or there has to be a large number of samples at each site. Smooths not always able to model the occasional extreme event.	The sampling design is unbalanced, irregular and unevenly spaced, there are missing values and a large number of samples at each site. Sufficient historical data is needed to generate indicator limits. Sampling at key times (highs/lows) throughout the period of interest will improve the model.

2.3. Forming Overall Water Quality Indices

A water quality index can be developed using a combination of non-compliance and amplitude scores across the site/(s) for a number of indicators of interest. The combining of scores depends on the structure of the water quality index and how many sites and indicators are used. We assume that data is collected for multiple indicators at a number of sites and the indicators are grouped into at least two water quality categories.

Two initial methods for combining the non-compliance and amplitude scores for site *i*, indicator *j* and water quality category *k* are:

$$S_{j,k} = \sum_{i=1}^{n^{(j,k)}} w_i \times \frac{(NC_{i,j,k} + D_{i,j,k})}{2} \quad (13)$$

$$S_{j,k} = \sum_{i=1}^{n^{(j,k)}} w_i \times \sqrt{NC_{i,j,k} \times D_{i,j,k}} \quad (14)$$

where w_i is a weighting for site *i* where the sum of the weights is equal to 1. If each site is equally important than w_i can be replaced by $1/n^{(j,k)}$. If either *D*5 or *D*6 is used as the amplitude score then due to these amplitude scores being conditional on exceeding the guideline (non-compliance) and the implied multiplicative nature a geometric mean of the non-compliance and amplitude scores would be more appropriate.

A squared transformation of Equation 13 results in:

$$S_{j,k} = \sum_{i=1}^{n^{(j,k)}} w_i \times NC_{i,j,k} \times D_{i,j,k} \quad . \quad (15)$$

The combining of scores this way results in a score that is less than or equivalent to the minimum of the two scores. It appears then that an extreme score in *NC* or *D* may not be well reflected by the water quality index of Equation 14.

When *D* contains a non-compliance and amplitude component within the score (such as *D*2 or even maybe *D*5 and *D*6) then the score could be simply calculated as:

$$S_{j,k} = \sum_{i=1}^{n^{(j,k)}} w_i \times D_{i,j,k} \quad . \quad (16)$$

Some slight modifications to Equation 13 give alternative equations for calculating a water quality index:

$$S_{j,k} = \sqrt{\sum_{i=1}^{n} \frac{NC_{i,j,k}}{n_j} \times \sum_{i=1}^{n} \frac{D_{i,j,k}}{n_j}} \quad (17)$$

$$S_{j,k} = \sqrt{\sum_{i=1}^{n} \frac{NC_{i,j,k} \times D_{i,j,k}}{n_j}} \tag{18}$$

where n_j is the number of sites within indicator j. The choice between Equation 13, 16 or 17 should relate to simplicity, understanding and usefulness of the score. Once a single score has been calculated it may be useful to be able for managers/researchers interested in understanding the scores to be able to trace and justify why the result was obtained. It makes sense therefore that this occurs from the highest summary down in the order of: final score, indicator, site and then *NC* or *D*. For this reason Equations 13 or 17 seem to be more appropriate. We choose to use Equation 17 for the case study in next section.

To combine the scores across indicators and water quality categories the final score is calculated as:

$$RCscore = \sum_{k=1}^{K} w_k \sum_{j=1}^{n_k} w_{kj} S_{j,k} \tag{19}$$

where k is the number of water quality categories, $\sum_{k=1}^{K} w_k = 1$ and $\sum_{j=1}^{n_k} w_{j,k} = 1$.

If each indicator and each water quality category are of equal importance then:

$$w_{j,k} = \frac{1}{n_k} \text{ and } w_k = \frac{1}{K}$$

where n_k is the number of indicators in water quality category k.

3. An Application

Many of the reservoirs within South-East Queensland (SE Qld), Australia are used for drinking water, recreation, industry and agriculture. Understanding the conditions on the lake is vital to the management of the reservoirs to ensure healthy water ecosystems for future generations.

The methods for calculating water quality indices were developed for an annual assessment of the reservoirs within SE Qld. They allow the reservoirs to be closely monitored throughout the year and provide necessary information for the ongoing management of the reservoirs.

The monitoring data collected by the Queensland Bulk Water Supply Authority (trading as Seqwater) is often unbalanced in design. Monitoring occurs for many water quality indicators on a monthly/fortnightly basis, however extra samples may be taken for event monitoring, for calibration/validation or checking of results and for specific projects/studies on the reservoir. As major events (algal blooms, rainfall and substantial runoff events) are more likely to occur during summer this may mean increases in the number of samples over this time thereby creating an unbalanced design. The number of sites that monitor each indicator may vary over time and may also differ between indicators. The sites used for calculating the water quality index should be representative of the dam and should remain fairly consistent throughout time in order for comparisons across years to be valid.

Increased variability in many water quality indicators occurs during the Australian spring (September - November) and summer (December - February) with peaks in concentrations and levels also often occurring during summer. For this reason we define our reporting year to be July to June so the summer period is not split between reporting years.

Preliminary analyses identified 16 water quality indicators within three water quality categories that will be used in developing a water quality index assessment for the health of the reservoirs. The first category (Water Quality) includes: surface and lake floor concentrations of dissolved oxygen (DO), surface pH, surface turbidity, surface filtered reactive phosphorus concentrations (surface FRP), surface total phosphorus (TP), surface dissolved nitrogen (DiN) and surface total nitrogen (TN). The second category (Toxicants and Pathogens; TandP) contain the surface algal ratio (total cyanophytes to total algal counts), surface toxic species ratio (proportion of *anabaena circinalis*, *microcystis aeruginosa* and *cylindrospermopsis raciborskii* to total cyanophyte counts), total toxic species at the surface and e-coli surface concentrations. The Biological category includes surface chlorophyll-a, and dissolved manganese (DiMang), ammonia nitrogen (NH_3-N) and FRP lake floor concentrations. In total there are 16 indicators split between three water quality categories. These 16 indicators are not all monitoring at the same number of sites (see Table 3). However, we believe it is more important to keep the number of sites consistent between consecutive reporting years to enable appropriate comparisons across years. The removal or addition of sites in some years may make it difficult to determine if changes are due to the addition/removal of sites or changes in the overall water quality on the dam.

Table 3. The number of sites monitoring each of the 16 water quality indicators used for the water quality index calculations

Indicator	Number of sites
DO (surface)	14
DO (floor)	12
pH (surface)	14
Turbidity (surface)	14
FRP (surface)	13
TP (surface)	13
DiN (surface)	13
TN (surface)	13
Algal ratio (surface)	4
Toxic ratio (surface)	14
Total toxic species (surface)	14
E-coli (surface)	14
Chlorophyll-a (surface)	4
DiMang (floor)	11
NH_3-N (floor)	11
FRP (floor)	11

3.1. Non-Compliance

The non-compliance scores have been calculated for FRP (at the lake floor) during the 2006/07 reporting year on one of the local reservoirs in SE Qld to compare the computational differences between methods (see Table 4).

Table 4. Calculation of non-compliance scores for Filtered Reactive Phosphorus samples taken at the site floor for numerous sites across a local SE Qld reservoir during 2006/2007. The scores are based on calculations using non-compliance methods 1 to 4 (NC1, NC2, NC3 and NC4)

Site	Number of samples above the guideline	Number of samples taken	Non-compliance score using Method 1 (NC1)	Average monthly proportion Method 2 (NC2)	Non-compliance using Method 3 (NC3)	Non-compliance score using Method 4 (NC4)
1	4	13	0.3077	0.3333	0.3781	0.4274
2	1	12	0.0833	0.0833	0.1068	0.2164
3	4	12	0.3333	0.3333	0.3507	0.3425
4	1	12	0.0833	0.0833	0.1370	0.2712
5	0	12	0.0000	0.0000	0.0000	0.1589
6	5	12	0.4167	0.4167	0.4301	0.3863
7	0	10	0.0000	0.0000	0.0000	0.0000
8	4	12	0.3333	0.3333	0.3370	0.4356
9	0	12	0.0000	0.0000	0.0000	0.1589
10	0	6	0.0000	0.0000	0.0000	0.1644
11	2	12	0.1667	0.1667	0.1699	0.2110

The first three non-compliance methods (*NC*1, *NC*2 and *NC3*) have very similar scores. This is because the sampling design for this particular water quality indicator is not greatly unbalanced. However, the fourth non-compliance scores are higher than the other non-compliance scores for each site. Values which observe a 0% non-compliance (score = 0) obtain scores greater than 0 for *NC*4. This occurs because we fit a single smooth function across sites with a mean-shift for each site. The averaging of the function across sites means some sites may have fits that indicate worse/better quality then what may have been observed. To fit a separate curve for each site would require a greater number of samples to be taken. It is for this reason that *NC*3 was chosen as the preferred method for calculating non-compliance over *NC*4. Scores *NC*1 and *NC*2 are not suited to unbalanced sampling designs, so *NC*3 also has the preference over these two scores. Non-compliance is therefore calculated using method *NC*3.

3.2. Amplitude

Greater differences exist between the different amplitude scores than the non-compliance scores (see Table 5). For example site 8 calculates a very large score using method *D*1 but very small scores using methods *D*2, *D*5 and *D*6. This particular site has a single sample that exceeded the recommended guideline. This particular value was close to the limit for that month and so received a large amplitude score. Scores *D*2, *D*5 and *D*6 do not give as much weight to this single value and so the amplitude score is very small. Amplitude scores *D*5 and *D*6 are very similar in terms of the ranking of amplitude scores from each site. This makes sense as the methodology for these two scoring procedures are similar. Amplitude score *D*4 is very similarly ranked to both *D*5 and *D*6 even though the amplitude scores themselves are very different. The other three scoring procedures give very different ranks and differing amplitude scores. Since the linear interpolation method is chosen for the non-compliance component (*NC*3) we also choose the linear interpolation method (*D*5) for the amplitude component.

Table 5. Calculation of amplitude scores using Methods 1 to 7 for Filtered Reactive Phosphorus at the site floor of 11 sites on a local SE Qld reservoir during 2006/2007

Site	No. of values >G	Distance Score 1 (D1)	Distance Score 2 (D2)	Distance Score 3 (D3)	Distance Score 4 (D4)	Distance Score 5 (D5)	Distance Score 6 (D6)
1	4	0.6906	0.3077	0.8557	0.9635	0.3867	0.2296
2	1	0.8571	0.0833	0.4925	0.4925	0.0307	0.0316
3	4	0.3820	0.3118	0.5821	0.6375	0.1970	0.1339
4	1	0.6316	0.0833	0.7164	0.7164	0.0678	0.0666
5	0	0	0	0	0	0	0.0124
6	5	0.2258	0.3727	0.5323	0.5605	0.2661	0.1871
7	0	0	0	0	0	0	0
8	4	0.6892	0.3118	0.8010	0.9237	0.3276	0.2377
9	0	0	0	0	0	0	0.0124
10	0	0	0	0	0	0	0.0131
11	2	0.2218	0.1179	0.2587	0.2052	0.0188	0.0290

3.3. Final Water Quality Index

As the amplitude score is conditional on non-compliance (conditional probability) it makes sense that the two scores are combined using a multiplicative operation. For this reason we calculate the average of the two scores using the geometric mean or transformation of a geometric mean rather than the arithmetic mean. We use Equation 17 as the means of calculating a score for each indicator. However we also need to combine indicator scores if we want to obtain a single score for the reservoir. We follow Equation 18 and assume that each water quality category is of equal importance and that the water quality indicators within each category are of equal importance:

$$RCscore = \frac{1}{K}\sum_{k=1}^{K}\sum_{j=1}^{n_k}\frac{1}{n_k}S_{j,k} \quad . \tag{20}$$

This means that greater emphasis is not placed on this particular category even though there are twice as many indicators in this category than in each of the other two categories.

The average non-compliance and amplitude scores, category scores and final water quality index for the reservoir in SE Qld are shown in Table 6. The water quality index assigned for the year is 0.3825 where a score of 0 implies that all indicators were fully compliant and a score of 1 implies all indicators were fully non-compliant. The Toxicant/Pathogen category is the category of greater water quality. Dissolved oxygen (bottom), surface pH, surface total phosphorus and surface total nitrogen are the indicators in the water quality category that indicate very poor water quality. For the Biological category chlorophyll-a surface concentrations indicate poor water quality followed closely by ammonia nitrogen levels near the lake floor. For most of these indicators this appears to be due to a high non-compliance component. Many sites must have observations that exceed the recommended guideline levels.

Table 6. Report Card Scores for a local SE Qld reservoir during 2006/2007

	Indicator	No. of sites	$\overline{NC}$	$\overline{D}$	$S_{j,k}$	Water Quality Index Score	Report Card Score
Water Quality	DO Surface	14	0.5105	0.1286	0.2760	0.4604	0.3825
	DO Bottom	12	0.8832	0.5318	0.6943		
	pH Surface	14	0.9303	0.5547	0.7273		
	Turbidity Surface	14	0.0261	0.1940	0.1488		
	FRP Surface	13	0.0264	0.0059	0.0325		
	Total Phosphorus Surface	13	1.0000	0.5051	0.7107		
	DiN Surface	13	0.4249	0.1455	0.3397		
	Total Nitrogen Surface	13	1.0000	0.5678	0.7535		
Toxicants / Pathogens	Algal ratio Surface	4	0.7753	0.3919	0.5565	0.2203	
	Toxic Species ratio Surface	14	0.7826	0.0749	0.2429		
	Total Toxic Species Surface	14	0.0000	0.0000	0.0000		
	E. Coli Surface	14	0.0325	0.0343	0.0816		
Biological	Chlorophyll a Surface	4	0.9438	0.4451	0.6503	0.4668	
	Dissolved Manganese Bottom	11	0.4699	0.3331	0.4502		
	Ammonia Nitrogen Bottom	11	0.8048	0.3482	0.5632		
	FRP Bottom	11	0.1736	0.1177	0.2036		

CONCLUSION

This Chapter details the development of water quality indices using the combination of two components: non-compliance and amplitude and within the context of monitoring designs being balanced or unbalanced.

Non-compliance measures how often values exceed a recommended guideline during a given period and amplitude measures departure from the guideline. A number of calculation methods were developed for each of these components, with recommendations in here based on the spatio-temporal monitoring context of the monitoring undertaken. The purpose of the water quality index and the sampling design of the monitored indicators will determine which method is most appropriate.

Many water quality indices assume regular and balanced sampling designs. In the case scenario for a local reservoir within SE Qld the design is often unbalanced or missing data and many of the documented water quality indices are inappropriate and would give biased results. We have proposed methods that could be used when the sampling design is unbalanced and missing values may occur. Some of these methods reduce bias that is present when certain sub-periods are over-represented or missing values.

The impact of this bias has been investigated for the 2006/2007 year (Table 7). Three scoring methods are compared. The first scoring method uses $NC1$ and $D1$, the second uses $NC3$ and $D5$ while the third uses $NC4$ and $D6$. The differences between the scoring methods 1 and 3 are compared to the adopted scoring method (method 2: $NC3$ and $D5$). Results for the non-compliance scores for each site and indicator combination differ by up to 0.25 for scoring method 1 and 2 but up to 0.44 between scoring methods 3 and 2. Much larger differences exist between the amplitude methods with differences in score of at most 0.7331 for each site and indicator combination. By the time the scores are multiplied and averaged across sites to derive an indicator score the difference in scores is at most 0.1271. The WQI scores differ by only 0.0256 and the final score differs by 0.03 at most. This reduction in differences between methods may be due in part to the averaging across sites, indicators and water quality groupings. This averaging is necessary if we are to adequately represent the overall condition of the lake. To avoid the overall water quality being influenced by a single site or indicator with poor water quality we take an average across a number of sites and indicators. This provides a more complete picture of the overall water quality. A poor score then indicates poor water quality at a number of sites and indicators and the reverse is also true.

The water quality indices developed within this Chapter are designed for data containing spatial and temporal information (i.e. when there are multiple samples taken during a given period of interest at a number of different sites). The presence of seasonality is an important consideration in the derivation of indicator limits. Many water quality variables show strong peaks or increased levels during the Australian summer. An extreme in summer may be much higher than a corresponding extreme in winter. The limits should therefore reflect changes with respect to season and so we model the seasonal limits using a non-parametric quantile regression.

Table 7. Comparison of scoring differences using Score1: NC1 and D1, Score 2: NC3 and D5 and Score 3: NC4 and D6

Scale		NC for each site and indicator	D for each site and indicator	NC * D for each site and indicator	Indicator score	WQI score	Final Score
Score 1 – Score 2	Minimum	-0.2541	-0.7331	-0.2432	-0.0673	0.0255	
	Q1	-0.0679	0.0000	0.0000	-0.0047		
	Median	0.0000	0.0313	0.0039	0.0416	0.0256	0.0325
	Q3	0.0000	0.1280	0.0583	0.0708		
	Maximum	0.1021	0.8264	0.3947	0.1271	0.0464	
Score 3 – Score 2	Minimum	-0.1984	-0.7065	-0.1006	-0.0509	-0.0182	
	Q1	0.0000	-0.0165	-0.0061	-0.0120		
	Median	0.0000	0.0000	0.0000	0.0000	-0.0015	-0.0028
	Q3	0.0904	0.0009	0.0131	0.0130		
	Maximum	0.4432	0.1232	0.1726	0.0298	0.0114	

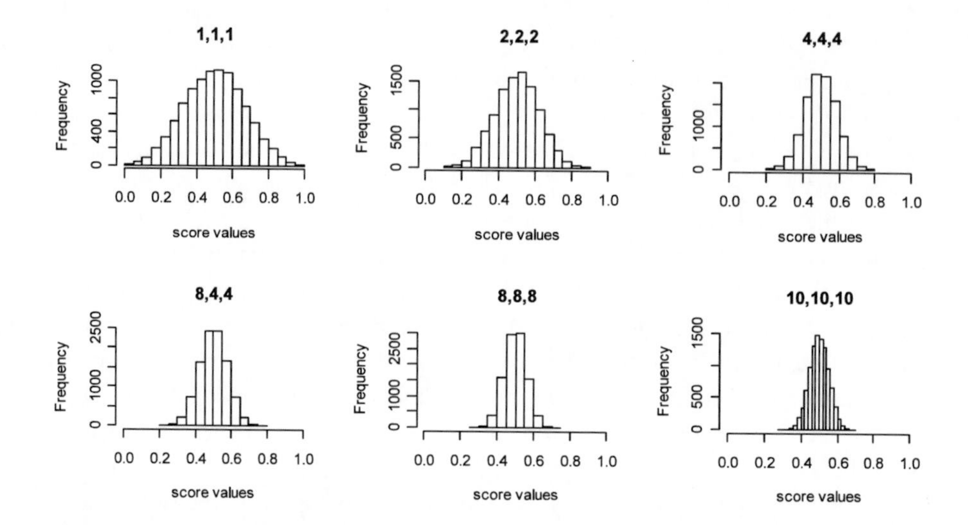

Figure 10. Distributions of final report card scores based on 3 categories with (a) 1 indicator in each category (b) 2 indicators in each category, (c) 4 indicators in each category, (d) 8 indicators in one category and 4 indicators in each of the other two categories (e) 8 indicators in each category and (f) 10 indicators in each category. Each indicator score was randomly sampled from a uniform distribution with range 0 to 1. The scores were combined using Equation 19 and the process simulated 10000 times to create the distributions above.

The non-compliance component of the water quality index is of at least equal if not greater importance than the amplitude component. The recommended guidelines used for the non-compliance scores are often safe levels beyond which serious public health or environmental problems may arise. If guidelines continue to remain non-compliant for large proportions of the year then management and preventative action may need to be implemented so reoccurrence doesn't occur in following years. Amplitudes $D5$ and $D6$ already incorporate aspects of non-compliance into the score and so do not need to be combined with a non-compliance score. However, we emphasize the importance of the non-compliance score by combining the amplitude score ($D5$) with the corresponding non-compliance score ($NC3$). The combination of these two scores provides a water quality index that captures the non-compliance and amplitude while incorporating spatial, temporal and seasonal information in the scores. The methods *NC*3 and *D*5 reduce bias that may result from using an unbalanced sampling design or from missing values within the data. There is also room for further research into the likelihood of scores for each water quality index when they contain a varied number of indicators. As the number of indicators in each index increases the scores are more likely to be closer to the mean 0.5, with much less chance of obtaining scores close to 0 or 1 (Figure 10). The indicator scores were randomly sampled from a uniform distribution with range of 0 to 1 to reflect the possible range of scores under an independence assumption. This follows the central limit theorem but creates some difficulties in the interpretation of scores or converting the scores to report card grades (A, B, C, D, Fail). We avoid these difficulties by simulating some worst case scenarios for each of the grades based on the number of indicators in each of the categories. These simulated cut-offs are then used to determine the range of scores for a particular grade. In this way we account for the effect of the number of indicators in each category within our calculations.

ACKNOWLEDGMENTS

This work is based on Lennox and Wang [2008] which can be consulted for further information. The work has been funded by the Queensland Bulk Water Supply Authority (trading as Seqwater) and the Commonwealth Scientific and Industrial Research Organisation. We also wish to thank Seqwater staff, Noel Cressie and Abdel El-Sharaawi for their valuable comments/suggestions on this work.

REFERENCES

Bartels, R. and Conn, A. (1980). Linearly Constrained Discrete *L_1* Problems, *ACM Transaction on Mathematical Software* 6, 594–608.

Canadian Council of Ministers of the Environment. 2001. Canadian water quality guidelines for the protection of aquatic life: CCME Water Quality Index 1.0, Technical report. In *Canadian environmental quality guidelines,* 1999; Canadian Council of Ministers of the Environment; Winnipeg.

Cude, C. G. (2001). Oregon water quality index: a tool for evaluating water quality management effectiveness. *Journal of the American Water Resources Association* 37(1), 125-137.

EHMP. (2007). Ecosystem Health Monitoring Program 2005-06 Annual Technical Report. South East Queensland Healthy Waterways Partnership, Brisbane.

He, X., and Ng, P. (1999). COBS: Qualitatively Constrained Smoothing via Linear Programming; *Computational Statistics* 14, 315–337.

Kannel, P. R., Lee, S., Lee, Y. S., Kanel S. R., and Khan, S. P. (2007). Application of water quality indices and dissolved oxygen as indicators for river water classification and urban impact assessment. *Environmental Monitoring Assessment* 132, 93-110.

Koenker, R. W., and D'Orey, V. (1987). Algorithm AS 299: Computing regression quantiles. *Applied Statistics* 36(3), 383-393.

Koenker, R. W., Ng, P., and Portnoy, S. (1994). Quantile Smoothing Splines. *Biometrika* 81(4), 673-680.

Koenker, R., and Ng, P. (1996). A Remark on Bartels and Conn's Linearly Constrained L1Algorithm, *ACM Transaction on Mathematical Software,* 22, 493–495.

Koenker, R., and Ng, P. (2005). Inequality Constrained Quantile Regression, *Sankhya, The Indian Journal of Statistics* 67, 418–440.

Lennox, S. M., and Wang, Y-G. (2008). Report Card Tool for Water Quality Monitoring at Wivenhoe, Somerset and North-Pine Storages. *CSIRO Mathematical and Information Sciences Report* 08/108.

Morgan , R. P., Kline, K. M, and Cushman, S. F. (2006). Relationships among nutrients, chloride and biological indices in urban Maryland streams. *Urban Ecosystems,* 10, 153-166.

Ng, P. (1996). An Algorithm for Quantile Smoothing Splines, *Computational Statistics and Data Analysis,* 22, 99–118.

Ng, P., and Maechler, M. (2008). COBS – Constrained B-splines (Sparse matrix based). R package version 1.1-5. http://wiki.r-roject.org/rwiki/doku.php?id=packages:cran:cobs

Said, A., Stevens, D. K., and Sehlke, G. (2004). An Innovative Index for Evaluating Water Quality in Streams. *Environmental Assessment,* 34(3), 405-414.

Sargaonkar, A., and Deshpande, V. (2003). Development of an Overall Index of Pollution for Surface Water Based on a General Classification Scheme in Indian Context. *Environmental Monitoring and Assessment,* 89, 43-67.

Sarkar, C. andAbbasi, A. (2006). Qualidex – A New Software for Generating Water Quality Indice. *Environmental Monitoring and Assessment,* 119, 201-231.

Smith, M. J., and Storey, A. W. (2001). Design and Implementation of Baseline Monitoring (DIBM3): Developing an Ecosystem Health Monitoring Program for Rivers and Streams in Southeast Queensland. Report to the South East Queensland Regional Water Quality Management Strategy, Brisbane.

Swamee, P. K., and Tyagi, A. (2000). Describing Water Quality with Aggregate Index. *Journal of Environmental Engineering,* 126(5, 451-455.

Queensland. (2006). Queensland water quality guidelines 2006. *Environmental Protection Agency,* Brisbane (accessed online 24th September 2008: http://www.epa.qld.gov.au/publications?id=1414).

Wood, S. N. (2006). Generalized Additive Models: An introduction with R; Chapman and Hall/CRC Press.

Yu, K., Lu, Z., and Stander, J. (2003). Quantile regression: applications and current research areas. *The Statistician,* 52, 331-350.

In: Water Quality
Editor: You-Gan Wang

ISBN: 978-1-62417-111-6

Chapter 2

ESTIMATES OF LIKELIHOOD AND RISK ASSOCIATED WITH SYDNEY DRINKING WATER SUPPLY FROM RESERVOIRS, LOCAL DAMS AND FEED RIVERS

Ross Sparks[1], Gordon J. Sutton[1,2], Peter Toscas[1] and Rod Mc Innes[3]

[1]CSIRO Mathematical, Informatics and Statistics, Australia
[2]School of Chemistry, University of New South Wales, Australia
[3]Sydney Catchment Authority, Australia

ABSTRACT

As the main supplier of potable water to Sydney, Sydney Catchment Authority aims to supply water that is in accordance with Australian drinking water quality standards. The standard specifies thresholds for acceptable ranges for over 80 analytes and attributes. When the quality thresholds are exceeded, the water needs to be treated, so that the treated water complies with the standard. The total expected cost of treating the water is the risk to be calculated.

The main objective of the chapter is to describe a water quality risk assessment process that is free from sampling biases. The risk assessment process involves modelling each analyte based on historical data so that estimates can be obtained for the likelihood of exceeding the water quality thresholds. These likelihood estimates are then combined with associated estimated water treatment costs to produce an annual water treatment cost estimate, or equivalently, an annual risk estimate.

Costs of exceeding thresholds are in their early stages of development and, at best, costs have been developed for each threshold exceedance separately. Some analytes have one-sided minor and major thresholds, while others have two-sided thresholds. Risk values (equal to the integral of exceedance likelihood estimates times the treatment cost associated with such exceedances) have been explicitly calculated for 2006.

A future aim is to repeat this process for all past years as a way of monitoring the year-to-year variation in risk for each analyte. Since the process is to be to be repeated for each year - the chapter works towards providing a robust approach to estimating the risk, which is:

1. Flexible in distributional assumptions; and
2. Capable of selecting variables that are important for estimating risk.

The application datasets come from the Sydney Catchment Authority and comprise approximately 80 analytes sampled at 35 different locations. Given the number of analyte-site combinations and the aim of yearly model fitting, the modelling process needs to be repeatable, so the process was fully automated using R scripts. Although the real-risk assessment problem is multivariate and spatio-temporal in nature, the associated cost profiles were not available in this form. The chapter includes an outline of a process that could be followed in the future if such cost profiles become routinely available. The cost profiles could be used to estimate the annual risk of all analyte exceedances, taking into account interactions. This requires a completely different way of thinking about the problem and requires a detailed understanding of variation and associated costs.

INTRODUCTION

The importance of water quality and its impact on health is well documented in the literature (e.g., see Abera et.al. [2011], Fawell and Nieuwenhuijsen [2008], Krewski *et al.* [2004], Hales *et al.* [2003], Urbansky and Schock [1999], Yang [1998]). There is evidence that poor water quality is linked to increased risk of cancer (e.g., see Morris [1995]). In addition, the concerns about the stringency of standard safe water thresholds are documented in Grinsven *et al.* [2006], Cross *et al.* [2004] and Pavlov *et al.* [2004] and therefore the risks are not well understood. This indicates the importance of collecting the data, controlling the variation in analytes in the drinking water system, and improving the understanding and management of the risks associated with water quality. The key purpose of the water quality risk assessment in this paper is to provide risk profiles for each water quality hazard group. This will assist in managing hazards by applying risk controls to hazard events in the supply systems with the highest priority (e.g., see Rosen *et al.* [2010]). There are several approaches to risk management and some of these are compared in Chowdhury *et al.* [2009]. In this chapter, we take a traditional approach to risk assessment and estimate the likelihood and consequences in terms of operational costs to calculate risk. In addition, the risk assessment process will also be used to develop a water quality management framework. Monitoring for deterioration in catchment water quality after adjusting for changes in climate is briefly discussed in the Conclusion Section of the Chapter. The purpose of the statistical analysis in this chapter is to apply statistically robust data analysis methods for the calculation of the risk assessment. The following general steps are carried out on each analyte to achieve this objective:

1. *Step 1: Develop model*
   Use all the available data to develop a model that can predict the analyte values throughout the year without sampling bias. This is done as follows:

   a. Clean the data – making certain all measures are recorded using the same unit of measurements – discarding all entries that are obvious errors.
   b. Decide on a set of parametric distributions to select from, with the aim of describing how measurements vary in distribution over time.

c. Decide on explanatory variables to consider in the models.
d. Fit models for each distribution, based on the explanatory variables that can predict temporal changes in the parameters of the distribution and that can interpolate the daily trend in the distribution of values for the target year.
e. Decide which model, and hence distribution, best describes how measurements vary about the trend.

2 *Step 2: Estimate Likelihood*
Use the model for the selected distribution to predict the trend in the analyte's values throughout the target year. Estimate the probability of exceeding the analyte's threshold for each day using the predicted distributions. Find the sum of the daily probabilities exceeding the threshold for a year. This estimates the expected number of days per annum that the analyte exceeds the threshold.

3 *Step 3: Estimate Risk*
Estimate the risk (expected value of likelihood multiplied by consequence) of poor water quality for the particular analyte as follows:

a. Establish (potentially multiple) poor quality thresholds that define different consequence levels and establish the associated consequences (treatment and operating costs to SCA (Sydney Catrchment Authority)).
b. Estimate the daily probabilities of exceeding each poor water quality threshold using the model for the selected distribution, as in Step 2.
c. Estimate the annual risk caused by variability in the analyte (sum over all days within a year the daily likelihoods for each consequence level multiplied by the respective consequences). This estimates the total expected consequence, or cost above normal operation, for the target year.

4 *Step 4: Estimate uncertainties*
Estimate the uncertainty in the estimates of likelihood and risk.

In more detail, empirical models are used to predict analyte values, particularly on days when no analyte value is measured. These models build equations linking the analyte value to a set of explanatory variables in a way that the equations can be used to reasonably accurately predict the analyte value given the explanatory variable values. More specifically, the equations define the distribution the analyte is expected to adhere to, as functions of the set of explanatory variables. The models have parameters assigned to each explanatory variable that describe the impact the explanatory variable has on predictions in the model. Methods designed to minimise the model error are used to estimate parameters. Model selection methods are used to decide on the most suitable distribution for the error structure in the model. The fitted models estimate the trend in mean values over the target year and the spread of values around that trend. These evolving mean values and volatility measures are used to:

- Estimate the likelihood – as the estimated number of days the water quality threshold will be exceeded for the target year;

- Estimate the risk – as the total estimated cost to SCA due to variable analyte values (for each analyte and supply site combination); and
- Estimate the uncertainty in the risk and likelihood estimates.

We start by looking at estimating the likelihood.

ESTIMATING THE LIKELIHOOD USING THE FITTED MODEL

In this section we develop estimates for the likelihood of exceeding water quality thresholds. We assume that a model has been fitted for the analyte of interest and that the model defines the probability density function of the analyte at each time point. In the SCA case study, gamlss models (Rigby and Stasinopoulos [2005], Stasinopoulos, Rigby and Akantziliotou [2006], and Stasinopoulos Rigby [2007]) were developed for each analyte, with models selected from the gamlss family of distributions: normal, log-normal, Inverse Gaussian and Zero Adjusted Inverse Gaussian. Discussion of the modelling for the likelihood is left to the model formulation section. For the sake of illustration, in this section we use a time-varying Gaussian distribution, having time-varying mean and standard deviation parameters, rather than a generic distribution. The method is applicable to any defined and calculable probability distribution.

Let the density for the analyte be denoted by $f(x_t / \mu_t, \sigma_t)$. The average likelihood (or average probability) of exceeding either the upper or lower threshold is then estimated as:

$$\overline{p} = \int_0^1 (1 - F(c_u(t) / \mu_t, \sigma_t) + F(c_L(t) / \mu_t, \sigma_t))dt$$

where the time period (t ranging from 0 to 1) represents one year, and

$$F(c / \mu_t, \sigma_t) = \int_0^c f(x / \mu_t, \sigma_t)dx$$

is the probability of the analyte being below threshold c at time t.

If there is only an upper threshold then this average is calculated as:

$$\overline{p}_u = \int_0^1 (1 - F(c_u / \mu_t, \sigma_t))dt.$$

The above integral is estimated by breaking time into daily intervals and summing over days. The average likelihood is estimated by first deciding on a time of day at which to calculate the likelihood. Estimates of the mean and standard deviation (denoted $\hat{\mu}_t$ and $\hat{\sigma}_t$, respectively) then give an estimate of the likelihood of exceeding a threshold on the day represented by time t of the target year, at the decided time of the day, as:

$$\hat{p}_t = 1 - F(c_u(t) / \hat{\mu}_t, \hat{\sigma}_t) + F(c_L(t) / \hat{\mu}_t, \hat{\sigma}_t)$$

$\bar{p}$ is then estimated by averaging these estimated daily likelihoods over each day of the year. Similarly, an estimate of

$$\hat{p}_{ut} = 1 - F(c_u(t) / \hat{\mu}_t, \hat{\sigma}_t)$$

can be found when there is only an upper threshold.

The number of days in the year that will exceed one of the thresholds is estimated by 365.25 days times the estimate of the daily likelihood value for either both thresholds (upper and lower) or the upper threshold only.

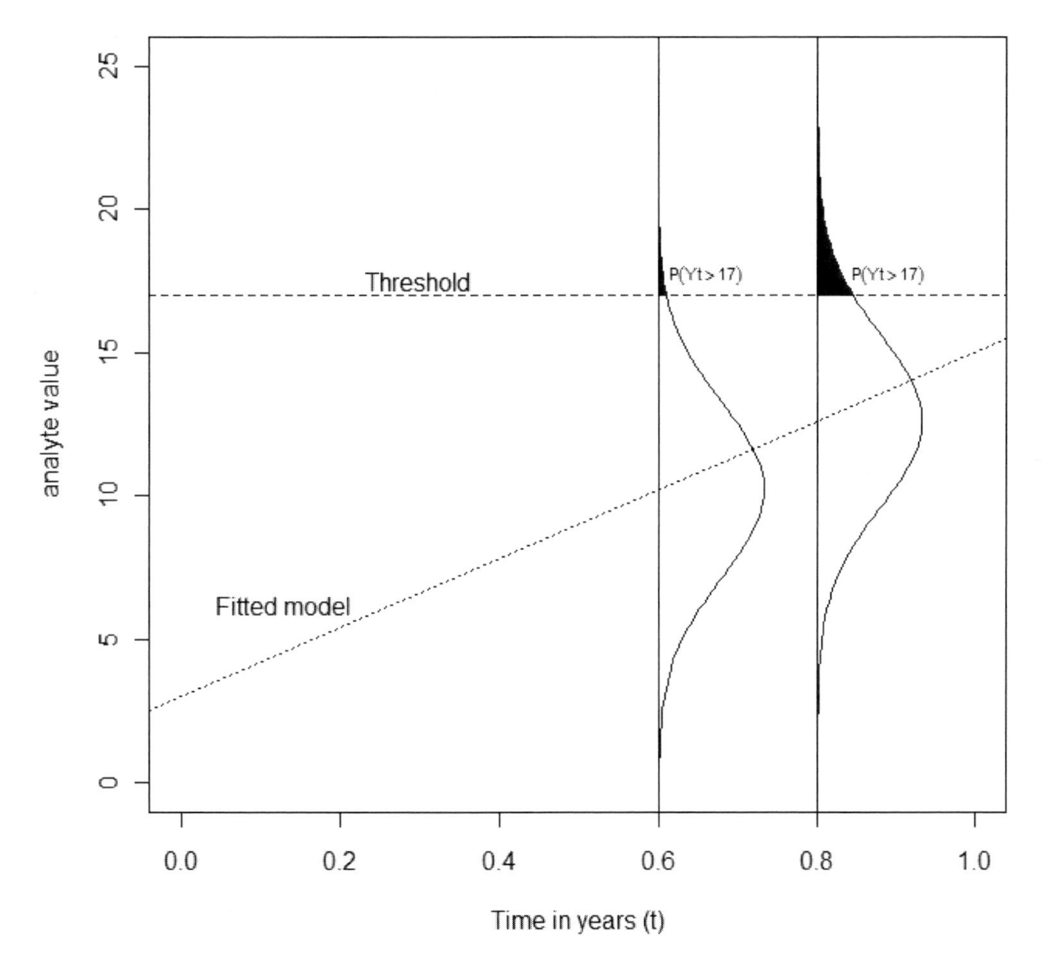

Figure 1. Simple illustration of the relationship between the fitted models and estimating probability of exceeding the threshold.

A Graphical Representation of the Likelihood Estimation Process

Figure 1 is used to demonstrate how the daily likelihood values are estimated for two days represented by t=0.6 and t=0.8 during the target year. A simple fitted model has been used comprising a linear trend term and a constant standard deviation for the volatility, giving:

$$\mu_t = 3 + 12t$$

$$\sigma_t = 3$$

where *t* is the fraction of the target year.

The distribution of analyte values about the mean, μ_t, is assumed to be normally distributed and the upper threshold has been set as 17. The density functions used to estimate the probability of exceeding the threshold at times t=0.6 and t=0.8 of the year are the bell shape curves plotted in Figure 1.

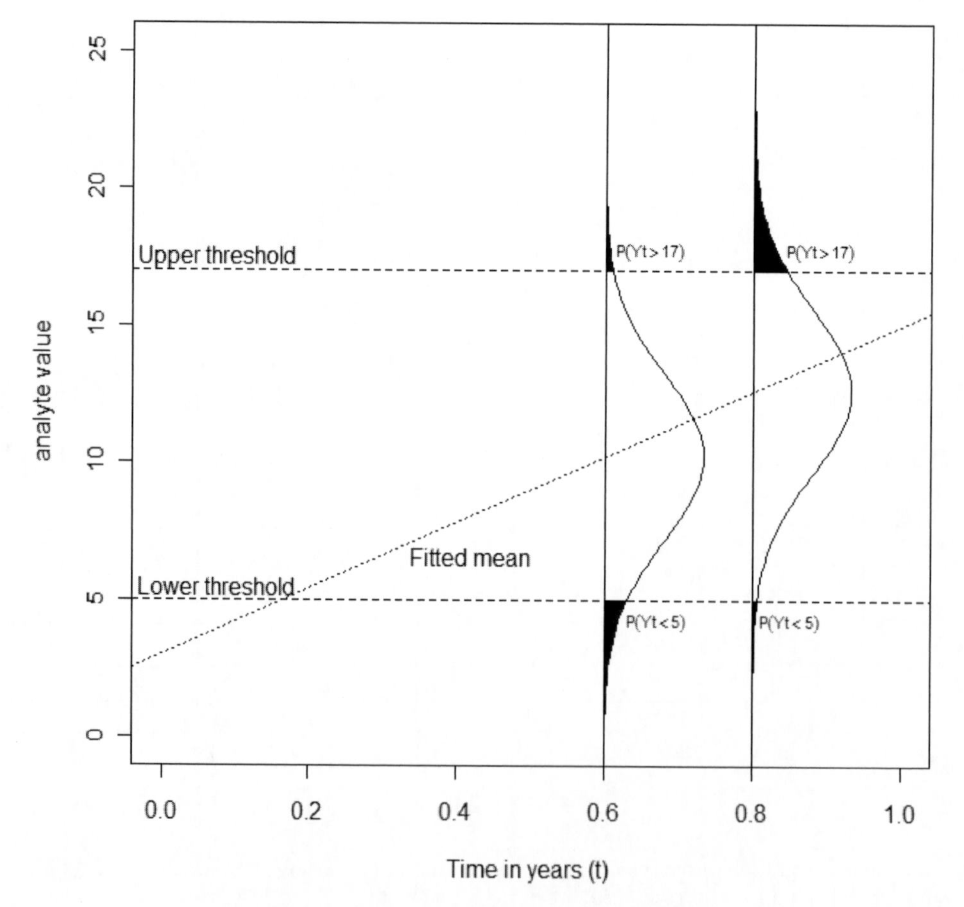

Figure 2. Examples of daily likelihood values – probability of exceeding an upper and lower threshold.

The probabilities $p_{0,6}=P(Y_t>17|t=0.6)$ and $p_{0,8}=P(Y_t>17|t=0.8)$ are equal to the areas shaded between the density estimates (bell shape curves) and the vertical line drawn at t=0.6 and t=0.8, respectively. The area is found for each time t using the normal distribution function, with mean μ_t and standard deviation σ_t, giving the probability that the random analyte value taken from the density at time t is greater than the threshold 17.

The probability of the analyte value exceeding the threshold for each day of the year is summed over all days in a year.

See Figure 2 for a graphical representation of estimating daily probabilities of exceeding either an upper or a lower threshold. The two examples of the probabilities of exceeding either of the thresholds illustrate that at times both tails of the distribution need to be considered for calculating daily likelihood values.

ESTIMATING THE UNCERTAINTY IN THE LIKELIHOOD ESTIMATES

The approach to estimating uncertainty in the likelihood estimates is developed for the situation where there is an upper and a lower threshold, and the upper threshold problem follows as a simplification of the results.

The likelihood is taken as the average of

$$\hat{p}_t = 1 - F(c_u(t) / \hat{\mu}_t, \hat{\sigma}_t) + F(c_L(t) / \hat{\mu}_t, \hat{\sigma}_t)$$

over days t = 1, 2, .., 365 (or 366) in the target year, where $\hat{\mu}_t$ and $\hat{\sigma}_t$ are estimates of the mean and standard deviation for day t, estimated from the selected gamlss regression model. We note that in the preceding section that introduces the likelihood, t is instead a continuous variable on [0,1], however here it is treated as discrete. This section develops a process for estimating how inaccurate $\hat{p}_t$ is relative to the true value, p_t, where

$$p_t = 1 - F(c_u(t) / \mu_t, \sigma_t) + F(c_L(t) / \mu_t, \sigma_t)$$

Moreover, we are interested in establishing the uncertainty in the likelihood estimate for the sum of $\hat{p}_t$ over all days within the target year.

The parameters of the time varying distributions, μ_t and σ_t, are estimated using regression models with time varying covariates, such as flows. Uncertainties in risk estimates are generated by the uncertainty in the fitted model, i.e., the uncertainty in the estimated regression parameters in the gamlss model (Rigby and Stasinopoulos [2005], Stasinopoulos, Rigby and Akantziliotou [2006], and Stasinopoulos Rigby [2007]).

Both the sequential estimates of the parameters ($\hat{\mu}_t$ and $\hat{\sigma}_t$) and the covariates are highly autocorrelated across days, and this autocorrelation influences the uncertainty in likelihood estimates. Perhaps the best way to cope with such situations is to develop a hierarchical Bayesian approach to estimating the uncertainty by establishing the posterior distribution of the likelihood estimate. An Markov Chain Monte Carlo (MCMC) approach could be used to estimate this posterior probability.

We did not follow this approach because we anticipated that it would take too long to apply to all analyte-site combinations. The alternative followed in this chapter is to establish reasonable estimates of the standard errors (standard deviations) of the estimated likelihood values.

At times it was necessary to make some simplifying assumptions. The first important assumption is that we assume all covariates are measured without error or with negligible error (i.e., near zero uncertainty in the measurement process).

The estimates $\hat{p}_t$ are established using the same fitted models. However, flow and storage variables are highly autocorrelated, and therefore $\hat{p}_t$ values are highly autocorrelated. Consequently, independence of consecutive daily estimates $\hat{p}_t$ cannot be assumed. We find the uncertainty in the sum of $\hat{p}_t$ under the assumption that $\hat{p}_t$ is an unbiased estimate of p_t for all t, i.e. $E[\hat{p}_t] = p_t$. This is a relatively strong assumption in practice because the thresholds that define p_t are often in the tails of the fitted distributions and the tails are less accurately estimated than the high density regions. Based on this assumption, the variance in the estimated sum probability of exceeding a threshold is given by:

$$\begin{aligned} Var(\textstyle\sum_{t=1}^{365}\hat{p}_t) &= E\left(\left[\textstyle\sum_{t=1}^{365}(\hat{p}_t - E[\hat{p}_t])\right]^2\right) \\ &= E\left(\textstyle\sum_{t=1}^{365}(\hat{p}_t - p_t)\textstyle\sum_{t=1}^{365}(\hat{p}_t - p_t)\right) \\ &= \textstyle\sum_{t=1}^{365}\textstyle\sum_{\tau=1}^{365}E[(\hat{p}_t - p_t)(\hat{p}_\tau - p_\tau)]. \end{aligned}$$

We now assume that $\hat{p}_t$ is estimated from a gamlss model. From Cogley [1999] and Thiebaux and Zwiers [1984], the effective sample size used to fit the model, n_e, is given by:

$$n_e = m/(1 + 2\textstyle\sum_{j=1}^{m-1}(1 - j/m)\hat{\rho}_{\Delta t}(j)),$$

where m is the number of observations used to fit the model and $\hat{\rho}_{\Delta t}(j)$ is the estimated autocorrelation in the model residuals. The R function effectiveSize, in the R library coda, is used to estimate n_e. The autocovariance of $\hat{p}_t$ can then be estimated as:

$$E[(\hat{p}_t - p_t)(\hat{p}_\tau - p_\tau)] \approx \sqrt{p_t(1-p_t)}\sqrt{p_\tau(1-p_\tau)}\rho(t-\tau)/n_e,$$

where $\rho(t-\tau)$ is the autocorrelation between the estimates of the probability of exceeding the threshold on days t and τ, so that

$$Var(\sum_{t=1}^{365}\hat{p}_t) \approx \sum_{t=1}^{365}\sum_{\tau=1}^{365}\sqrt{p_t(1-p_t)}\sqrt{p_\tau(1-p_\tau)}\rho(t-\tau)/n_e.$$

What is missing in this equation is some goodness of fit measure for the gamlss model, and this could be estimated using the bootstrap approach or MCMC. Bootstrap and MCMC are both computationally intensive, so for the sake of simplicity the model uncertainty part of the variance is ignored.

Therefore, an estimate of $E[(\hat{p}_t - p_t)(\hat{p}_\tau - p_\tau)]$is given by:

$$\sqrt{\hat{p}_t(1-\hat{p}_t)}\sqrt{\hat{p}_\tau(1-\hat{p}_\tau)}\ \hat{\rho}(t-\tau)/n_e,$$

where $\hat{\rho}(t-\tau)$ is approximated by the sample autocorrelation for variable $\hat{p}_t/\sqrt{\hat{p}_t(1-\hat{p}_t)} = \sqrt{\hat{p}_t/(1-\hat{p}_t)}$. This then gives the estimated variance of the annual likelihood estimate as:

$$\hat{V}ar(\sum_{t=1}^{365}\hat{p}_t) = \sum_{t=1}^{365}\sum_{\tau=1}^{365}\hat{\rho}(t-\tau)\sqrt{\hat{p}_t(1-\hat{p}_t)}\sqrt{\hat{p}_\tau(1-\hat{p}_\tau)}/n_e.$$

Simulation studies found this estimated variance to be fairly volatile, but the alternative MCMC approach was considered untenable given the computational cost of the model selection and retrospective surveillance aspects of the work.

Very roughly, we expect approximately 70% of the total probabilities to be within the bounds

$$\sum_{t=1}^{365}p_t \pm \sqrt{\hat{V}ar(\sum_{t=1}^{365}\hat{p}_t)}.$$

ESTIMATING THE COST RISK ASSOCIATED WITH EXCEEDING UPPER THRESHOLDS

Using a step-function: For each analyte-site combination, let the increase in drinking water supply and operating costs be $c_1 < c_2 < \ldots < c_m$ when analyte measurement boundaries $b_1 < b_2 < \ldots < b_m$ are exceeded. In addition, assume that the probability of the analyte at a site on day *t* exceeding b_j is given by:

$$p_{tj} = 1 - \int_o^{b_j} f(x_t/\mu_t,\sigma_t)dx_t,$$

For day *t* the expected increase in supply and operating costs is given by:

$$EC_t = 0 \times (1 - p_{t1}) + c_1 \times (p_{t1} - p_{t2}) + c_2 \times (p_{t2} - p_{t3}) + ... + c_m \times p_{tm}.$$

These can be thought of as costs above the normal running costs associated with adverse variation for the respective analyte being considered. The annual expected cost due to variable analyte values is given by:

$$EC = \sum_{t=1}^{365} EC_t$$

The individual expected daily costs are estimated by:

$$\hat{E}C_t = 0 \times (1 - \hat{p}_{t1}) + c_1 \times (\hat{p}_{t1} - \hat{p}_{t2}) + c_2 \times (\hat{p}_{t2} - \hat{p}_{t3}) + ... + c_m \times \hat{p}_{tm},$$

and the expected annual cost is estimated by:

$$\hat{E}C = \sum_{t=1}^{365} \hat{E}C_t.$$

$\hat{E}C$ is taken as the increase to the cost risk associated with analyte variability over the year.

Using a Double-Sided Two Staged (Minor or Major) Step-Profile Cost Function with an Escalating Cost for Consecutive Major Exceedances

For each analyte-site combination, define the analyte measurement boundaries $0 = b_0 < b_1 < ... < b_m < b_{m+1} = \infty$. Now define analyte measurement levels L0, L1,..., Lm so that when an analyte measurement lies between measurement boundaries b_j and b_{j+1} it is in level Lj.

This gives the probability of an analyte measurement being in level Lj at a site on day t as $p_{tj} - p_{tj+1}$, where p_{tj} is defined, as before, as:

$$p_{tj} = 1 - \int_o^{b_j} f(x_t / \mu_t, \sigma_t) dx_t$$

We now extend the cost structure to allow both lower and upper quality thresholds and also consecutive events. Denote the analyte measurement at the site of interest on day t as y_t.

Let L0 and Lm represent major events, then define a new cost level for day t that applies when both y_t and y_{t-1} are major events (of the same sort).

So, define the increases in drinking water supply and operating costs on day t as:

c_{c0}, for y_t and y_{t-1} lying in level L0 (consecutive major events);

c_0, for y_t lying in level L0 and y_{t-1} not (major event);

$c_1..c_{m-1}$, for y_t lying in levels L1, L2,…,Lm-1 .

c_m, for y_t lying in level Lm and y_{t-1} not (major event); and

c_{cm}, for y_t and y_{t-1} lying in level Lm (consecutive major events).

Note that Lj is the region where $b_j < y < b_{j+1}$, j=0,…,m.

Figure 3 provides an example with m=4. Although this figure considers the application that excludes changing costs for consecutive events, it does plant the seed for considering consecutive events.

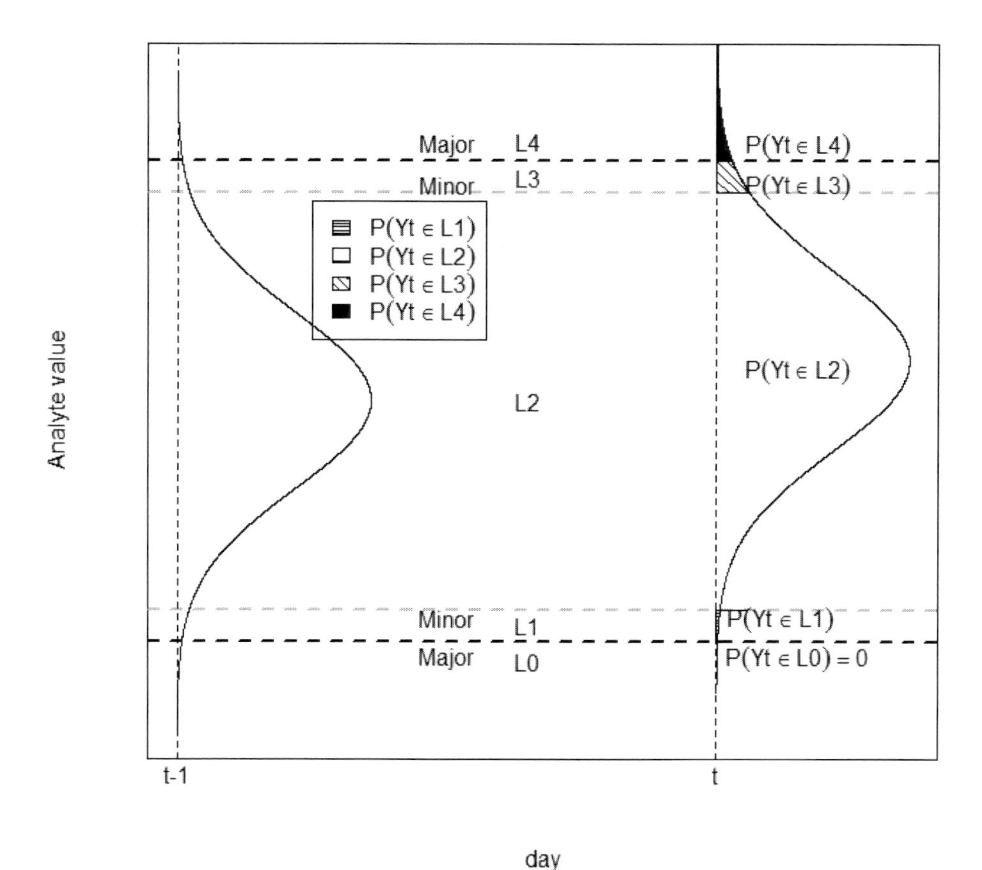

Figure 3. Demonstrating the minor and major exceedance thresholds for day t.

The following table gives the cost incurred on day *t*, as a function of the analyte levels on day *t* and day (*t*-1), for the general case that there are thresholds on both the low and the high sides and the second of two consecutive major events incurs an extra cost (i.e., c_{c0} and c_{c4}). For illustration the number of boundaries is restricted to 4, $b_0 < b_1 < \ldots < b_5$, where $b_0 = 0$ and b_5 is infinite.

		Level on day (t-1) (Yesterday)				
		L0	L1	L2	L3	L4
Level on day t (Today)	L0	c_{c0}	c_0	c_0	c_0	c_0
	L1	c_1	c_1	c_1	c_1	c_1
	L2	c_2	c_2	c_2	c_2	c_2
	L3	c_3	c_3	c_3	c_3	c_3
	L4	c_4	c_4	c_4	c_4	c_{c4}

The next table gives the costs incurred on day *t*, for the same case as in the previous table, along with the corresponding probabilities of occurrence, assuming the model residuals are uncorrelated.

	Cost	Probability	Interpretation
y_t and y_{t-1} in L0	c_{c0}	$(1-p_{t1})(1-p_{(t-1)1})$	consecutive major events
y_t in L0, y_{t-1} not L0	c_0	$(1-p_{t1})(p_{(t-1)1})$	major event
y_t in L1	c_1	$p_{t1}-p_{t2}$	minor event
y_t in L2	c_2	$p_{t2}-p_{t3}$	in control
y_t in L3	c_3	$p_{t3}-p_{t4}$	minor event
y_t in L4, y_{t-1} not L4	c_4	$p_{t4}(1-p_{(t-1)4})$	major event
y_t and y_{t-1} in L4	c_{c4}	$p_{t4}p_{(t-1)4}$	consecutive major events

By setting the appropriate costs to zero and appropriate consecutive analyte measurement boundaries equal, this structure can accommodate cases with just lower thresholds or just upper thresholds as well as the double sided case. Similarly, the cost of two consecutive days with major events need not be implemented.

The following formula gives the expected cost incurred on day t for the double sided case, including consecutive major events:

$$EC_t = \sum_i (\text{cost}_\text{i})(\text{probability}_\text{ti})$$
$$= c_{c0}(1-p_{t1})(p_{(t-1)1}) + c_0(1-p_{t1})(p_{(t-1)1}) + c_1(p_{t1}-p_{t2}) + c_2(p_{t2}-p_{t3}) +$$
$$c_3(p_{t3}-p_{t4}) + c_4 p_{t4}(1-p_{(t-1)4}) + c_{c4} p_{t4} p_{(t-1)4}$$

The expected cost, or risk, for the year 2006 is then given as:

$$EC_{t\in 2006} = \sum_{t=1}^{365} EC_t,$$

where t = 1 represents January 1, 2006. Note that to calculate $EC_{t\in 2006}$, the expected cost over 2006, p_{t1} and p_{t4} will be needed for 31 Dec 2005.

Estimating the Uncertainty in the Risk Estimator, Excluding Altered Costs for Consecutive Exceedances

The challenge in this section is to come up with a relatively simple way of estimating the uncertainty for the annual risk estimator defined as the total cost for the year caused by variable quality in analyte values at a supply site.

Assume there are m thresholds. Define p_{ti} as before and set $p_{t1} = 1$ and $p_{tm+1} = 0$. Then, write the estimated daily cost as:

$$EC_t = c_1 \times (\hat{p}_{t1} - \hat{p}_{t2}) + c_2 \times (\hat{p}_{t2} - \hat{p}_{t3}) + \cdots + c_m \times (\hat{p}_{tm} - \hat{p}_{tm+1})$$
$$= c_1 \times \hat{q}_{t1} + c_2 \times \hat{q}_{t2} + \cdots + c_m \times \hat{q}_{tm}$$
$$= c^t \hat{q}_t$$

where $c^T = [c_1 \quad c_2 \quad c_3 \quad \ldots \quad c_m]$; $\hat{q}_t^T = [\hat{q}_{t1}^T \quad \hat{q}_{t2}^T \quad \ldots \quad \ldots \quad \hat{q}_{tm}^T]$; and $\hat{q}_{ti} = \hat{p}_{ti} - \hat{p}_{ti+1}$

Then the estimated total annual cost is given by:

$$\sum_{t=1}^{365} \hat{E}C_t = \sum_{t=1}^{365} c^T \hat{q}_t = (1^T \otimes c^T)\hat{q},$$

where $\hat{q}^T = [\hat{q}_1^T \quad \hat{q}_2^T \quad \hat{q}_3^T \quad \ldots \quad \hat{q}_{365}^T]$ and $\otimes$ is the Kronecker product.

We note that y_t is being represented as a multinomial with levels L1, ..., Lm and with probabilities $q_{t1}, \ldots, q_{tm}$, such that $\sum_{i=1}^{m} q_{ti} = 1$, for all t, and $E[q_{ti} q_{tj}] = 0$ for $i \neq j$.

The variance of the estimated daily cost is given by:

$$\operatorname{var}(\hat{E}C_t) = c^T \Sigma_t c,$$

where Σ_t is the covariance matrix of $\hat{q}_t$, given by:

$$\Sigma_t = \begin{bmatrix} q_{t1}(1-q_{t1}) & -q_{t1}q_{t2} & . & . & -q_{t1}q_{tm} \\ -q_{t1}q_{t2} & q_{t2}(1-q_{t2}) & . & . & -q_{t2}q_{tm} \\ . & . & . & & . \\ . & . & & . & . \\ -q_{t1}q_{tm} & -q_{t2}q_{tm} & . & . & q_{tm}(1-q_{tm}) \end{bmatrix} / n_e.$$

The matrix Σ_t is estimated by replacing q_{tj} by $\hat{q}_{tj}$ in the equation for Σ_t above. This gives an estimate for the variance of daily costs as:

$$\hat{\operatorname{var}}(\hat{E}C_t) = (\sum_{j=1}^{m} c_j^{\ 2} \times \hat{q}_{tj}(1-\hat{q}_{tj}) - 2\sum_{j=1}^{m-1}\sum_{k=j+1}^{m} c_j c_k \hat{q}_{tj}\hat{q}_{tk}) / n_e.$$

Note that this only estimates the uncertainty for the cost on day *t*, not for the total cost for the year.

To obtain an estimate for the annual expected cost, it is assumed that there is autocorrelation between consecutive estimates of the likelihoods, i.e.

$$\operatorname{cov}(\hat{q}_{tk}, \hat{q}_{\tau k}) = \rho(t-\tau)\sqrt{q_{tk}(1-q_{tk})}\sqrt{q_{\tau k}(1-q_{\tau k})} / n_e,$$

where $\rho(t-\tau)$ is the autocorrelation between estimate $\sqrt{\hat{q}_t/(1-\hat{q}_t)}$ and $\sqrt{\hat{q}_\tau/(1-\hat{q}_\tau)}$. Therefore estimating the standard deviation of the estimated annual expected cost is much more complicated than estimating the standard deviation for daily risk estimates and a few assumptions are needed to simplify the process. Two methods are now presented for estimating this standard deviation, each based on different assumptions about the autocorrelations.

Method 1: We assume that the distribution of p_t is invariant of time *t*, such that $\hat{q}_{tj}$ are roughly constant over time. In addition, we assume that we can approximate the covariance matrix of $\hat{q}$ by $R \otimes \Sigma_{m\times m}$, where R is the 365x365 matrix with ones down the main diagonal and the time-invariant autocorrelations of $\hat{q}_{tj}$ on the jth off diagonal for all *j*; and $\Sigma_{m\times m}$ is the time-invariant covariance matrix for all q_t.

As such the variance of the total estimated cost for the year is given by:

$$\operatorname{var}(\sum_{t=1}^{365} \hat{E}C_t) = (1^T \otimes c^T)(R \otimes \Sigma_{m\times m})(1 \otimes c) = 1^T R 1 \times c^T \Sigma_{m\times m} c,$$

where 1^T is the 1x365 matrix of 1's.

The time-invariant autocorrelations in R are approximated by averaging the estimated autocorrelations of q_{tj} over j and $\Sigma_{m\times m}$ is approximated by the time-average sample-estimated covariance matrix of q_t. Treating each q_t as comprising the probabilities of a multinomial distribution, $\Sigma_{m\times m}$ is approximated by:

$$\Sigma_{m\times m} = \sum_{t=1}^{365} \begin{bmatrix} \hat{q}_{t1}(1-\hat{q}_{t1}) & -\hat{q}_{t1}\hat{q}_{t2} & . & . & -\hat{q}_{t1}\hat{q}_{tm} \\ -\hat{q}_{t1}\hat{q}_{t2} & \hat{q}_{t2}(1-\hat{q}_{t2}) & . & . & -\hat{q}_{t2}\hat{q}_{tm} \\ . & . & . & & . \\ . & . & & . & . \\ -\hat{q}_{t1}\hat{q}_{tm} & -\hat{q}_{t2}\hat{q}_{tm} & . & . & \hat{q}_{tm}(1-\hat{q}_{tm}) \end{bmatrix} / n_e .$$

Method 2: We now allow the distribution of q_t to vary with time and assume that the covariance matrix of q can be written:

$$\Sigma_T = \begin{bmatrix} \Sigma_1 & \rho(1)\Sigma_{12} & \rho(2)\Sigma_{13} & & \rho(364)\Sigma_{1,365} \\ \rho(1)\Sigma_{21} & \Sigma_2 & \rho(1)\Sigma_{23} & & \rho(363)\Sigma_{2,365} \\ \rho(2)\Sigma_{32} & \rho(1)\Sigma_{32} & \Sigma_3 & & \rho(362)\Sigma_{3,365} \\ . & . & & & \\ . & . & & & \\ . & . & & & \\ \rho(364)\Sigma_{365,1} & \rho(363)\Sigma_{365,2} & \rho(362)\Sigma_{365,3} & & \Sigma_{365} \end{bmatrix}, \quad (1)$$

where Σ_t on the diagonal are defined previously; and $\rho(|t-\tau)|)\Sigma_{t\tau})$ on the off-diagonals are the covariances between $\hat{q}_t$ and $\hat{q}_\tau$. The (i^{th} , j^{th}) element of $\rho(|t-\tau)|)\Sigma_{t\tau})$ is assumed to be given by: $\{-\rho(|t-\tau|)p_{ti}p_{\tau j}\sqrt{p_{ti}(1-p_{ti})}\sqrt{p_{\tau j}(1-p_{\tau j})}\}$ when $j \neq i$ and $\{-\rho(|t-\tau|)\sqrt{p_{ti}(1-p_{ti})}\sqrt{p_{\tau i}(1-p_{\tau i})}\}$ when $j = i$.

An estimate for Σ_T, denoted $\hat{\Sigma}_T$, is obtained by replacing each variable by its estimate. That is, p_{tj} by $\hat{p}_{tj}$ and $\rho(t-\tau)$ by $\hat{\rho}(t-\tau)$. $\hat{\rho}(t-\tau)$ is approximated as the average over all *j* of the sample autocorrelation of $\hat{q}_{tj}$ at time difference $(t-\tau)$ days.

The variance estimate of the annual cost estimate is then given by:

$$\hat{\text{var}}(\sum_{t=1}^{365}\hat{E}C_t) = (1^T \otimes c^T)\hat{\Sigma}_T(1 \otimes c).$$

The uncertainty in the total costs estimate $\sum_{t=1}^{365} \hat{E}C_t$ is given by the estimated standard error:

$$\sqrt{\hat{\text{v}}\text{ar}(\sum_{t=1}^{365} \hat{E}C_t)}$$

A similar approach can be used for a two stage (minor or major) additional cost step function.

Assessing Trends in the Likelihood and Risk

We decided to calculate the average likelihood and risk for each annual period, rather than averaging over the whole history, and then presenting the trend in these annual likelihoods and risks over the whole history. The advantage of this approach is it allows SCA to monitor trends in the annual risk over time in a way that is linked to their associated budgeted costs. In addition, annual averages remove the seasonal influence on the likelihood and risk scores. Understanding the variation in annual likelihoods and risks across years is essential for managing the increased costs associated with these risks.

Future analyses should look at evaluating whether SCA's risk management strategy is improving from year to year. This calls for a different monitoring strategy to the one referred to in this chapter. This is much more difficult because the variation in risk that is beyond the control of SCA, such as weather changes, need to be removed before the management performance can be assessed for those aspects (hazard events) that are within the control of SCA.

Models for Estimating the Likelihood

Models were developed for estimating the likelihood of analyte values exceeding upper or lower thresholds for all days within a year. These models needed to account for seasonal variation, flow characteristics and other explanatory variables that were measured daily. These models were used to predict analyte values during periods where measurements are not made.

Having daily measurements for each analyte would have been ideal for modelling. However, only a few analytes were measured daily (e.g., pH). Another, less ideal, scenario would have been having the sampled days sufficiently frequent. If the sampled days were representative of the period of study it would provide enough accuracy to the likelihood of threshold exceedance estimates. However, this too was seldom true, which made assessing the annual risk of exceeding thresholds at a site problematic. Therefore, the purpose of building a model was to have the ability to interpolate analyte values even for days when the analyte had not been measured.

Potential explanatory variables: Operating models where established by finding the set of explanatory variables that could best predict the analyte value on any day. Water temperature was a potential explanatory variable for many analytes, but was not used because of missing data problems. Therefore harmonics for within a day and across the year was used as a surrogate for temperature. The following explanatory variables were considered:

- Daily and annual harmonics
- Flow variables – mean flow, maximum flow and flow velocity at representative sites for each water catchment
- Storage level at the specified sites
- Depth
- Depth interactions with harmonics and/or flows.

Model selection: The model selection process was based on the generalized additive modelling approach using the gamlss library of functions in R-code (Stasinopoulos, Rigby and Akantziliotou [2006]). The gamlss library is a collection of functions to fit Generalized Additive Models for Location, Scale and Shape using the R package. Generalized Additive Models for Location, Scale and Shape (gamlss) were introduced as a way of overcoming some of the limitations associated with Generalized Linear Models (GLM) and Generalized Additive Models (GAM) (e.g., see Hastie and Tibshirani [1990]). In gamlss, the exponential family distribution assumption for the response variable (y) is relaxed and replaced by a general distribution family, including highly skewed and/or kurtotic distributions. The following subset of the available gamlss distributions was used for model selection and was considered sufficiently versatile to model the analytes:

1. Normal
2. Log-normal
3. Inverse Gaussian
4. Zero adjusted Inverse Gaussian where the low detects or zeros are fitted using a logistic regression model and the remaining values are fitted using the Inverse Gaussian distribution.

The models were fitted using maximum (penalised) likelihood estimation. This is implemented in the *gamlss* package in R; see Stasinopoulos *et al.* [2006].

The EM algorithm and/or MCMC algorithm (e.g., see Sparks *et al.* [2011]) was used to deal with partial missing information where analyte values have exceeded their measurement limits (e.g., the measured values are only known to be below 10).

Variable selection process: For each model, a forward selection approach was used to include variables that were significant in a step-wise fashion, one explanatory variable at a time. A first explanatory variable was selected that best explained the analyte's behaviour on its own. Subsequent explanatory variables were included if and when they:

a) reduced the model's AIC (Akaike Information Criterion) when included; and
b) reduced the model's AIC more than the other remaining explanatory variables.

Model selection (variables and distribution) of every analyte-site combination proved too onerous a task to be carried out independently in a hands-on manual fashion, because of the number of combinations of site and analytes. Therefore the process was automated so that it was repeatable for different time periods. Several rules needed to be developed (such as small observation sizes limit the complexity of the model selected) so that reasonable models would be selected in the automatic model selection process.

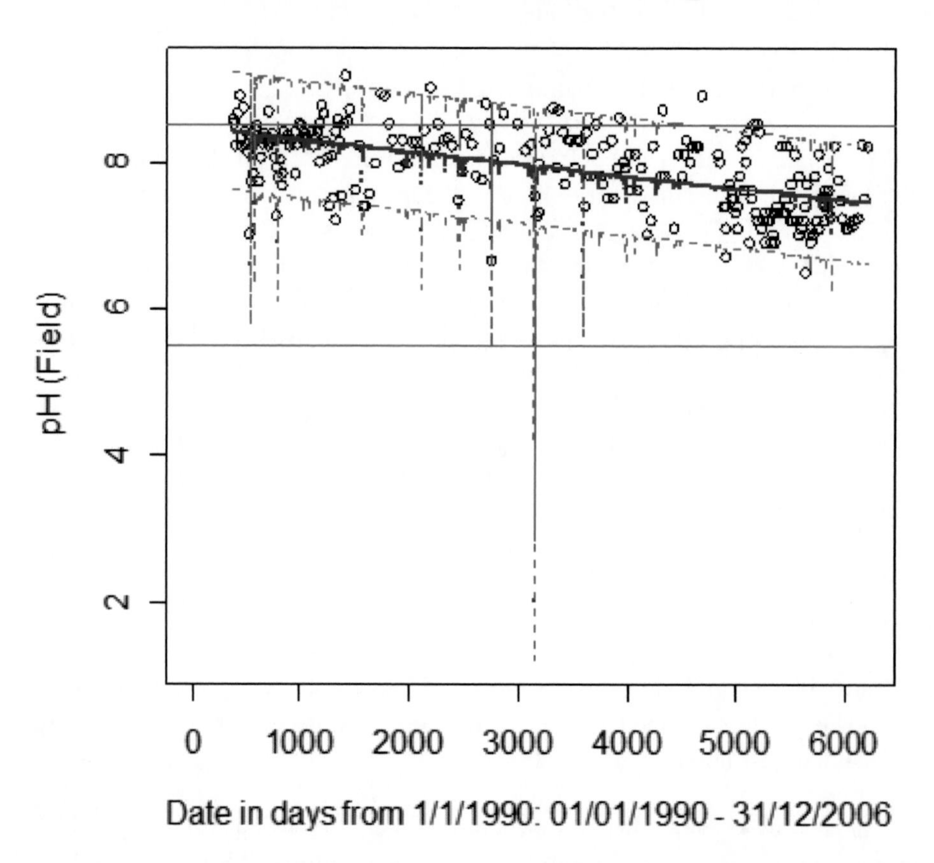

Figure 4. Time series plot of pH at Wollondilly River for the full period from 1990 to the end of 2006.

An example of model fit is presented in Figure 4 for pH for 1991 to 2006. The black circles are the measured values, the blue points are the model fitted values and the green lines are the prediction bounds. The red lines are the quality thresholds for pH. Notice that Sydney Catchment Authority was able to deliver within the threshold limits by the end of 2006, while at the end of 1990 it was not. That is, there has been a steady reduction from high alkalinity in 1990 to a more acceptable range in 2006. In addition, the pH values are fairly predictable meaning that they are largely in-control. Notice the model does on occasions produce strange predictions when extrapolating beyond the range of flows experienced in the past dataset. These can be seen by the prediction intervals dramatically dropping in values in Figure 4. In the future, the historical data used to fit the models will include more extreme events and thus extrapolation will become less likely. Therefore, having a long historical dataset is essential to avoiding this issue in estimating the risks of exceeding thresholds.

CONSEQUENCE - COST ESTIMATES

The cost of a water quality exceedance is viewed from a water consumer perspective – that is, the consequence is either a loss of consumer welfare, e.g., health or aesthetic (taste, odour), or alternatively, the opportunity cost of mitigation actions at multiple barrier exceedances in the water supply network to reduce the risk of this consequence to the consumer occurring. Ideally, sufficient analysis would be undertaken to ensure that the selection between and within these alternatives is optimised so that the least cost consequence is selected in all cases. In this application, the available information has not allowed more than a rudimentary optimisation. Further development of this technique should focus on this aspect. However, with this caveat, the consequence of a given breach in a particular analyte's thresholds (usually three thresholds have been set for each analyte - minor, major and emergency) would vary according to:

- The location of the testing station – the further upstream the greater the probability that this threshold will not be breached at downstream locations, and that therefore there will be no consumer losses or mitigation action required to avoid them.
- The size and location of the community being supplied with water from a particular source (the representative monitoring station).
    - There are strong economies of scale for water supply for lower level breaches of water quality thresholds – water supplies to large communities can have typical contaminants treated for lower unit cost, thus significantly reducing the costs of addressing a given low level water quality event.
    - Simultaneously, there are strong diseconomies of scale for water supply for higher level breaches, particularly when breaches are uncorrelated between water sources. To explain, for breaches in small water supply systems (e.g. less than 100 connections), when other water sources are not in breach, alternative supplies can be trucked in at a cost of around \$7 per KL, or less than \$5,000 per day total. Whereas, for breaches in large water quality systems, logistic and congestion costs may make alternative supplies uneconomic. In these situations, least cost short term solutions such as boil water alerts have been estimated (Sloane Cook and King [2004] and Jaguar Consulting Pty Ltd [2004]) to cost from \$78M to \$350M in 2004 dollars, or roughly an increase of \$0.90 to \$3 Per KL supplied, depending on the extent of consumer costs (time lost, income, health etc.) included. Some of this high cost is caused by the mitigation not being able to address this risk in full, and so costly consequences to health can ensue. Harrington *et al.* [1991] details how pathogen contamination can lead to such high consumer costs. Note that it is the volume of contaminated water and the subsequent impact on consumers that drives total cost, and that therefore creates the diseconomy. (For such systems, longer term solutions such as diversifying

supplies may be required, which would have minimum costs of the order of $1/KL (transfers) to $3 per KL (desalination). However, long term capital solutions to address infrequent water quality events can result in poorly utilised capital, with sometimes exorbitantly high unit costs.

- The type of response by consumers.
    - For example, even when a boil water alert is in place, many consumers will not deem the risk great enough to comply, while others will be ignorant of the alert. In a situation where there are no known clinical consequences (e.g. Sydney in 1998), these people will not bear any costs. However, where there are, and people may suffer chronic or acute ill health, they will bear significant loss of income and direct or indirect health care costs (e.g., increase in insurance premiums).
    - Alternatively, water quality perceptions can be changed by a well publicised breach, even where there is no health or aesthetic consequence. Consumers will increase averting behaviours such as purchasing bottled water, which can be very costly (e.g., thousand fold increase from $2 per kL to greater than $2 per L), even though the total quantity drunk is small.
- The nature of the water supply
    - The correlation between water quality in different parts of the network is a major driver of cost. All things being equal, spatially uncorrelated water quality events in the bulk water supply system are lower cost, with fewer consumers impacted, if the system allows water to be supplied from uncontaminated sources.
    - The frequency of the breach – where breaches are unusual, there will be one-off incident costs (incident management, ad hoc testing, overtime, etc.). These costs can be in the order of $50,000 to $750,000 per event even where it is later established a breach did not cause impact to consumers (SCA [2008]). Where breaches are frequent, lower unit-cost-per-incident responses can be put in place (e.g., putting increased monitoring on routine programs rather than ad hoc delivery). See Miller *et al.* [2004], Pg. 44ff for a discussion of consequence scoring for raw water sources.

There are a wide range of considerations that apply in modelling costs, and these become more complex the greater the level of accuracy that is required. It has not been possible to cover all issues, but the following pragmatic approach has been applied.

Costing Approach

The approach taken is to give all analytes a cost, even when this cost cannot be accurately estimated. This approach is analogous to a Bayesian prior, and can be refined in subsequent development.

Analytes have been grouped according to the types of mitigation cost that they would normally incur, and an approximate cost allocated to that group. All costs are divided into three groupings: the cost of exceeding a minor threshold (labelled 1st), the cost of a single day exceedance of a major threshold (2nd) and the cost when consecutive days exceed the major threshold (3rd).

Note that all costs are on an incremental basis. That is, the cost does not include any baseline or routine monitoring or treatment, only the costs incremental (additional) to this baseline cost.

The key groups were:

Analytes with pathogenic impact but difficult to treat:

- Giardia and Cryptosporidium per event at each threshold 1st \$10,000, 2nd \$100,000, 3rd \$4,680,000. The huge cost increase at the higher thresholds is based on boil water costs to community and disease burden (see Sloane Cook and King [2004] and Jaguar Consulting [2004]). As noted above, these studies estimated total incident cost for these kinds of events in the tens and hundreds of millions, which gives very high daily and consecutive day costs for incidents lasting for a few months at the maximum.

Analytes treatable by disinfection:

- E. Coli as indicator of treatable pathogen presence. Cost per event at each threshold 1st \$1,000, 2nd \$15,000, 3rd \$10,000. This cost curve assumes incremental disinfectant and monitoring costs. For pathogens, the fact that the routine operational monitoring is frequent, means that the incremental monitoring cost for an incident is low).
- For other treatable pathogens, the incremental cost is assumed to be \$1,000 at first threshold only reflecting treatment, as monitoring of indicator pathogens would be covered by the other item.

Analytes not readily[1] treatable such as pesticides:

- Cost per event at each threshold 1st - \$5,000, 2nd \$10,000, 3rd \$30,000 which includes intensive monitoring and incident management to determine appropriate responses (e.g., Water source management, or changing farm or forest pesticide practices). These costs are based on the lower bound historical incident management cost cited above \$50,000. This is incremental monitoring only, but non-routine (cf. e.g., E.coli).
- True colour – No costs for any thresholds. Penalties related to this item were removed from the Bulk Water Supply Agreement with SCA's major customers in 2003.

---

[1] Some pesticides may be removed in small systems (cf. cyanobacteria) using activated carbon, e.g. US EPA. [2000]. However, in Australian coastal urban water systems, these treatments are not typically available in plant infrastructure because influent is normally high quality. Thus, activated carbon would have high incremental cost, .

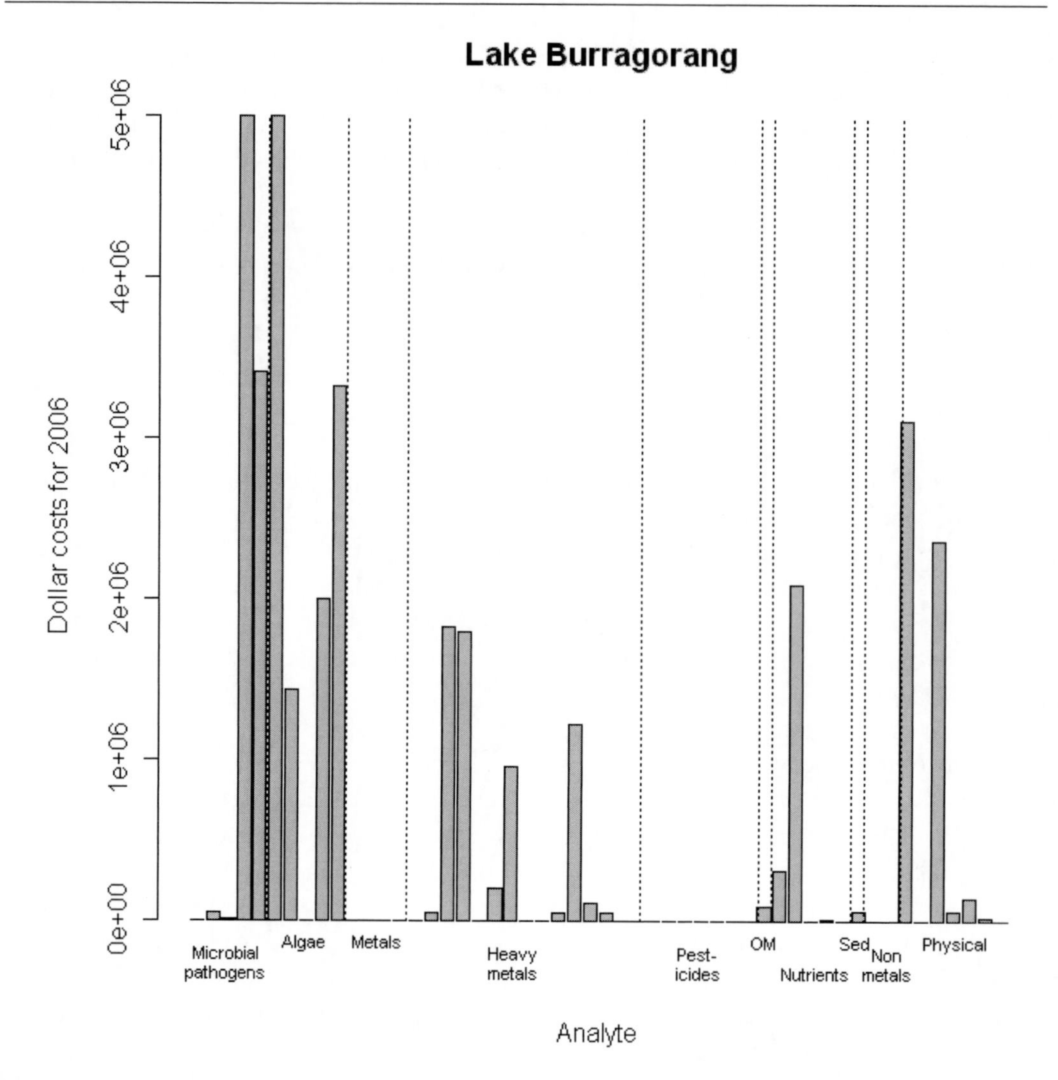

Figure 5. The estimated costs for each analyte at the major drinking water supply point of Lake Burragorang Dam Wall (OM denotes Organic matter and Sed denotes sediment (turbidity)).

Analytes with biological impact:

- Microcystin LR equiv, Toxic Cyanobacterial Count (cells/mL), Toxic Cyanobacterial biovolume (mm3/L) 1st $100,000, 2nd $200,000 and 3rd $1,000,000 involving treatment costs.
- This costing is based on applying activated carbon treatment if available, or where not, as in a major incident response, on the opportunity cost of loss of water use to consumers. No health impact has been assumed.
- Areal Standard Unit (algae) 1st $50,000, 2nd $150,000, 3rd $200,000, where the first threshold cost only includes extra monitoring and management costs. These relate to incident costs. The second and third thresholds include incremental treatment costs for algae, but do not assume any loss of supply to consumers (cf. toxic algae above).

- Chlorophyll–a – 1st $50,000. These relate to incident costs only, as explained above for the Areal Standard unit measure.

Analytes with chemical impact:

- This includes various forms of nitrogen, e.g., nitrate or ammonia. For these analytes, a single 1st of $10,000 was set, based on incident costings.

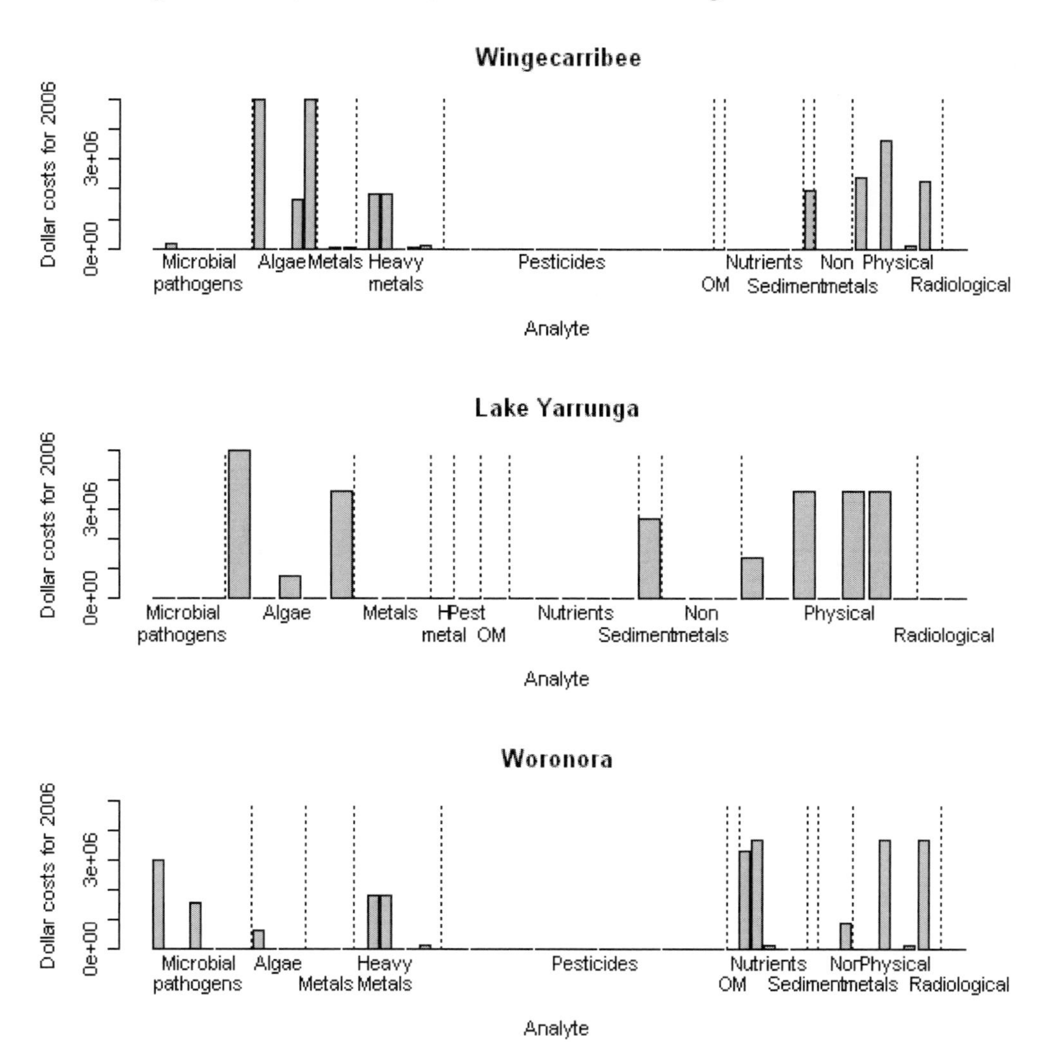

Figure 6. The estimated costs for each analyte at other places in the drinking water supply chain.

RESULTS

The results of applying the above theory are now presented. The annual risk estimates are provided for 2006, with likelihoods and consequence estimates being based on the interpolated values from each analyte's gamlss model, which were fitted using known flows for 2006, annual and daily harmonics and time as explanatory variables.

Figures 5 to 7 present the annual risk estimates for the SCA supply sites as barplots, with estimates truncated at 5 million dollars. The vertical dashed lines separate the analytes into their higher-level groupings, such as Microbial pathogens and Algae, as labelled. There are too many analytes to label the bars at the individual analyte level.

Instead, the analytes with high estimated annual risk are summarised in Table 1. Note that not all analytes are measured at each site.

Table 1. Summary of analytes with high estimated annual risk, by location

Water Quality Hazard	Analyte	Burragorang	Wingercarri bee	Lake Yarrunga	Woronora	Lake Prospect	Illawara	Nepean
mircrobial pathogens	Cryptosporidim	x						
	Giardia						x	
algae	areal standard unit	x	x	x	x	x	x	x
	toxic cyanobacterial	x	x	x				
heavy metals	barium		x		x			
	boron		x		x			
Organic	Total organic carbon	x						
Nutrients	nitrogen oxidised	x						x
sediments	turbidity		x	x				
Non-metals	sulphate				x			
	chlorine free						x	
physical	alkalinity		x	x		x		
	total hardness		x	x	x		x	x
	dissolved oxygen	x			x		x	x
	temperature	x		x				

The sites investigated are Burragorang, Wingecarribee, Lake Yarrunga, Woronora, Lake Prospect, Illawarra and Lake Nepean. The full list of the analytes is recorded in Table 2 in Appendix A together with their threshold values. The source of the threshold values are given in the last column of Table 2 in Appendix A.

Figure 5 presents the results for the supply site: Lake Burragorang @ 500m u/s Dam Wall (DWA2). The dam wall, which is 500 metres downstream from the Lake Burragorang measurement site, is the major source of drinking water supplied to Sydney.

Figures 6 and 7 present the results for upstream supply sites. These annual risk (cost) estimates are not assumed accurate in an absolute sense. Instead, they should be interpreted in a relative sense, with relatively high values being considered indicative of high-risk areas.

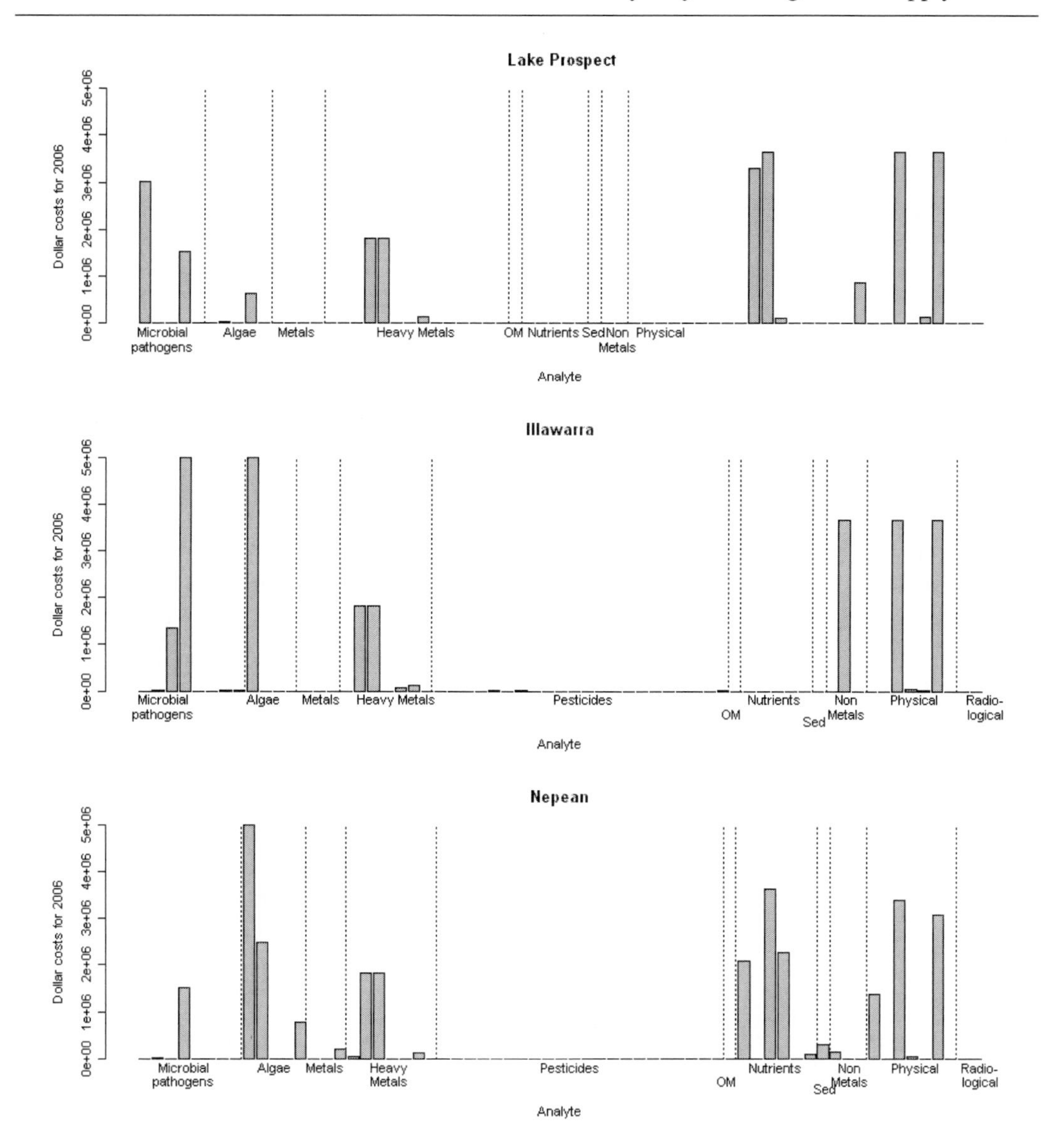

Figure 7. The estimated costs for each analyte at other places in the drinking water supply chain.

Table 1 presents the analytes posing high estimated annual risk for each supply site. There is a high risk from algae, especially areal standard unit algae, across all supply sites. This is in part due to the high cost of occasional blooms, but this analysis highlights a need to investigate the causes of the high algae levels and highlights that there are potentially large benefits to be gained from investing in preventative measures, as part of the water catchment management.

Exact interpretation and consideration of the biology is beyond the scope of this chapter and is left to experts in water ecology.

The other area of common high risk is the physical higher-level water quality hazard grouping. The analysis indicates that alkalinity and hardness both tend to have a high estimated risk (cost) and that it would be prudent to investigate possible management methods. These may be difficult of manage, considering the geology of the catchment area.

FUTURE RISK ASSESSMENT CHALLENGES

Estimating the costs associated with past events and using these to estimate future costs is made difficult by the multivariate nature of these events. Generally, analyte values are correlated with time, and therefore often several thresholds are exceeded simultaneously. In this chapter, cost estimates have been derived univariately, which is largely because the cost thresholds were assumed univariate. If water treatment costs are triggered by univariate threshold exceedances, then these costs may reflect accurately the nature of the operational costs. However, the true cost to society is unlikely to be reflected in univariate rectangular cost boundaries.

The challenge of defining thresholds that account for the multivariate nature of the problem is enormous. However, if true society risk estimates are required then working towards this objective is necessary. An initial approach to this end would be to approximate the cost boundaries by simple hyper-ellipsoidal lines and estimate the likelihoods under the assumption that analyte values are normally distributed, so that distributions of quadratic forms could be used to estimate the likelihood of such exceedances.

The current approach, at best, provides users with the analyte grouping that are more likely to exceed univariate quality thresholds. This is useful in that it helps decide which analytes need to be more closely monitored, however, the risk estimates derived in this chapter are not an accurate reflection of the operational costs resulting from poor water quality, and are at best interpreted in a relative sense.

Most catchment managers are interested in whether they are improving the condition of their catchment. However, this can only be efficiently achieved after correcting for changes in climate and environmental conditions from year to year. Risk adjusted control charts are used in health surveillance (e.g., Steiner *et al.* [2000]) and provide a methodology for achieving this aim. These charts would first correct for the changes in climate and the environment that are beyond the managers' control by removing their influence on the analyte trends. Examples for climate might be temperature, rainfall, flows, etc, and for the environment, might be the number of wildlife in the catchment, the number of livestock, percentage trees, percentage of land under agriculture, etc. Adjusted scores, derived from the adjusted trends, could then be interpreted as measures of improvement in the catchment that are attributable to the managers' actions.

CONCLUSION

This chapter examines the Sydney Catchment Authority's operating risk relating to treatment costs and costs of managing poor drinking water outbreaks for individual analytes. The risk for each analyte at each site was calculated by combining a fitted time-dependent model for the analyte that gave the distribution of analyte values for each day, with a cost profile for the analyte exceeding certain thresholds.

The analysis was performed on previously collected analyte measurements that were made irregularly and often infrequently, which posed modelling challenges. The modelling was performed using gamlss models. Estimates of the uncertainties in the likelihood and annual risk estimates were derived that accounted for autocorrelation in the analytes.

Some conceptual changes could improve the cost estimates. One improvement would be to consider consequences outside the Australian Drinking Water Guidelines, such as the health and other consequences for the customers of drinking the water or to consider the aquatic environmental concerns. Another possible improvement would be to treat risk as multivariate by attempting to estimate the true cost of multivariate hazard events rather than approximate it as the sum of multiple univariate hazard events. Both these objectives offer potential future research opportunities.

REFERENCES

Abera, S., Zeyinudin, A., Kebede, B., Deribew, A., Ali, S. and Zemene, E. (2011). Bacteriological analysis of drinking water sources. *African Journal of Mibrobiology Research.* 5: 2638-2641.

Chowdhury, S., Champagne, P. and McLellan, P.J. (2009). Uncertainty characterization approaches for risk assessment of DPBs in drinking water: a review. *J. Environ. Manage.* 90:1680-1691.

Cogley, JG (1999). Effective sample size for glacier mass balance. *Geografiska Annaler.* 81; 497-507.

Cross, A., Cantor, K.P., Reif, J.S., Lynch, C.F., Ward, M.H. (2004). Pancreatic cancer and drinking water and dietary sources of nitrate and nitrite. *American Journal of Epidemiology.* 159: 693-701.

Hales, S., Black, W., Skelly, C., Salmond, C., and Weinstein, P. (2003). Social deprivation and public health risks of community drinking water supplies in New Zealand. *J. Epidemiol. Community Health.* 2003;57:581-583 doi:10.1136/jech.57.8.581.

Harrington, W. Krupnick, A. J. Spofford W. O. (1991) Economics and episodic disease: the benefits of preventing a giardiasis outbreak Resources for the Future. Quality of the Environment Division.

Fawell, J and Nieuwenhuijsen, M.J. (2008), Contaminants in drinking water: Environmental pollution and health. *British Medical Bulletin.* 68 (1): 199-208.

Jaguar Consulting Pty Ltd (2004) Drinking Water Quality Regulatory Framework For Victoria Regulatory Impact Statement For The Safe Drinking Water Regulations, for Department of Human Services, Victoria, September.

Krewski, D., Balbus, J., Butler-Jones, D., Haas, C., Isaac-Renton, J., Roberts, K. and Sinclair, M. (2004). Managing the mirobiological risks of drinking water. *Journal of Toxicology and Environmental Health Part A.* 67: 1591-1617.

Miller, R, Guice, J and Deere, D. (2004) Risk Assessment for Drinking Water Sources CRC for Water Quality and Treatment, Research Report No 78 CRC for Water Quality and Treatment

Morris, R.D. (1995). Drinking water and cancer. *Environ. Health Perspect.* 103: 225-231.

Pavlov, D., de Wet, C.M.E., Grabow, W.O.K., Ehlers, M.M. (2004). Potentially pathogenic features of hterotrophic plate countbecteria isolated from treated and untreated drinking water. *International Journal of Food Microbiology.* 92: 275-287.

Rigby, R. A. and Stasinopoulos D. M. (2005). Generalized additive models for location, scale and shape,(with discussion), *Appl. Statist.*, 54, part 3, pp 507-554.

Rosen, L., Lindhe, A., Bergstedt, O., Norberg, T. and Pettersson, T.J.R. (2010). Comparing risk-reduction measures to reach water safety targets using an integrated fault tree model. *Water Science and Technology Water Supply.* 10: 428-436.

Sloane Cook and King Pty Ltd (2004) Prospect Raw Water Pumping Station Economic Appraisal Final Draft, NSW Department Of Commerce, May.

Sparks, RS, Sutton, G. and Toscas, P. (2011). "Modeling inverse Gaussian random variables when there are missing data and low detection limits. *Advances in Decision Sciences.* doi:10.1155/2011/571768.

Stasinopoulos D. M., Rigby R.A. and Akantziliotou C. (2006) Instructions on how to use the GAMLSS package in R. Accompanying documentation in the current GAMLSS help files, (see also http://www.gamlss.org/).

Stasinopoulos D. M. Rigby R.A. (2007) Generalized additive models for location scale and shape (GAMLSS) in R.*Journal of Statistical Software*, Vol. 23, Issue 7, Dec 2007, http://www.jstatsoft.org/v23/i07.

Steiner, S. H., Cook, R.J., Farewell, V.T., Treasure, T. (2000). Monitoring surgical performance using risk-adjusted cumulative sum charts. *Biostatistics.* 1: 441-452.

Thiebaux, HJ and Talkner, P (1984). On the interpretation and estimation of effective sample size. *J. Climate Appl. Meteorol.*, 23; 800-811.

Urbansky, E.T. and Schock, M.R. (1999). Issues in managing the risks associated with perchlorate in drinking water. *Journal of Environment Management.* 56: 79-95.

US EPA (2000). Summary of Pesticide Removal/Transformation Efficiencies from Various Drinking Water Treatment Processes - Prepared for the Committee to Advise on Reassessment And Transition (CARAT) October 3

van Grinsven, H.J., Ward, M.H., Benjamin, N., and de Kok , T.M. (2006). Does the evidence about health risks associated with nitrate ingestion warrant an increase of nitrate standard for drinking water? *Environmental Health.* 5:26. DOI: 10.1186/1476-069X-5-26.

Yang, C. (1998). Calcium and magnesium in drinking water and risk of death from cerebrovascular disease. *Stroke.* 29: 411-414.

ATTACHMENT A– ANALYTES FOR ASSESSMENT, FOLLOWED BY AN EXAMPLE DATA SET

Table 2. Analytes to be assessed under each water quality hazard group

Water Quality Hazard Group	Analyte	Concentration Limit for Assessment	Guideline
Microbial Pathogens	*E. coli* (cfu/100mL)	500	BWSA
	Coliforms Total (cfu/100mL)	1000	BWQIRP
	Enterococci (cfu/100mL)	35	ANZECC
	Cryptosporidium (oocysts/100L)	1	BWQIRP
	Giardia (cysts/100L)	1	BWQIRP
	Adenovirus	1	ADWG
	Enterovirus	1	ADWG
	F-RNA Phage	1	ADWG

Water Quality Hazard Group	Analyte	*Concentration Limit for Assessment*	Guideline
	Reovirus (Reoviridae)	1	ADWG
Algae	Areal Standard Unit (algae)	500	BWSA
	Chlorophyll a (µg/L)	5	ANZECC
	Microcystin LR equiv (µg/L)	1.3	BWQIRP
	Toxic Cyanobacterial Biovolume (mm^3/L)	0.4	BWQIRP
	Toxicigenic Cynobacterial Count (cells/mL)	5000	BWQIRP
Metals	Aluminium (acid soluble) (mg/L)	0.2	ADWG
	Aluminium Total (mg/L)	1.0	BWQIRP
	Iron Total (mg/L)	0.3	ADWG
	Manganese Total (mg/L)	0.1	ADWG
	Sodium Total (mg/L)	180	ADWG
Heavy Metals	Total Antimony (mg/L)	0.003	ADWG
	Total Arsenic (mg/L)	0.007	ADWG
	Total Barium (mg/L)	0.7	ADWG
	Total Boron (mg/L)	4	ADWG
	Total Cadmium(mg/L)	0.002	ADWG
	Total Chromium (mg/L)	0.05	ADWG
	Total Copper (mg/L)	2.0	ADWG
	Total Lead (mg/L)	0.01	ADWG
	Total Mercury (mg/L)	0.001	ADWG
	Total Molybdenum (mg/L)	0.05	ADWG
	Total Nickel (mg/L)	0.02	ADWG
	Total Selenium (mg/L)	0.01	ADWG
	Total Silver (mg/L)	0.1	ADWG
	Total Uranium (mg/L)	0.02	ADWG
	Total Zinc (µg/L)	8	ANZECC
Pesticides	Aldrin (µg/L)	0.01	ADWG
	Amitrole (µg/L)	1	ADWG
	Atrazine (µg/L)	0.1	ADWG
	Chlordane (µg/L)	0.01	ADWG
	Chlorpyrifos (µg/L)	0.1	ADWG
	Clopyralid (µg/L)	1000	ADWG
	2,4 Dichlorophenooxyacetic acid (µg/L)	0.1	ADWG
	DDT (µg/L)	0.06	ADWG
	Dieldrin (µg/L)	0.01	ADWG
	Diquat (µg/L)	0.5	ADWG
	Diuron (µg/L)	30	ADWG
	Total Endosulfan (µg/L)	0.05	ADWG
	Heptachlor (µg/L)	0.05	ADWG

Table 2. (Continued)

Water Quality Hazard Group	Analyte	Concentration Limit for Assessment	Guideline
	Hexazinone (μg/L)	2	ADWG
	Lindane (μg/L)	0.05	ADWG
	Molinate (μg/L)	0.5	ADWG
	Paraquat (μg/L)	1	ADWG
	Picloram (μg/L)	300	ADWG
	Propiconazole (μg/L)	0.1	ADWG
	Temephos (μg/L)	300	ADWG
	Triclopyr (μg/L)	10	ADWG
	2,4,5-T	0.05	ADWG
Organic Matter	Total Organic Carbon (mg/L)	10	SCA Science and Research
Nutrients	Ammonia (as NH3) (μS/cm)	0.5	ADWG
	Ammonium (mg/L)	0.01	ANZECC
	Nitrate as N (mg/L)	50	ANZECC
	Nitrogen Oxidised (mg/L) (NOx)	0.01	ANZECC
	Nitrogen Total (mg/L)	0.35	ANZECC
	Phosphorus Filterable (mg/L)	0.005	ANZECC
	Phosphorus Total (mg/L)	0.01	ANZECC
Sediment	Turbidity Field (NTU)	40	BWQIRP
	Total Dissolved Solids (mg/L)	1000	ANZECC
Non-metals	Chloride (mg/L)	250	ADWG
	Chlorine Free (mg/L)	0.6	ADWG
	Cyanide (mg/L)	0.08	ADWG
	Iodide (mg/L)	0.001	BWQIRP
	Sulphate (mg/L)	250	ADWG
Non-metals	Chloride (mg/L)	250	ADWG
	Chlorine Free (mg/L)	0.6	ADWG
	Cyanide (mg/L)	0.08	ADWG
	Iodide (mg/L)	0.001	BWQIRP
	Sulphate (mg/L)	250	ADWG
Physical	Alkalinity (mg as CaCO3/L)	15 - 60	BWSA
	Conductivity Field (μS/cm)	350	ANZECC
	Dissolved Oxygen (%Sat)	90-110	ANZECC
	pH (Field)	5.5 – 8.5	BWQIRP
	Temperature (Deg C)	5.0-25	BWSA
	Total Hardness (mgCaCO3/L)	25 – 70	BWSA

Water Quality Hazard Group	Analyte	Concentration Limit for Assessment	Guideline
	True Colour at 400nm	40	BWQIRP
Radiological	Gross alpha emitters (Bq/L)	0.5	ADWG
	Gross beta emitters (Bq/L)	0.5	ADWG

BWSA = Bulk Water Supply Agreement; ADWG = Australian Drinking Water Guidelines; ANZECC = ANZECC Water Quality Guidelines; BWQIRP = Bulk Water Quality Incident Response Plan.

In: Water Quality
Editor: You-Gan Wang

ISBN: 978-1-62417-111-6

Chapter 3

THREE-DIMENSIONAL NUMERICAL MODELING OF WATER QUALITY AND SEDIMENT-ASSOCIATED PROCESSES IN NATURAL LAKES

Xiaobo Chao*[*] *and Yafei Jia
National Center for Computational Hydroscience and Engineering,
The University of Mississippi, US

ABSTRACT

This chapter presents the development and application of a three-dimensional water quality model for predicting the distributions of nutrients, phytoplankton, dissolved oxygen, etc., in natural lakes. In this model, the computational domain was divided into two parts: the water column and the bed sediment layer, and the water quality processes in these two domains were considered. Three major sediment-associated water quality processes were simulated, including the effect of sediment on the light intensity for the growth of phytoplankton, the adsorption-desorption of nutrients by sediment and the release of nutrients from bed sediment layer. This model was first verified using analytical solutions for the transport of non-conservative substances in open channel flow, and then calibrated and validated by the field measurements conducted in a natural oxbow lake in Mississippi. The simulated concentrations of water quality constituents were generally in good agreement with field observations. This study shows that there are strong interactions between sediment and water quality constituents.

INTRODUCTION

Sediment is a major nonpoint source pollutant. It may be transported into surface water bodies from agricultural lands and watersheds through runoff. These sediments could be associated with nutrients, pesticides, and other pollutants, and greatly affect the surface water qualities. Therefore, sediment has been listed as the most common pollutant in rivers,

[*] E-mail address: chao@ncche.olemiss.edu.

streams, lakes, and reservoirs by the US Environmental Protection Agency (USEPA). Sediment creates turbidity in water bodies, reducing the light intensity in water columns, which is one of the most important factors affecting the phytoplankton growth. In water bodies, nutrients can interact with sediment particles through the processes of adsorption and desorption. In addition, nutrients in bed sediments may be released into the water column. Those sediment-associated processes play important roles in water quality interaction systems.

Numerical modeling is a very effective approach to study water quality constituents in surface water bodies. In recent years, some well-established three dimensional models, such as WASP6[1], CE-QUAL-ICM[2], Delft3D-WAQ[3], MIKE3_WQ[4], MOHID[5], EFDC [6], etc., have been used to simulate water quality constituents in rivers, lakes, and coastal waters. These models generally cover basic physical, chemical and biological processes of aquatic ecosystems. However, only a few are capable of simulating the effects of sediment on the water quality. In WASP6, CE-QUAL-ICM, and MIKE3-WQ, the effect of suspended sediment (SS) concentration on the growth of phytoplankton was not taken into account. The processes of adsorption-desorption were not simulated in MIKE3-WQ and MOHID. In WASP6, CE-QUAL-ICM and EFDC, the adsorption-desorption of nutrients by sediment were described using a simple linear isotherm. In the above mentioned models, the release rate of nutrients from the bed sediment was determined based on the concentration gradient across the water-sediment interface; however, the effects of pH and dissolved oxygen concentration on the release rate were not considered. In view of the limited understanding of the sediment-associated water quality processes, the accuracy of simulations using the aforementioned models has room for improvement.

The development and application of a three-dimensional model for simulating the water quality constituents in natural lakes were presented. The computational domain was divided into two parts: the water column and the bed sediment layer. This model was decoupled with a three-dimensional free surface hydrodynamics model CCHE3D [7], and the major sediment-associated processes were simulated, including the effect of sediment on light penetration, the adsorption-desorption of nutrients by sediment and the release of nutrients from bed sediment.

This model was first verified by a mathematic solution consisting of the movement of a non-conservative tracer in a prismatic channel with uniform flow, and the numerical results agreed well with the analytical solutions. Then it was applied to a real case to validate its capability for simulating the concentrations of phytoplankton and nutrients in Deep Hollow Lake, a small oxbow lake in the Mississippi Delta. The concentrations of water quality constituents obtained from the numerical model were generally in good agreement with observations.

This chapter presents detailed technical information on the modeling of water quality and sediment-associated processes in natural lakes. The general water quality processes considered in the numerical model, including the phytoplankton kinetics, nitrogen cycle, phosphorus cycle, dissolved oxygen balance, and processes in benthic sediment layer are described first, and then the sediment-associated water quality processes are presented. The detailed information on the development, verification, and application of a three-dimensional water quality model are given in the following three Sections. The discussion and conclusion are described in the last two Sections.

WATER QUALITY PROCESSES

Water quality is scaled by a combination of many chemical variables, which interact based on laws of chemistry and bio-chemistry under natural conditions. The interactions of water quality constituents in water column and sediment layer are shown in Figure 1.

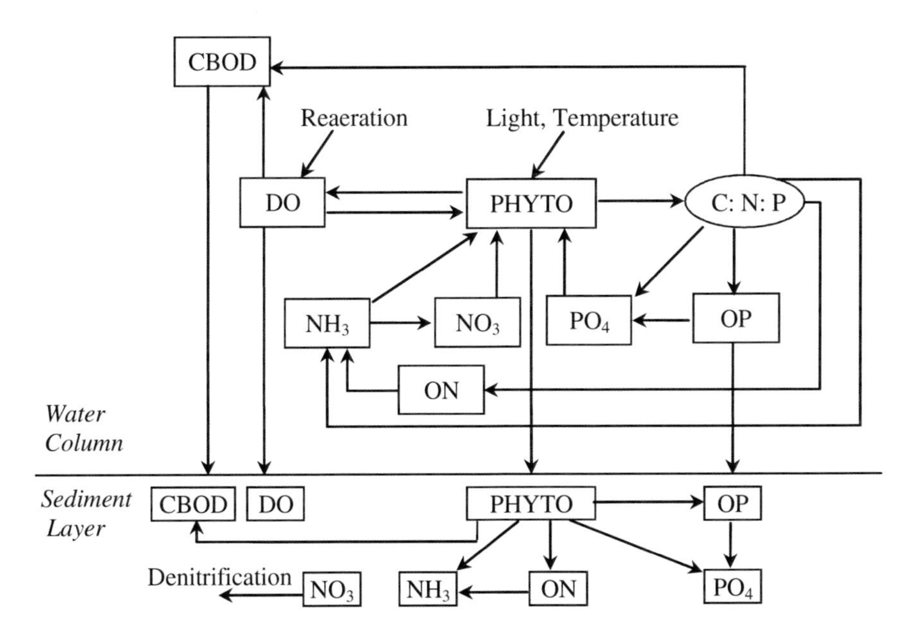

Figure 1. Interactions of water quality constituents in water column and sediment layer.

In the water column, four biochemical processes were considered: the phytoplankton kinetics, nitrogen cycle, phosphorus cycle, and dissolved oxygen (DO) balance.

The conceptual framework for the eutrophication kinetics was mainly based on the WASP6 model [1]. Eight state variables were involved in the interacting systems: ammonia nitrogen (NH_3), nitrate nitrogen (NO_3), phosphate (PO_4), phytoplankton (PHYTO), carbonaceous biochemical oxygen demand (CBOD), DO, organic nitrogen (ON), and organic phosphorus (OP).

In bed sediment layer, the decomposition of organic material releases nutrients to the sediment porous water and also results in the exertion of oxygen demand at the sediment-water interface. Similar to the water column, eight state variables are involved, including NH_3, NO_3, PO_4, PHYTO, CBOD, DO, ON, and OP.

Phytoplankton Kinetics

Phytoplankton are microscopic plant-like organisms that live in water environments, play a central role in aquatic ecosystem. Via photosynthesis, phytoplankton produce organic compounds from inorganic nutrients. They consume nutrients, including inorganic nitrogen, phosphate, silica, and carbon dioxide from the water and release oxygen as a by-product to the water. The source term S_m due to phytoplankton growth and reduction is calculated by

$$S_m = (G_p - D_p - P_{set})M \tag{1}$$

where M is the phytoplankton biomass; G_p is the growth rate of phytoplankton (day^{-1}); D_p is the reduction rate of phytoplankton (day^{-1}); and P_{set} is the effective phytoplankton settling rate(day^{-1}). In the model, the total chlorophyll is used as a simple measure of phytoplankton biomass. The relationship between the chlorophyll and phytoplankton biomass can be expressed as

$$M = \alpha C_{chl} \tag{2}$$

in which C_{chl} is the chlorophyll concentration (mg/l); α is the carbon to chlorophyll ratio.

The growth rate of phytoplankton is determined by the availability of nutrients, the intensity of light, and by the ambient temperature. The effects of each factor are considered to be multiplicative:

$$G_p = P_{mx} f_N f_I f_T \tag{3}$$

in which f_N, f_I and f_T are the limitation factors due to nutrient availability, light intensity, and temperature, respectively; P_{mx} is the maximum phytoplankton growth rate (day^{-1}).

The limitation factor f_N is calculated based on Michaelis-Menten Equation and Liebig's law of the minimum [1]:

$$f_N = \min\left(\frac{C_{NH3} + C_{NO3}}{C_{NH3} + C_{NO3} + K_{mN}}, \frac{C_{PO4}}{C_{PO4} + K_{mP}}\right) \tag{4}$$

in which C_{NH3}, C_{NO3} and C_{PO4} are the concentrations of NH_3, NO_3 and PO_4, respectively (mg/l); K_{mN} and K_{mP} are the half-saturation constants for nitrogen and phosphorus uptake, respectively(mg/l).

The limitation factor f_I is calculated by integrating the Steele equation over water depth and time [8]:

$$f_I = \frac{2.72 f_d}{K_e \Delta z}\left[\exp\left(-\frac{I_0}{I_m} e^{-K_e(z_d+\Delta z)}\right) - \exp\left(-\frac{I_0}{I_m} e^{-K_e \cdot z_d}\right)\right] \tag{5}$$

where f_d is the photoperiod; Δz is the vertical thickness of a computational element (m); z_d is the distance from the water surface to the top of a computational element in the water column (m); K_e is the total light attenuation coefficient (m^{-1}), it is determined by the pure water and the concentrations of chlorophyll and suspended sediment in the water column; I_0 is the daily light intensity at the water surface (ly/day); I_m is the saturation light intensity of phytoplankton (ly/day).

The field observations [2,6] show that there is an optimum temperature T_m (^oC) for the growth of phytoplankton. When the temperature T (^oC) is below T_m, the growth rate of phytoplankton increases as a function of T; while the growth rate decreases when the temperature T is above T_m. The limitation factor f_T can be calculated using formulas proposed by Cerco and Cole [2]:

$$f_T = \exp\left[-KTg_1(T - T_m)^2\right] \quad \text{when } T \le T_m \tag{6}$$

$$f_T = \exp\left[-KTg_2(T_m - T)^2\right] \quad \text{when } T > T_m \tag{7}$$

in which KTg_1 and KTg_2 are coefficients representing the effects of temperature on the phytoplankton growth below and above T_m, respectively.

Phytoplankton losses mainly include endogenous respiration, death and grazing by zooplankton. The reduction rate of phytoplankton is given as follows:

$$D_p = k_{pr}\theta_{pr}^{T-20} + k_{pd} + k_{pzg}C_{zoo}\theta_{pzg}^{T-20} \tag{8}$$

where k_{pr} and k_{pd} are the rates of endogenous respiration and death, respectively (day^{-1}); k_{pzg} is the zooplankton grazing rate (*l/mgC/day*); C_{zoo} is the concentration of zooplankton(*mgC/l*); θ_{pr} and θ_{pzg} are the temperature coefficients.

The effective phytoplankton settling rate P_{set} can be given as:

$$P_{set} = \frac{\omega_{s4}}{D_j} \tag{9}$$

where ω_{s4} is the settling velocity of phytoplankton (*m/day*) and D_j is the depth of element j (*m*).

Nitrogen Cycle

The major components of nitrogen cycle in aquatic environments include: ammonia nitrogen (NH_3), nitrate nitrogen (NO_3), phytoplankton (PHYTO) and organic nitrogen (ON).

NH_3 and NO_3 are consumed by phytoplankton for its growth. Due to physiological reasons, NH_3 is the preferred form of inorganic nitrogen for phytoplankton. Nitrogen is returned from the phytoplankton biomass pool to particulate and dissolved organic nitrogen pools as a result of phytoplankton death, endogenous respiration and zooplankton grazing. ON is converted to NH_3 at a temperature-dependent mineralization rate, and NH_3 is then converted to NO_3 at a temperature- and oxygen-dependent nitrification rate. In the absence of oxygen, NO_3 can be converted to nitrogen gas (denitrification) at a temperature-dependent rate. Nitrogen may interact with sediment through the processes of adsorption and desorption. Dissolved inorganic and organic nitrogen at the bed sediment layer may also be released to the water column under certain conditions.

In the numerical model, the kinetic source terms for NH_3, NO_3 and ON were calculated using formulas presented by Wool et al. [1].

Phosphorus Cycle

The major components of phosphorus cycle in aquatic environments includes: phosphate (PO_4), phytoplankton (PHYTO) and organic phosphorus (OP).

Phosphorus kinetics are basically similar to the nitrogen kinetics except there is no process analogous to denitrification. PO_4 is utilized by phytoplankton for its growth and is incorporated into phytoplankton biomass. Phosphorus is returned to the water column from dead or decaying phytoplankton biomass in the bed sediment. The various forms of OP undergo settling, hydrolysis and mineralization, and are converted to inorganic phosphorus at temperature-dependent rates. In addition, phosphorus may interact with sediment through the processes of adsorption and desorption. Dissolved inorganic and organic phosphorus at the bed sediment layer may also be released to the water column under certain conditions.

In the numerical model, the kinetic source terms for PO_4 and OP were calculated using formulas summarized by Wool et al. [1].

Dissolved Oxygen Balance

Dissolved oxygen (DO) is one of the most important parameters in water quality analysis and is used to measure the amount of oxygen available for biochemical activity in waters. Five water quality constituents, including ammonia nitrogen (NH_3), nitrate nitrogen (NO_3), phytoplankton (PHYTO), carbonaceous biochemical oxygen demand (CBOD) and DO, are involved in the DO processes. The decomposition of organic material in benthic sediment can greatly reduce the concentration of DO in the water column. In the model, sediment oxygen demand (SOD) is used to calculate the sink of oxygen due to this process.

The level of DO is increased due to the atmospheric reaeration and photosynthesis of aquatic plants; and decreased due to phytoplankton respiration, oxidation of CBOD, nitrification and sediment oxygen demand (SOD).

CBOD is a measure of oxygen demand consumed by bacteria in the oxidation of organic (carbonaceous) matters present in water. The major source of CBOD is the result of phytoplankton death. The loss of CBOD usually results from the settling of the particulate CBOD, oxidation of the carbonaceous materials and the denitrification reaction under low DO conditions.

The kinetic source terms for DO and CBOD can be calculated using formulas presented by Wool et al. [1] and Chapra [8].

Processes in Bed Sediment Layer

For simplicity, the bed sediment layer is represented using one layer in the model. Similar to the water column, eight state variables including NH_3, NO_3, PO_4, PHYTO, CBOD, DO, ON, and OP are involved in the bed sediment layer.

The decomposition of phytoplankton releases organic nitrogen and organic phosphorus based on the ratios of nitrogen/carbon and phosphorus/carbon, respectively. The ON and OP are converted to NH_3 and PO_4 at temperature-dependent decomposition rates. NH_3 and PO_4

can also be generated by the anaerobic decomposition of phytoplankton. In addition NO_3 can be converted to nitrogen gas at a temperature-dependent rate due to denitrification.

The anaerobic decomposition of phytoplankton releases CBOD based on the ratio of oxygen/carbon. Under the anoxic condition, the denitrification reaction provides a sink for CBOD. The decomposition reaction of benthic organic carbon causes reductions of CBOD and DO.

The water quality constituents in this layer may be exchanged with overlying water column due to the processes of diffusion, resuspension and settling.

Sediment-Associated Water Quality Processes

The field observations conducted by Stefan et al.[9] show that suspended sediment reduces light penetration in water column. Therefore it reduces the growth of phytoplankton by limiting the amount of solar energy available for photosynthesis. The processes of adsorption-desorption of nutrients by suspended sediment are also important for the fate and transport of nutrients. In addition, nutrients may exchange between bed sediment layer and water column due to diffusion. To study these complex processes, the effects of sediment on the phytoplankton and nutrients were considered, and three major sediment-associated water quality processes, including the effect of sediment on the growth of phytoplankton, the adsorption-desorption of nutrients by sediment and the release of nutrients from bed sediment layer, were taken into account.

Effect of Sediment on the Growth of Phytoplankton

The light attenuation coefficient K_e in Eq.(5) is determined by the effects of water, chlorophyll and SS, and can be expressed by the formula proposed by Stefan et al. [9]:

$$K_e = K_0 + K_{chl} + K_{ss} \tag{10}$$

where K_0 is the light attenuation by pure water (m^{-1}); K_{chl} is the attenuation by chlorophyll (m^{-1}); K_{ss} is the attenuation by SS (m^{-1}). Wool et al. [1] proposed a formula to calculate K_e based on the light attenuation due to pure water and chlorophyll without considering the effect of SS:

$$K_e = K_0 + 0.0088C_{chl} + 0.054C_{chl}^{0.67} \tag{11}$$

in which C_{chl} is the concentration of chlorophyll (mg/l).

In fact, SS increases both water surface reflectivity and light attenuation. Field measurements by Stefan et al. [9] showed that the attenuation by SS can be given by:

$$K_{ss} = \gamma S \tag{12}$$

where s is the concentration of suspended sediment; γ is the attenuation parameter due to suspended sediment. Combining Eqs. (11) and (12), Eq. (10) can be written as:

$$K_e = K_0 + 0.0088C_{chl} + 0.054C_{chl}^{0.67} + \gamma s \tag{13}$$

The parameter K_0 and γ can be obtained based on field measurements.

Processes of Adsorption-Desorption of Nutrients by Sediment

Mathematical Descriptions

Adsorption and desorption are important processes between dissolved nutrients and suspended sediment in the water column. Adsorption is a process in which dissolved nutrients become associated with suspended particles. Desorption is the reverse process and thus refers to the release of adsorbed nutrients from suspended sediment.

Therefore, nutrients can be directly transported in the flow, they can also be transferred from the dissolved phase to the particulate phase associated with sediment and then transported with sediment in the flow.

In water quality processes, the reaction rates for adsorption-desorption are much faster than that for the biological kinetics, so an equilibrium assumption can be made [1,2]. It is assumed that the interaction of the dissolved and particulate phases reach equilibrium instantaneously in response to nutrient and sediment inputs so as to redistribute nutrients between dissolved and solid-phase compartments.

Thus, the processes of adsorption-desorption are assumed to reach equilibrium at each time step in the numerical simulation.

In some models, the adsorption-desorption of nutrients by sediment is described by a linear isotherm, and the ratio of particulate and dissolved nutrient concentration is assumed as a constant [1, 10]. However, most experimental results show that the Langmuir equilibrium isotherm is a better representation of the relations between the dissolved and particulate nutrient concentrations [11,12,13].

It is assumed that the volume of the nutrient/water/sediment mixture solution is V_0, which is a constant before and after adsorption. C_0 is the total initial nutrient concentration at each time step, and s is the sediment concentration. The concentrations of dissolved and particulate nutrients are defined as:

$$C_d = \frac{M_d}{V_0} \tag{14}$$

$$C_p = \frac{M_p}{V_0} \tag{15}$$

where C_d and C_p are the concentrations of dissolved and particulate nutrients when the adsorption reaches equilibrium; M_d and M_p are the masses of dissolved and particulate nutrients, respectively.

The total initial amount of nutrients in the solution is same as when the adsorption reaches equilibrium, so the following equation can be obtained:

$$C_0 V_0 = C_d V_0 + C_p V_0 \tag{16}$$

$$C_0 = C_d + C_p \tag{17}$$

The equilibrium adsorption content (Q) is defined as:

$$Q = \frac{M_p}{M_s} = \frac{M_p}{sV_0} = \frac{C_p}{s} \tag{18}$$

where M_s is the mass of sediment in water column. In the proposed model, the Langmuir equation was adopted to calculate the adsorption and desorption rate [8,13]. The equilibrium adsorption content (Q) can be expressed as:

$$Q = \frac{Q_m K C_d}{1 + K C_d} \tag{19}$$

where Q_m is the maximum adsorption capacity; and K is the ratio of adsorption and desorption rate coefficients. Based on Eqs. (17) and (18), the following equations can be obtained:

$$C_p = sQ \tag{20}$$

$$C_d = C_0 - C_p = C_0 - sQ \tag{21}$$

By substituting Eqs.(18) and (21) into Eq.(19) and simplifying, a quadratic equation is obtained

$$C_p^2 - \left(C_0 + \frac{1}{K} + sQ_m \right) C_p + C_0 s Q_m = 0 \tag{22}$$

By using the quadratic formula, C_p can be solved as:

$$C_p = \frac{1}{2} \left[\left(C_0 + \frac{1}{K} + sQ_m \right) \pm \sqrt{\left(C_0 + \frac{1}{K} + sQ_m \right)^2 - 4 C_0 s Q_m} \right] \tag{23}$$

In order to determine the "+" or "–" sign to be used in front of the radical in Eq (23), this equation was rewritten as:

$$C_p = \frac{1}{2}\left[\left(C_0 + \frac{1}{K} + sQ_m\right) \pm \sqrt{\left(C_0 + \frac{1}{K} - sQ_m\right)^2 + \frac{4sQ_m}{K}}\right] \tag{24}$$

In Eq. (24), $\sqrt{\left(C_0 + \frac{1}{K} - sQ_m\right)^2 + \frac{4sQ_m}{K}} > \left(C_0 + \frac{1}{K} - sQ_m\right)$. If the "+" is adopted, then $C_p > C_0$.

Based on Eq. (21), it is known that C_p has to be less than C_0. So in Eq. (24), only the "–" sign is possible, and it becomes

$$C_p = \frac{1}{2}\left[\left(C_0 + \frac{1}{K} + sQ_m\right) - \sqrt{\left(C_0 + \frac{1}{K} - sQ_m\right)^2 + \frac{4sQ_m}{K}}\right] \tag{25}$$

Based on Eq (21), C_d can be expressed as:

$$C_d = \frac{1}{2}\left[\left(C_0 - \frac{1}{K} - sQ_m\right) + \sqrt{\left(C_0 + \frac{1}{K} - sQ_m\right)^2 + \frac{4sQ_m}{K}}\right] \tag{26}$$

Eqs. (25) and (26) are used to calculate the concentration of particulate and dissolved nutrients when the adsorption reaches equilibrium. The two equations show the concentrations of particulate and dissolved nutrients due to adsorption-desorption are determined by the total initial concentration of nutrients C_0, the adsorption constants K and Q_m, and the suspended sediment concentration s. As C_0 and s may vary with time, the ratio of C_p and C_d is not a constant.

Comparison with Experimental Measurements

A laboratory experiment was conducted by Bubba et al. [13] to study adsorption processes of ortho-phosphorus by sediment. Figure 2 shows the measured concentrations of dissolved and particulate phosphate at equilibrium. The results obtained based on the Langmuir equation (Eq.19) and linear isotherms with ratios for the particulate to dissolved phosphate ranges from 0.01 to 0.5 suggested by some researchers [1,2] were also plotted for comparison. It can be observed that the Langmuir equation is generally in good agreement with the full range of the data, but the linear isotherm appears to be valid only when the concentration of dissolved phosphate is very low, producing large errors outside this range.

Eqs. (25) and (26) were tested using this experimental measurement case. Based on the measured data, the maximum adsorption capacity Q_m and the Langmuir adsorption constant K were 0.7 l/mg and 5.1×10^{-3} mg P/mg, respectively.

At equilibrium adsorption conditions, the concentrations of SS, the particulate and dissolved phosphate under different initial concentrations of 2.5, 5, 10, 20, and 40 mg/l were measured.

Two Stations' data (Birkesig and Vestergard) were compared with computation results obtained by Eqs. (25) and (26). Figures 3a and 3b show that the computed concentrations of particulate and dissolved phosphate are generally in good agreement with measurements.

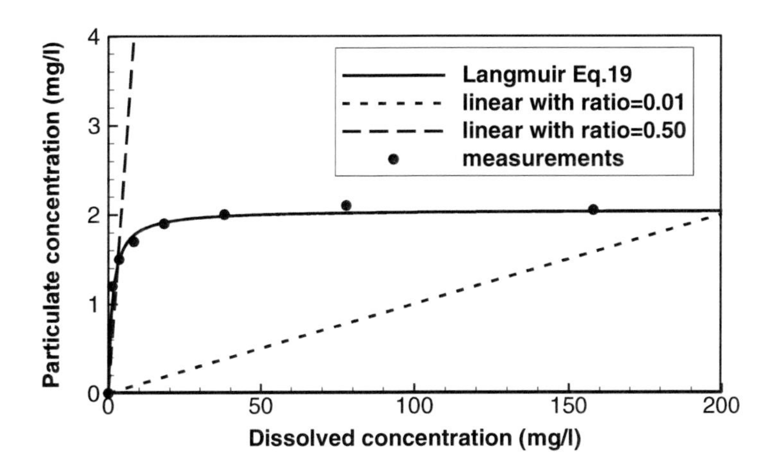

Figure 2. Relations between dissolved and particulate phosphate concnetrations obtained by Langmuir equation and linear assumption.

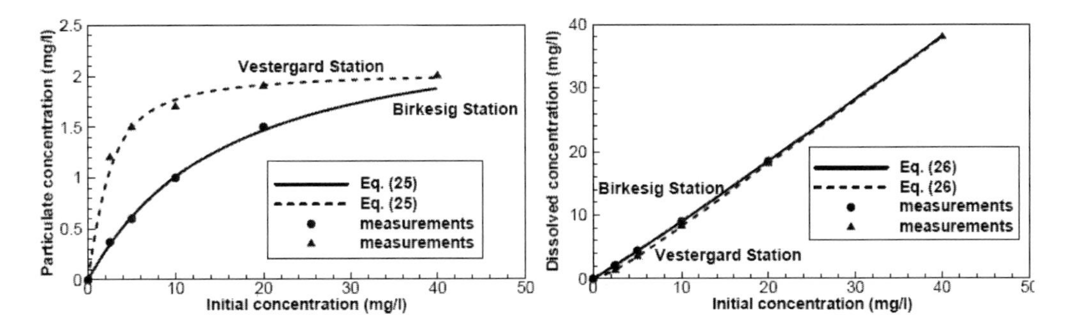

Figure 3. Concnetration of phosphate versus different initial concnetration at equilibrium.

Release of Nutrients from Bed Sediment

Mathematical Descriptions

The process of the decomposition of organic material in bed sediment can release nutrients to the sediment interstitial waters and remove oxygen from the overlying water. As a result, the bed release is important sources of nutrients in the water column.

In many numerical models, the release rate of nutrients from bed sediment is determined based on the concentration gradient across the water-sediment interface [1,2,4,5]. In fact, the bed release rate is also affected by pH, temperature and dissolved oxygen concentration [14,15]. Based on Romero [16], the bed release rate can be expressed as:

$$S_{diff} = \theta_{sed}^{T-20} S_c \left(\frac{K_{dos}}{K_{dos} + DO} + \frac{|pH - 7|}{K_{pHs} + |pH - 7|} \right) \tag{27}$$

where S_{diff} is the bed release rate (mg/m^2day); S_c is the diffusive flux of nutrients (mg/m^2day); K_{dos} (mg/l)and K_{pHs} are the values that regulate the release of nutrient according to the dissolved oxygen (DO) and pH in the bottom layer of the water column of depth Δz_b (m); θ_{sed} is the temperature coefficient. The diffusive flux S_c can be calculated using Fick's first law which expresses that the flux is directly proportional to the concentration gradient and the porosity of sediment [17,18]:

$$S_c = -\phi D_m \frac{dC}{dz} \approx \frac{\phi D_m}{\Delta z_b}(C_b - C_w) = k(C_b - C_w) \tag{28}$$

where D_m is the molecular diffusivity (m^2/day); ϕ is the porosity of sediment; Δz_b is the diffusive sub-layer thickness near the bed (m); k is the diffusive exchange coefficient at water-sediment interface (m/day); C_w and C_b are the concentration of nutrients in water and water-sediment interface, respectively.

Steinberger and Hondzo [19] investigated the factors affecting k and established an empirical relation for diffusional transfer of dissolved oxygen across the bed surface. Their studies show that k is governed by the Reynolds and Schmidt numbers. Based on their studies, k is expressed by

$$k = \begin{cases} 0.012 \dfrac{\phi D}{\Delta z} \left(\dfrac{U \Delta z}{\nu} \right)^{0.89} \left(\dfrac{\nu}{D} \right)^{0.33} & U \neq 0 \\ \dfrac{\phi D}{\Delta z} & U = 0 \end{cases} \tag{29}$$

where U = depth averaged velocity; and ν = kinematic viscosity. This formula takes the flow conditions into account for estimating the exchange coefficient k.

Comparison with Experimental Data

A laboratory experiment was conducted by Kim et al.[14] to study the effects of pH and DO on the phosphorus release rate at Jamsil Submerged Dam (JSD) Station in Han River. Eq.(27) was tested using the experiment measurements. Figures 4 and 5 show the comparisons between the computational results by Eq. (27) and experimental measurements.

In general, the effects of pH and DO on the phosphorus release rate are reasonably predicted by Eq. (27). The experimental and computational results show the release rate of phosphorus decreases with the increase of DO. Under acidic conditions, the release rate decreases as pH increases, while under basic conditions, the release rate increases as pH increases. These trends are similar to other experimental measurements [10,20,21].

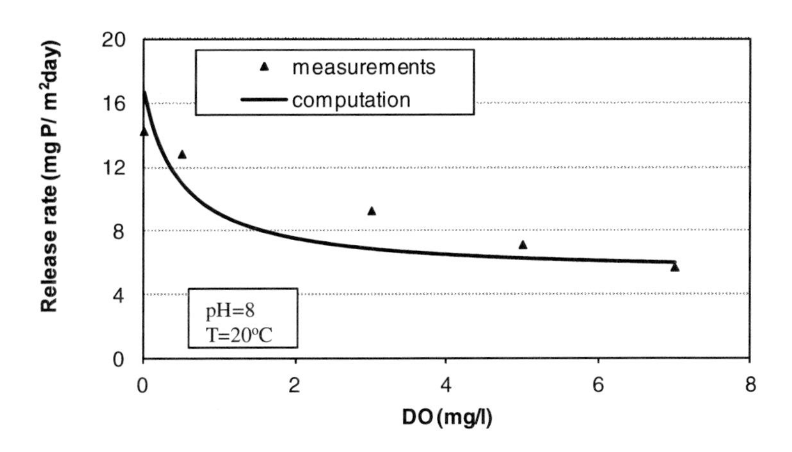

Figure 4. The effect of DO on the release rate of phosphorus at JSD Station in Han River.

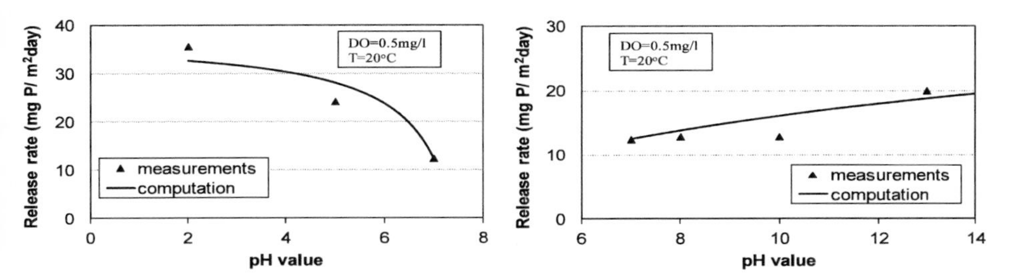

Figure 5. The effect of pH on the release rate of phosphorus at JSD Station in Han River.

NUMERICAL MODEL DEVELOPMENT

The proposed water quality model (CCHE3D_WQ) was developed based on CCHE3D hydrodynamic model, a three dimensional model developed at the National Center of Hydroscience and Engineering, the University of Mississippi [7]. CCHE3D is a three-dimensional model that can be used to simulate unsteady turbulent flows with irregular boundaries and free surfaces. It is a finite element model utilizing a special method based on the collocation approach called the efficient element method. This model is based on the 3D Reynolds-averaged Navier-Stokes equations. By applying the Boussinesq Approximation, the turbulent stress can be simulated by the turbulent viscosity and time-averaged velocity. There are several turbulence closure schemes available within the model for different purposes, including the parabolic eddy viscosity, mixing length, k–ε and nonlinear k–ε models. This model has been successfully applied to analyze wind-driven flow, turbulent buoyant flow, turbulent flow fields in scour holes and around a submerged training structure in a meander bend [22, 23,24].

Governing Equations

The governing equations of continuity and momentum of the three-dimensional unsteady hydrodynamic model can be written as follows:

$$\frac{\partial u_i}{\partial x_i} = 0 \tag{30}$$

$$\frac{\partial u_i}{\partial t} + u_j \frac{\partial u_i}{\partial x_j} = -\frac{1}{\rho}\frac{\partial p}{\partial x_i} + \frac{\partial}{\partial x_j}\left(\nu \frac{\partial u_i}{\partial x_j} - \overline{u_i' u_j'} \right) + f_i \tag{31}$$

where u_i $(i=1,2,3)$ = Reynolds-averaged flow velocities (u, v, w) in Cartesian coordinate system (x, y, z); t = time; ρ = water density; p = pressure; ν = fluid kinematic viscosity; $-\overline{u_i' u_j'}$ =Reynolds stress; and f_i = body force terms.

The free surface elevation (η_s) is computed using the following equation:

$$\frac{\partial \eta_s}{\partial t} + u_s \frac{\partial \eta_s}{\partial x} + v_s \frac{\partial \eta_s}{\partial y} - w_s = 0 \tag{32}$$

where u_s, v_s and w_s = surface velocities in x, y and z directions; η_s = water surface elevation.

In water column, each one of the water quality constituents can be expressed by the following mass transport equation:

$$\frac{\partial C_i}{\partial t} + \frac{\partial (uC_i)}{\partial x} + \frac{\partial (vC_i)}{\partial y} + \frac{\partial (wC_i)}{\partial z} = \frac{\partial}{\partial x}(D_x \frac{\partial C_i}{\partial x}) + \frac{\partial}{\partial y}(D_y \frac{\partial C_i}{\partial y}) + \frac{\partial}{\partial z}(D_z \frac{\partial C_i}{\partial z}) + \Sigma S_i \tag{33}$$

in which u, v, w are the water velocity components in x, y and z directions, respectively; C_i is the concentration of the ith water quality constituent; D_x, D_y and D_z are the diffusion coefficients in x, y and z directions, respectively; ΣS_i is the effective source term for the ith water quality constituent.

In bed sediment layer, the decomposition of organic material can have profound effects on the concentrations of oxygen and nutrients in the overlying waters. In this layer, the pore water may infiltrate in and out and thus induce additional nutrients transfer.

To simulate the water quality processes in this layer, the exchange between the sediment-water interface and the kinetic transformation of water quality constituent are considered. To simplify the model, the bed sediment layer is represented using one layer. In general, the depth of this layer is about 0.1 to 0.3 m. In the current version, only diffusive flux of water quality constituents is considered in this layer:

$$\frac{\partial C_j}{\partial t} = \Sigma S_j \tag{34}$$

where C_j =concentration of water quality constituents in sediment layer; ΣS_i is the effective source term, which includes biochemical kinetic transformation rate, bed release rate, and the settling and re-suspension terms.

Wind_Induced Eddy Viscosity

In a natural lake, wind stress is the important driving force for lake flow circulation. For simulating the wind driven flow, the distribution of vertical eddy viscosity is required. A parabolic eddy viscosity distribution was proposed by Tsanis [25] based on the assumption of a double logarithmic velocity profile. The vertical eddy viscosity was expressed as:

$$\nu_t = \frac{\lambda u_{*s}}{H}(z + z_b)(z_s + H - z) \tag{35}$$

in which λ= numerical parameter; z_b and z_s = characteristic lengths determined at bottom and surface, respectively; u_{*s} = surface shear velocity; H = water depth. To use this formula, three parameters, λ, z_b and z_s have to be determined. For some real cases with very small water depths, using this formula to calculate eddy viscosity may cause some problems.

Koutitas and O'Connor [26] proposed two formulas to calculate the eddy viscosity based on a one-equation turbulence model. Their formulas were:

$$\nu_t = \nu_{t,\max}\eta_z(2-\eta_z) \qquad (0 \le \eta_z \le 0.5) \tag{36}$$

$$\nu_t = \nu_{t,\max}(1-\eta_z)(5\eta_z - 1) \qquad (0.5 < \eta_z \le 1) \tag{37}$$

Where

$$\nu_{t,\max} = \lambda u_{*s} 0.105 H 0.3^{-0.25} = 0.142\lambda u_{*s} H \tag{38}$$

in which, λ = numerical parameter; η_z = non-dimensional elevation, $\eta_z = z/H$.

In this model, a new formula was proposed based on experimental measurements conducted in a laboratory flume with steady-state wind driven flow reported by Koutitas and O'Connor [26]. The form of eddy viscosity was similar to Koutitas and O'Connor's assumption (Eq. 36 and 37) and expressed as:

$$\nu_t = \nu_{t,\max} f(\eta_z) \tag{39}$$

Based on measured data, a formula is obtained to express the vertical eddy viscosity:

$$\nu_t = \nu_{t,\max}\eta_z(-3.24\eta_z^2 + 2.78\eta_z + 0.62) \tag{40}$$

A figure was plotted to compare the vertical distributions of eddy viscosity obtained from formulas provided by Tsanis, Koutitas, and the proposed model (Figure 7, next Section).

Boundary Conditions

Wind stress is generally the dominant driving force for flow currents in natural lakes. The wind shear stresses at the free surface are expressed by

$$\tau_{wx} = \rho_a C_d U_{wind} \sqrt{U_{wind}^2 + V_{wind}^2} \tag{41}$$

$$\tau_{wy} = \rho_a C_d V_{wind} \sqrt{U_{wind}^2 + V_{wind}^2} \tag{42}$$

where ρ_a = air density; U_{wind} and V_{wind} = wind velocity components at 10 m elevation in x and y directions, respectively. Although the drag coefficient C_d may vary with wind speed [26,27], for simplicity, many researchers assumed the drag coefficient was a constant on the order of 10^{-3} [28, 29]. In this model, C_d was set to 1.0×10^{-3}, and this value is applicable for simulating the wind driven flow in Deep Hollow Lake [30].

On the free surface, the normal gradient of mass concentration is set to be equal to zero. In the vicinity of the solid walls and bed, the gradients of flow properties are steep due to wall effects. The normal gradient of concentration on the wall is set to be zero.

At the inlet boundary, flow discharge and mass concentration are required; at the outlet, the water surface elevation is set as boundary condition.

Numerical Solution

The numerical model was developed based on the finite element method. Each element is a hexahedral with three levels of nine-node quadrilaterals, and the governing equations are discretized using a 27-node hexahedral (Figure 6).

Staggered grid is adopted in the model. The grid system in the horizontal plane is a structured conformal mesh generated on the boundary of the computational domain. In vertical direction, either uniform or non-uniform mesh lines are employed. In order to get more accurate results, the mesh lines are placed with finer resolution near the wall, bed and free surface.

The unsteady equations are solved by using the time marching scheme. A second-order upwinding scheme is adopted to eliminate oscillations due to advection.

In this model, a convective interpolation function is used for this purpose. This function is obtained by solving a linear and steady convection-diffusion equation analytically over a one-dimensional local element shown in Figure 6.

Although there are several other upwinding schemes, such as the first order upwinding, the second order upwind and Quick scheme, the convective interpolation function is selected in this model due to its simplicity for the implicit time marching scheme [7].

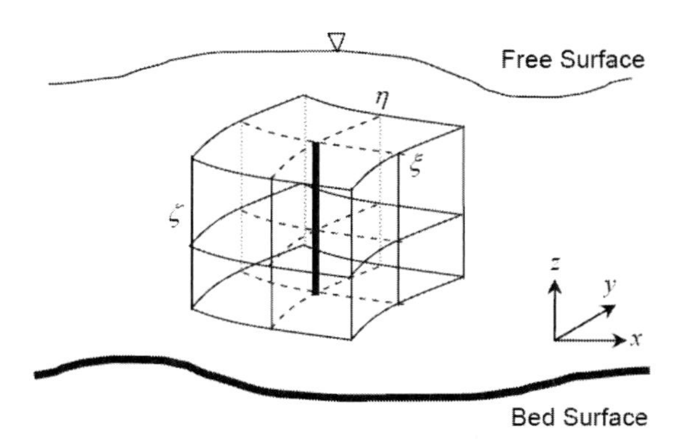

Figure 6. Coordinate system and computational element. (In Figure6, the x, y ,and z are axes of the Cartesian coordinate system and ξ , η and ζ are axes of a local system).

The velocity correction method is applied to solve the dynamic pressure and enforce mass conservation. Provisional velocities is solved first without the pressure term, and the final solution of the velocity is obtained by correcting the provisional velocities with the pressure solution [7, 22]. The system of the algebraic equations is solved using the Strongly Implicit Procedure (SIP) method. Flow fields, including water elevation, horizontal and vertical velocity components, and eddy viscosity parameters were computed by CCHE3D and set as input data. After getting the effective source terms ΣS_i , the concentration distribution of water quality constituents can be obtained by solving mass transport equations (33) numerically. The effective source term ΣS_i includes the kinetic transformation rate, external loads and sinks for water quality constituents. The kinetic transformation rate can be obtained by analyzing the complex processes of water quality constituents. Normally, the external loads and sinks come from the boundary, including inlet, outlet, benthic, water surface and atmospheric. In sediment layer, the concentration distribution of water quality constituents can be obtained by solving equations (34) numerically after obtaining the effective source terms ΣS_j .

Model Validation and Verification

Model Validation for Wind-Driven Flow

A laboratory test case was carried out in a wind-wave flume by Koutitas and O'Connor [26] to study the wind-driven flow.

The length and width of the measureing flume were 5 m and 0.2m, and the water depth was 0.31m. The vertical current profile near the middle section was measured. The detailed information of the experiment was described by Koutitas and O'Connor [26]. Figure 7 shows the vertical distributions of eddy viscosity obtained from experimental measurements and formulas provided by Tsanis (Eq.35), Koutitas (Eq. 36 and 37), and the proposed formula (Eq.40). This proposed formula can be used to calculate the eddy viscosity over the full range of water depth. At water surface ($z/H=1$), Eqs. (35) and (37) show the eddy viscosity is zero or very closed to zero.

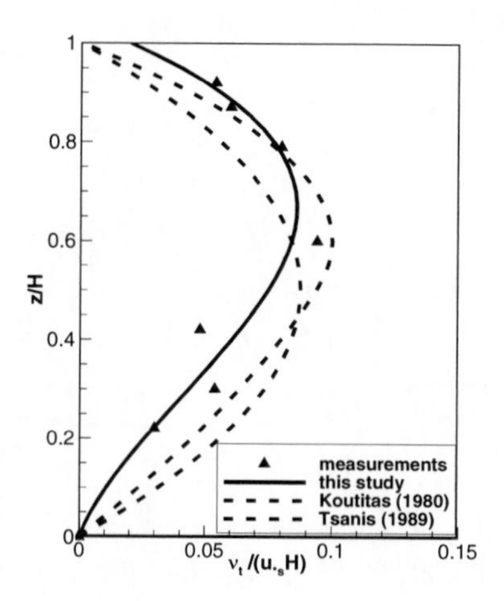

Figure 7. Comparison of vertical eddy viscosity formulas and experimental data.

The newly proposed formula shows the eddy viscosity at the water surface is about 16% of the maximum eddy viscosity $\nu_{t,\max}$.

The developed numerical model was applied to simulate the velocity profile of the wind driven flow for the experimental case. A non-uniform grid with 21 vertical nodes with fine resolution near surface and bed were used for model validation. This non-uniform grid was generated using a flexible and powerful two-direction stretching function EDS [31].

The parameters in the stretching function were set to E=1, D=0.5, and S=5.2. In the numerical modeling, both Eq. (35), and (40) were used for calculating the eddy viscosity, and the simulated vertical velocity profiles were compared with experimental measurements (Figure 8). Numerical results are generally in good agreement with experimental measurements. Near the water surface, the model overestimated the velocity using Tsanis's formula (Eq. 35) to calculate the eddy viscosity; while it underestimated the velocity by using the proposed formula (Eq. 40). Near the bottom, the simulation obtained by using the proposed formula gave better predictions.

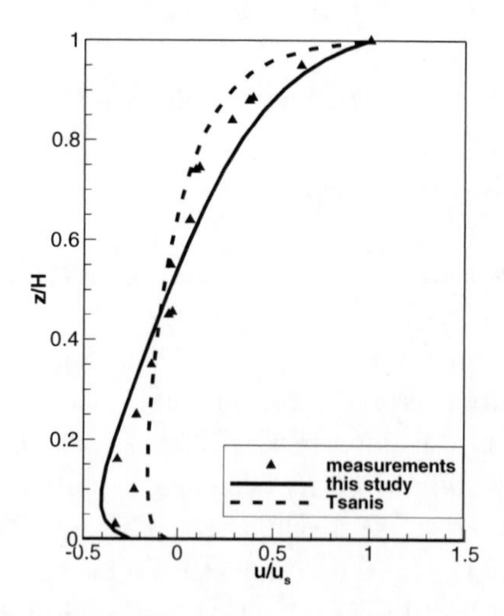

Figure 8. Normalized vertical velocity profile (u_s= surface velocity).

Model Verification for the Mass Transport Simulation

The proposed water quality model was tested against an analytical solution for predicting the concentrations of a non-conservative substance in a hypothetical one-dimensional river flow with constant depth and velocity.

A continuous source of a non-conservative substance was placed at the upstream end of a straight channel for a finite period of time, τ (Figure 9). Under the unsteady condition, the concentration of the substance throughout the river can be expressed as:

$$\frac{\partial C_s}{\partial t} + U\frac{\partial C_s}{\partial x} = D_x \frac{\partial^2 C_s}{\partial x^2} - K_d C_s \quad (43)$$

where U = velocity; C_s = concentration of substance; D_x = mixing coefficient; and K_d = decay rate. An analytical solution given by Chapra (1997) is:

$$C_s(x,t) = \frac{C_0}{2}\left[\exp\left(\frac{Ux}{2D_x}(1-\Gamma)\right) erfc\left(\frac{x-Ut\Gamma}{2\sqrt{D_x t}}\right) + \exp\left(\frac{Ux}{2D_x}(1+\Gamma)\right) erfc\left(\frac{x+Ut\Gamma}{2\sqrt{D_x t}}\right)\right] \quad (t<\tau) \quad (44)$$

$$C_s(x,t) = \frac{C_0}{2}\left\{\exp\left(\frac{Ux}{2D_x}(1-\Gamma)\right)\left[erfc\left(\frac{x-Ut\Gamma}{2\sqrt{D_x t}}\right) - erfc\left(\frac{x-U(t-\tau)\Gamma}{2\sqrt{D_x(t-\tau)}}\right)\right]\right.$$
$$\left. + \exp\left(\frac{Ux}{2D_x}(1+\Gamma)\right)\left[erfc\left(\frac{x+Ut\Gamma}{2\sqrt{D_x t}}\right) - erfc\left(\frac{x+U(t-\tau)\Gamma}{2\sqrt{D_x(t-\tau)}}\right)\right]\right\} \quad (t > \tau) \quad (45)$$

where $\Gamma = \sqrt{1+4f}$, and $f = \frac{K_d D_x}{U^2}$. For the river conditions shown in Figure 9, with a depth of 10 m, U = 0.03m/s, D_x = 30 m^2/s, τ = 6 hr, and the values of K_d = 0, 1.0/day and 2.0/day, respectively. Figure 10 shows the time series of concentration at the section x = 2000 m obtained by the numerical model and analytical solution. The maximum error is less than 2%.

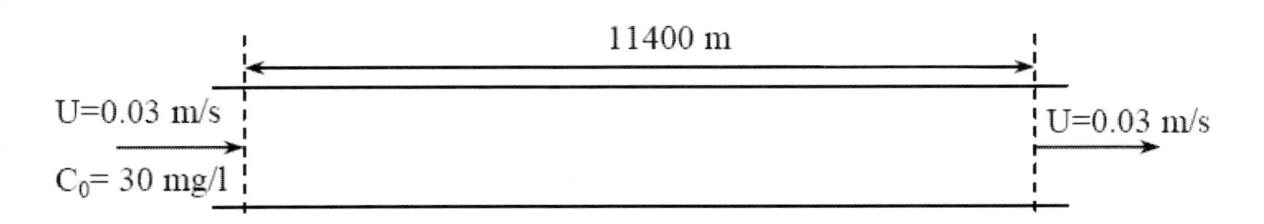

Figure 9. Test river for verification case.

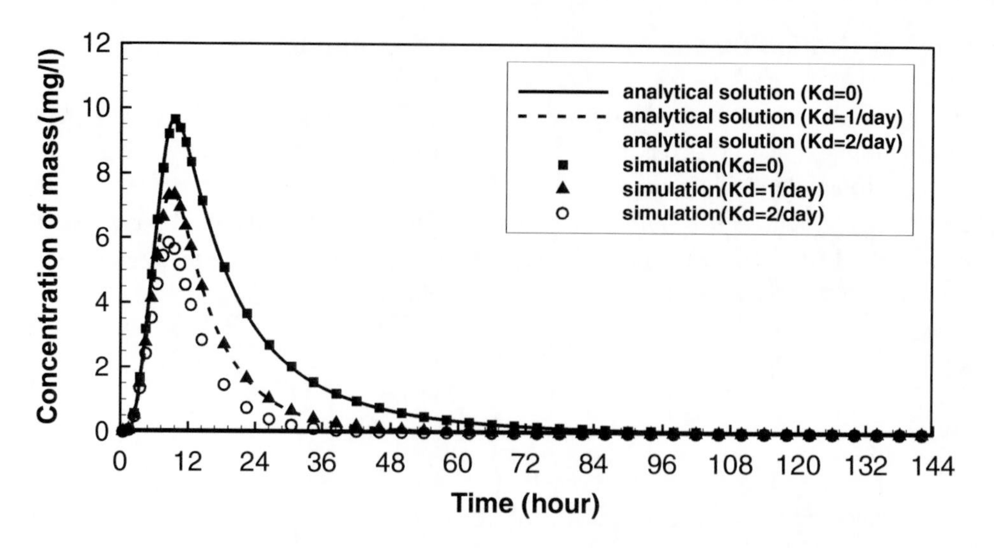

Figure 10. The time series of concentration at the section x = 2000m obtained from the numerical model and analytical solution.

MODEL APPLICATION TO DEEP HOLLOW LAKE

Study Area

Mississippi Delta is one of the most intensively farmed agricultural areas of the United States. The soils in this region are highly erodible, resulting in a large amount of sediment discharged into the water bodies.

Sediments are normally associated with many pollutants and greatly affect water quality and aquatic lives. The Mississippi Delta Management System Evaluation Area (MDMSEA) project is part of a national program designed to evaluate the impact of agricultural production on water quality and to develop best management practices (BMPs) to minimize adverse effects. Deep Hollow Lake, a small oxbow lake located in Leflore County, Mississippi, was selected as a study site for MDMSEA.

Figure 11 shows the study area of Deep Hollow Lake. It has a morphology typical of an oxbow lake, with a length of about 1 km and a width of about 100 m. Lake water depth ranges from 0.5 m to 2.6 m, with greatest depth in the middle. The lake receives runoff from a two square kilometer watershed that is heavily cultivated.

Weekly or biweekly samples of suspended sediment, nutrients, chlorophyll, bacteria, and other selected water quality variables were collected at Stations DH1, DH2 and DH3. Two of the major inflows, located at the Stations UL1 and UL2, were monitored for flow and water quality by the U.S. Geological Survey. The nutrient concentrations in Deep Hollow Lake are mainly dependent on the fertilizer loadings in the surrounding farmland and the quantity of runoff. Field measurements show that the concentrations of nitrate and ammonia in the lake are very low, while the concentration of phosphorus is relatively high in comparison with other areas of the United States. Suspended sediment concentrations are relatively high, exceeding published levels known to adversely impact fish growth and health [32].

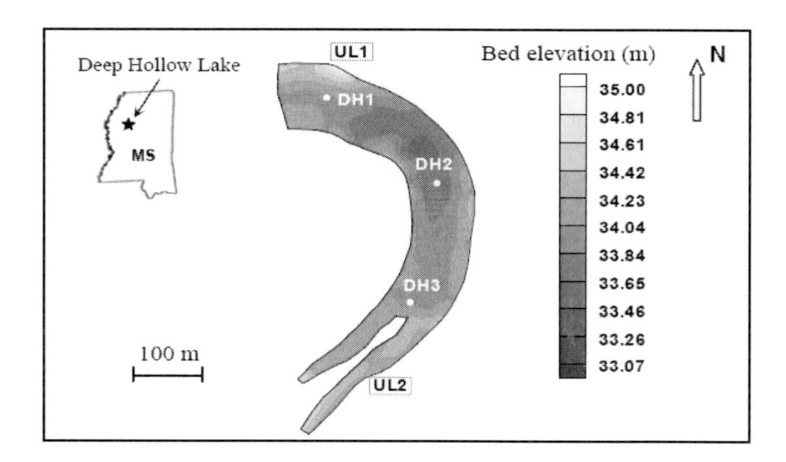

Figure 11. The study area of Deep Hollow Lake.

The water quality of the lake is sensitive to suspended sediment concentrations because photosynthetic activity is limited by elevated turbidity levels following runoff events. Researches in lakes similar to Deep Hollow have shown that the nutrients can be released from suspended and bed sediments [15]. So the sediment-associated water quality processes need to be considered in the water quality modeling.

Based on bathymetric data, the computational domain was discretized into a structured finite element mesh using the CCHE Mesh Generator[33].

In the horizontal plane, the irregular computational domain was represented by a mesh with 95×20 nodes. In the vertical direction, the domain was divided into 8 layers with finer spacing near the bed. The bed layer was represented by a 0.1 m single layer.

Light Attenuation Coefficient in Deep Hollow Lake

Light intensity is one of important factors for the growth of phytoplankton. In the numerical model, the effect of light intensity on the growth of phytoplankton is determined by the light limitation factor, which can be calculated based on the light attenuation coefficient, light intensity and water depth (Eq. 5).

As shown in Eq.(10), the light attenuation coefficient K_e is affected by water, concentrations of chlorophyll and SS. It has been observed that the light intensity in the water column decreases with the water depth, and the attenuation of light is proportional to the light intensity in the water:

$$\frac{dI}{dz_d} = -K_e I \tag{46}$$

in which I is the light intensity in the water column; z_d is the distance to water surface.

On the water surface, $z_d = 0$ and $I = I_0$, so

$$I = I_0 e^{(-K_e z_d)} \tag{47}$$

Eq.(47) can also be written as:

$$\ln\left(\frac{I}{I_0}\right) = -K_e z_d \quad (48)$$

So K_e can be obtained by linear fitting based on the measured I, I_0 and z_d.

A field measurement was conducted to measure the light intensity on the water surface and in the water column, and concentrations of suspended sediment and chlorophyll in Deep Hollow Lake. Based on the measured data, the light attenuation coefficient K_e, and the parameters K_0 and γ in Eq. (13) can be obtained. Light intensities were measured using a LICOR LI-250 light meter and a spherical quantum radiation sensor which measures photon flux from all directions underwater. This measurement is called Photosynthetic Photon Flux Fluence Rate (PPFFR). Units are in micromole per second per square meter. Data was collected at Deep Hollow Lake approximately every two weeks from January to March, 2004 (weather permitted). Light radiation was measured during conditions of unobstructed sunlight (no clouds) at three sites: DH1, DH2, and DH3 Stations (shown in Figure 11) between the hours of 10:00 AM and 2:00 PM. At each station, light intensities were measured in air, at water surface and approximately every 10 cm in the water body until the measurement unit touched the lake bottom. In addition, the concentrations of SS and chlorophyll at the same sites were also measured. Total 20 sets of data were obtained.

Figure 12 shows the light intensity of three stations (DH1, DH2 and DH3) on water surface and in the water column measured on January 20, 2004. It can be observed that light intensity decreases quickly with the increase of depth in the water column due to the effects of water, sediment and phytoplankton. The light can penetrate about 1/3 of the total water depth. Figures 13a, 13b and 13c show the relations between $\ln(I/I_0)$ and distance to the water surface z_d at DH1, DH2 and DH3 stations on Jan 20, 2004. It is obviously there is a linear relation between $\ln(I/I_0)$ and z_d, and the slope of the line is the measured light attenuation coefficient.

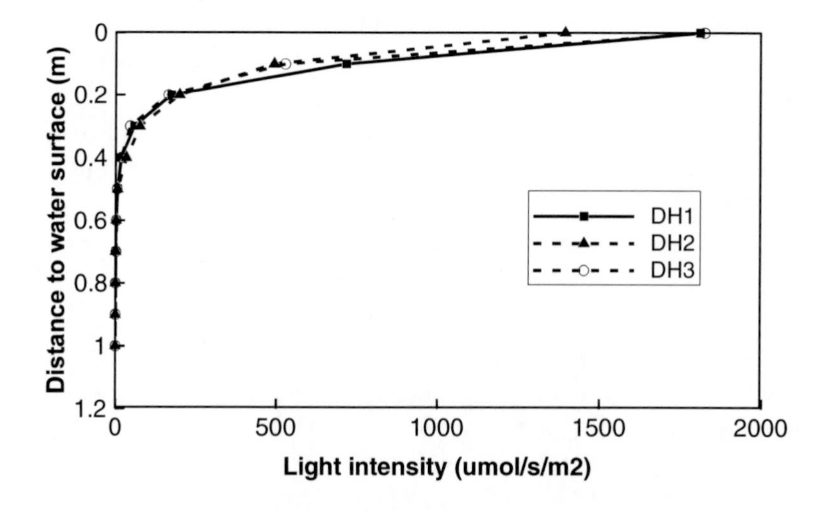

Figure 12. Measured light intensity in water at DH1, DH2 and DH3 Stations (Jan. 20, 2004).

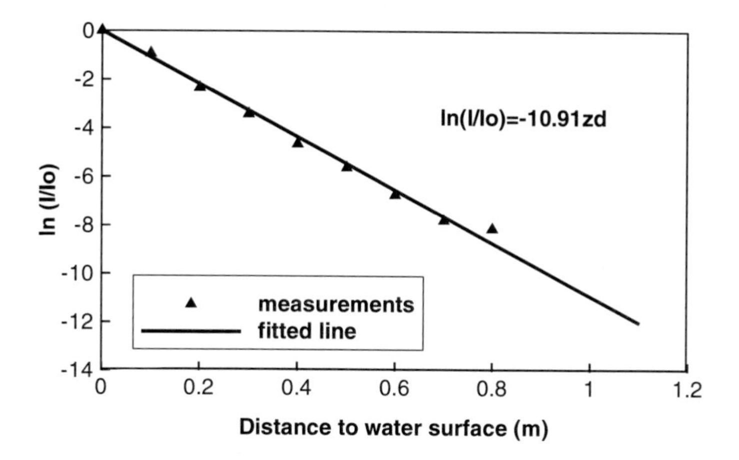

Figure 13a. Relations between $\ln(I/I_0)$ and distance to the water surface z at DH1 Station.

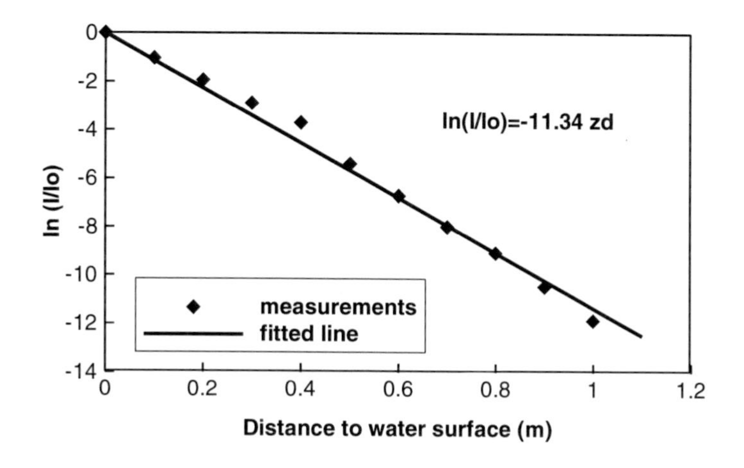

Figure 13b. Relations between $\ln(I/I_0)$ and distance to the water surface z at DH2 Station.

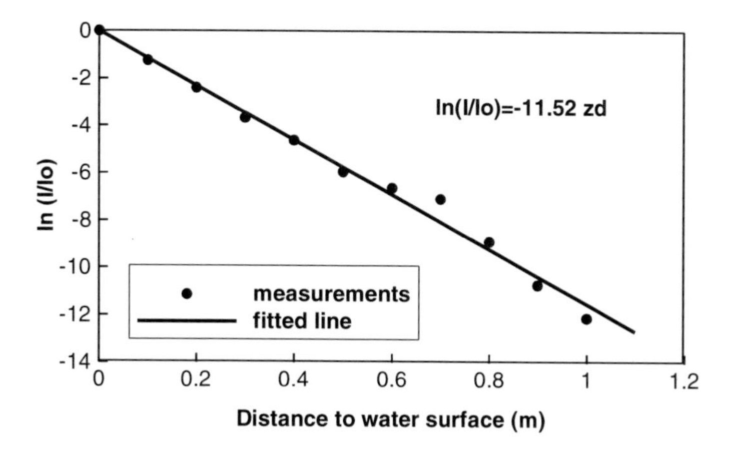

Figure 13c. Relations between $\ln(I/I_0)$ and distance to the water surface z at DH3 Station.

The Eq. (13) can also be written as

$$K_e - (0.0088C_{chl} + 0.054C_{chl}^{0.67}) = \gamma s + K_0 \quad (49)$$

Set $f(K_e) = K_e - (0.0088C_{chl} + 0.054C_{chl}^{0.67})$,so

$$f(K_e) = K_e - (0.0088C_{chl} + 0.054C_{chl}^{0.67}) = \gamma s + K_0 \quad (50)$$

Based on all sets of field measured data, a regression line, with the slope γ and intersection K_0 equal to 0.0452 and 1.2, respectively, was obtained (Figure14). So Eq. (13) can be expressed as

$$K_e = 1.2 + 0.0088C_{chl} + 0.054C_{chl}^{0.67} + 0.0452s \quad (51)$$

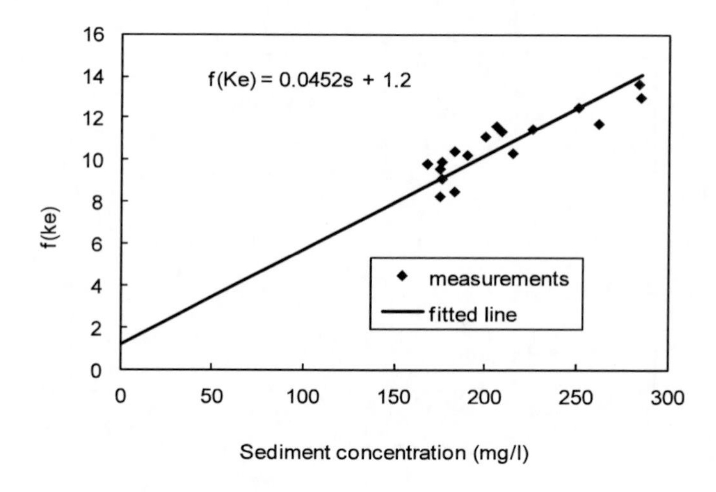

Figure 14. $f(K_e)$ versus suspended sediment concentration s.

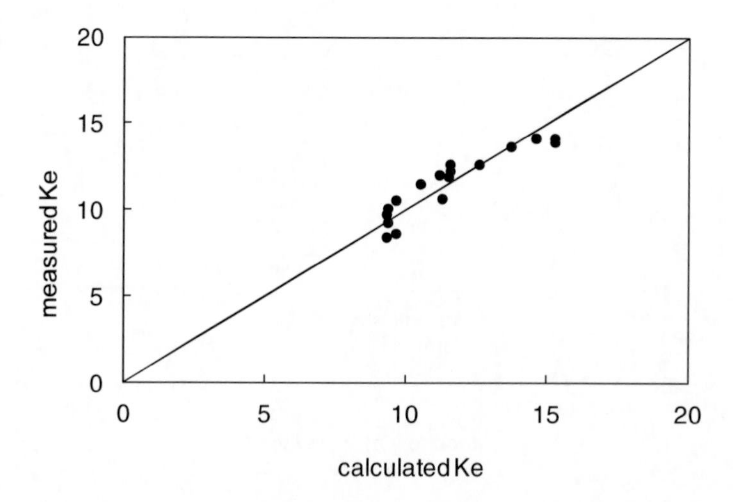

Figure 15.Comparison of the light attenuation coefficient obtained from field measurements and Eq. (51).

Based on the field measurements in Deep Hollow Lake, the value of □ was equal to 0.0452. This value was close to the value of 0.043 obtained by Stefan et al. [9] for a nearby lake. Eq. (51) was adopted to calculate the light attenuation coefficient in Deep Hollow Lake. Figure 15 compares the light attenuation coefficient obtained from field measurements and calculations from Eq. (51). Good agreement was obtained with an r2 value of 0.86.

Model Application

In Deep Hollow Lake, wind stress is the most important driving force for lake flow circulations. The CCHE3D hydrodynamic model was first calibrated using field measurements obtained from Deep Hollow Lake using an Acoustic Doppler Current Profiler (ADCP), and then it was applied to simulate the flow fields during the selected simulation period.

Figure16 shows the comparison of simulated flow velocities with field measurements conducted on November 12, 2003. In general, numerical model provides good predictions for flow fields.

The proposed water quality model CCHE3D_WQ was calibrated using weekly/biweekly field measured data obtained in the lake between April to June, 1999. In this period, the flow fields including water surface elevation, velocity, eddy viscosity, etc., were obtained from the CCHE3D hydrodynamic model. After obtaining the flow fields, the concentrations of water quality constituents can be simulated using the proposed CCHE3D_WQ model.

Figure 17 shows the measured suspended sediment at DH1 Station in 1999. The SS concentration is relatively high in spring, fall, and winter, so the sediment-associated water quality processes play important roles in the lake aquatic ecosystem. The above mentioned three major sediment-associated processes were integrated into the CCHE3D_WQ model to simulate the water quality constituents in the lake.

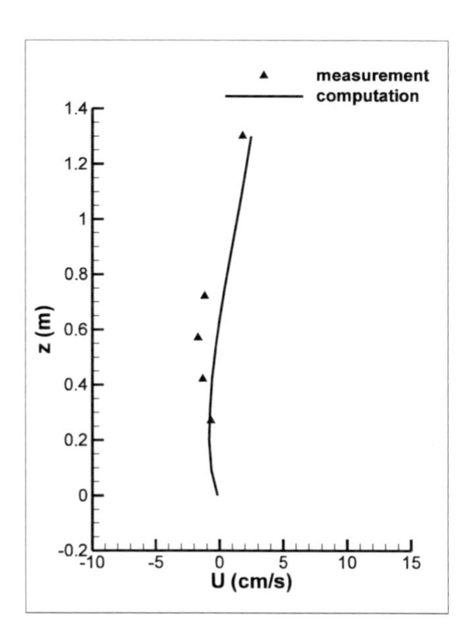

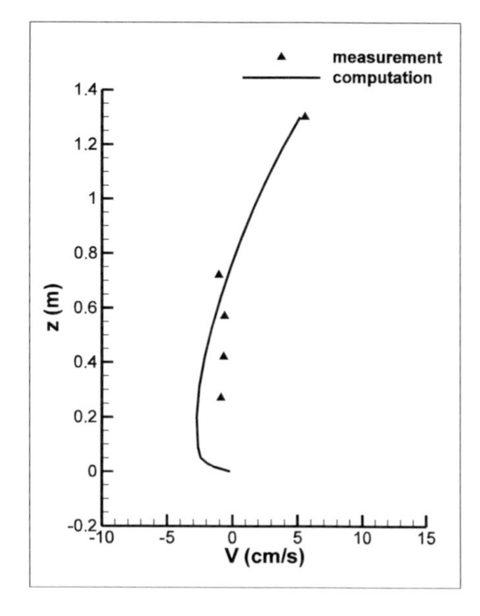

Figure 16. Observed and simulated velocities at Station DH1 (Nov. 12, 2003).

The light attenuation coefficient K_e was calculated using Eq.(51) by considering the effects of water, chlorophyll and suspended sediment. The adsorption and desorption of nutrients by sediment were described using Langmuir equation, and the concentrations of dissolved and particulate nutrient at equilibrium were calculated using Eqs. (25) and (26). In Deep Hollow Lake, the concentrations of ammonium and nitrate were very low, so the adsorption and desorption of ammonium and nitrate from sediment were expected to be insignificant and were not incorporated.

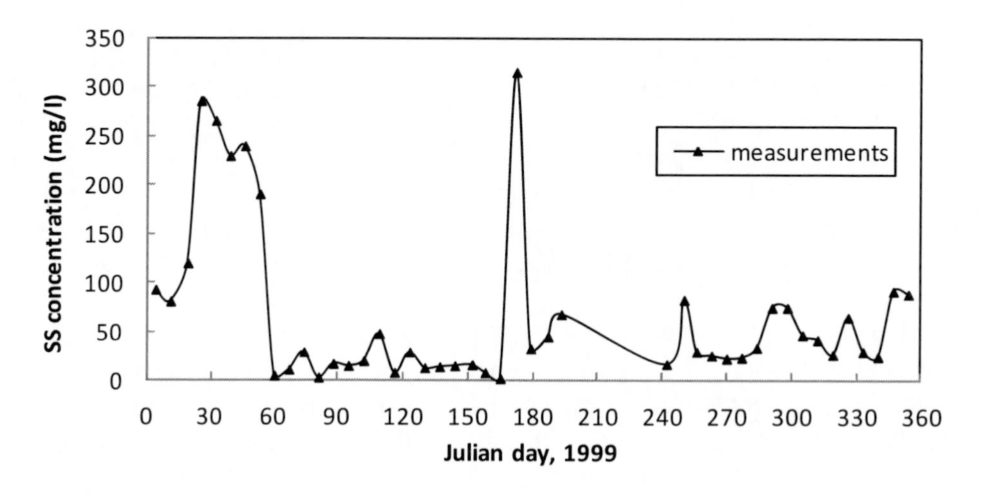

Figure 17. Measured SS concentration at Station DH1.

Dissolved nutrients may be released to the water column from bed sediments. In the numerical simulation, the bed release rate of ammonium, nitrate, organic nitrogen, phosphate, and organic phosphorus were calculated using Eq. (27). For calibration runs, parameters in the water quality model were adjusted repeatedly to obtain a reasonable reproduction of the field data. The model parameters were obtained either directly from experiments and field measurements, or by others [1,12,13,34,35,36]. Table 1 shows some calibrated parameters used in the model to simulate the water quality constituents in Deep Hollow Lake. All adopted parameter values are in the range reported in the literatures [1,2]. In the water quality model, the measured water temperature, surface light intensity, wind field, SS concentration, and pH were set as input data. The measured depth averaged concentrations of nutrients and chlorophyll were used for model calibration.

Figure 18 shows the comparison of the simulated chlorophyll concentrations with field measurements. In this figure, the simulated results at different layers were plotted for comparison. Figure 19 and 20 show the model calibration results of nitrogen and phosphorus at Station DH1. In general, the model provided reasonable reproduction of patterns and acceptable magnitudes for water quality constituents. The mean values of the model results are generally in good agreement with the field observations. However, the ability of the model to reproduce temporal variations in field measurements was not as good as the mean values. Those differences may arise due to the fact that measurements occurred weekly while the time step for the simulation was 1 hour.

Table 1. Calibrated values of parameters for the water quality model applied to Deep Hollow Lake

Parameter definition	Symbol	Value	Units
Maximum phytoplankton growth rate	P_{mx}	2.0	day^{-1}
Background light attenuation coefficient	K_0	1.2	m^{-1}
Light attenuation parameter related to SS	γ	0.0452	*l/mg/m*
Saturation light intensity of phytoplankton	I_m	300	*ly/day*
Half-saturation constant for nitrogen in phytoplankton growth	K_{mN}	0.01	*mg/l*
Half-saturation constant for phosphorus in phytoplankton growth	K_{mP}	0.001	*mg/l*
Effect coefficient of temperature below optimal temperature on growth	KTg_1	0.006	*none*
Effect coefficient of temperature above optimal temperature on growth	KTg_2	0.008	*none*
Phytoplankton endogenous respiration rate	k_{pr}	0.125	day^{-1}
Phytoplankton mortality rate	k_{pd}	0.02	day^{-1}
Temperature correction coefficient	θ_{pr}	1.068	
Settling velocity of phytoplankton	ω_{s4}	0.01	*m/day*
Diffusive exchange coefficient of nitrogen at water-sediment interface	k_n	0.01	*m/day*
Control value of nitrogen release via *DO*	K_{ndos}	0.5	mg/l
Ratio of adsorption and desorption rate coefficients for phosphorus	K	0.7	*l/mg*
Maximum adsorption capacity of phosphorus	Q_m	0.0051	*mg* P/*mgSS*
Diffusive exchange coefficient of phosphorus at water-sediment interface	k_p	0.01	mg/m^2day
Temperature coefficient	θ_{sed}	1.05	none
Control value of phosphorus release via DO	K_{pdos}	0.5	*mg/l*
Control value of phosphorus release via pH	K_{pH}	18	*none*

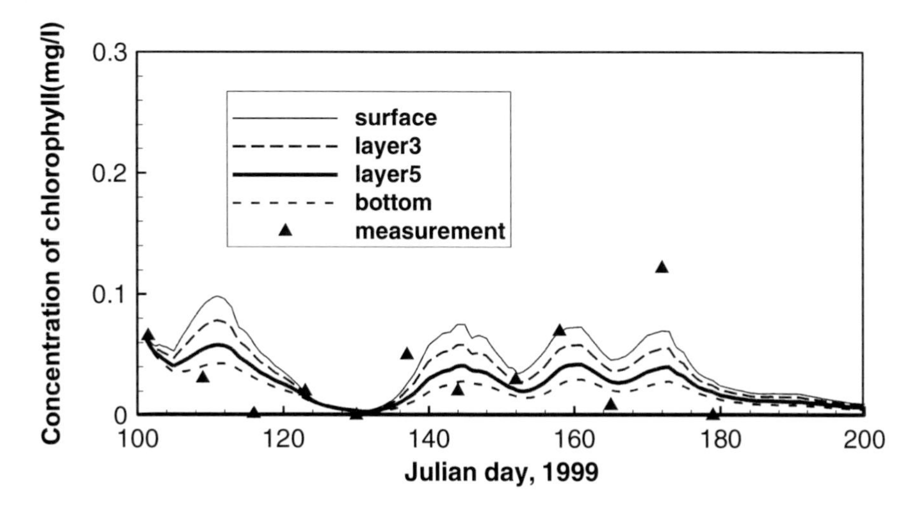

Figure 18. The concentration of chlorophyll at Station DH1 for calibration.

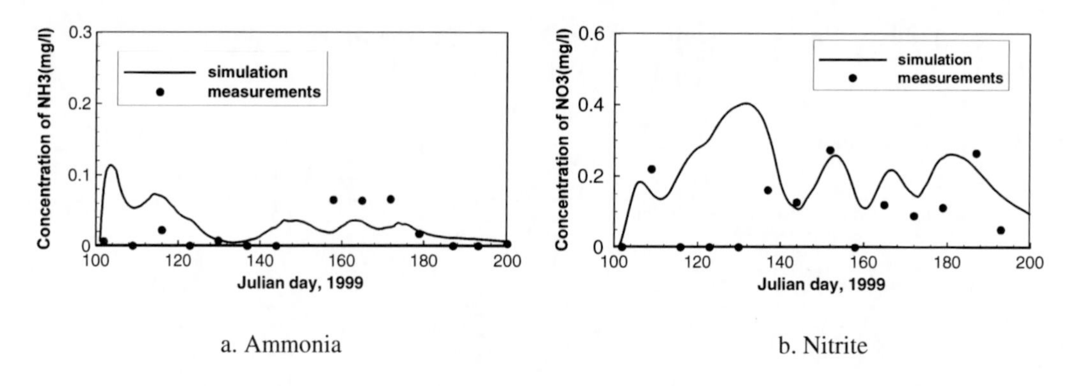

a. Ammonia b. Nitrite

Figure 19. The concentration of nitrogen at Station DH1for calibration.

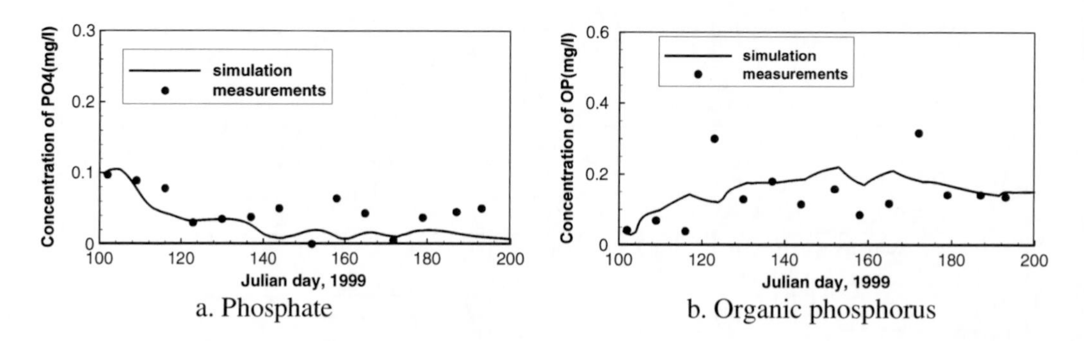

a. Phosphate b. Organic phosphorus

Figure 20. The concentration of phosphorus at Station DH1for calibration.

In addition, the water quality processes in the lake system are likely more complicated than those processes considered in the numerical model.

The period from September to December 1999 was chosen for model validation. After obtaining the flow fields from CCHE3D hydrodynamic model, the measured boundary conditions, weather data and SS concentrations, CCHE3D_WQ model was used to simulate the concentrations of water quality constituents. Parameter values in the water quality model were same as those calibrated values. Figures 21 and 22 show the simulated and measured concentrations of chlorophyll and nitrogen. For the validation run, trends and quantities of concentration of chlorophyll and nitrogen obtained from the numerical model are generally in agreement with the observations. Figure 23 shows the simulated and observed concentrations of ortho-phosphorus and total organic phosphorus, respectively. Without considering the sediment-associated processes of adsorption, desorption and bed release, the model overestimated ortho-phosphorus concentration and underestimated organic phosphorus. After considering those processes, the root mean square error (RMSE) of ortho-phosphorus concentration was reduced from 0.029 to 0.016 mg/l, or reduced 45%; for organic phosphorus RMSE was reduced from 0.051 to 0.037 mg/l, or reduced 28%.

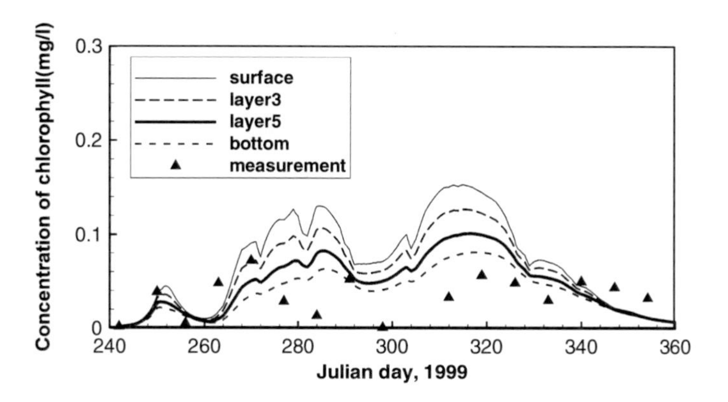

Figure 21. The concentration of chlorophyll at Station DH1 for validation.

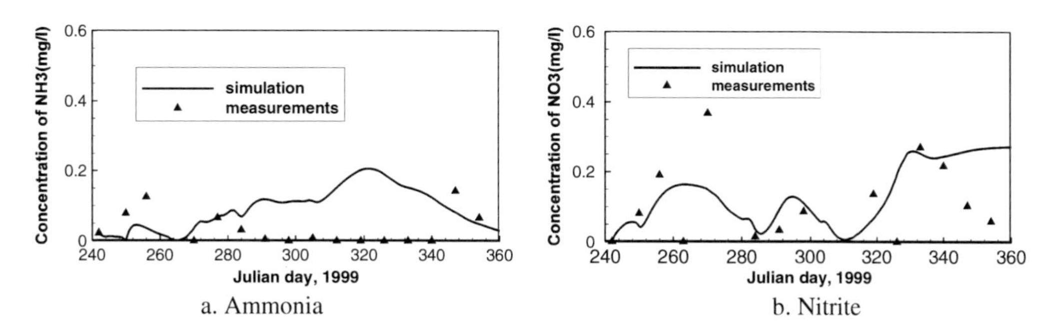

a. Ammonia

b. Nitrite

Figure 22. The concentration of nitrogen at Station DH1for validation.

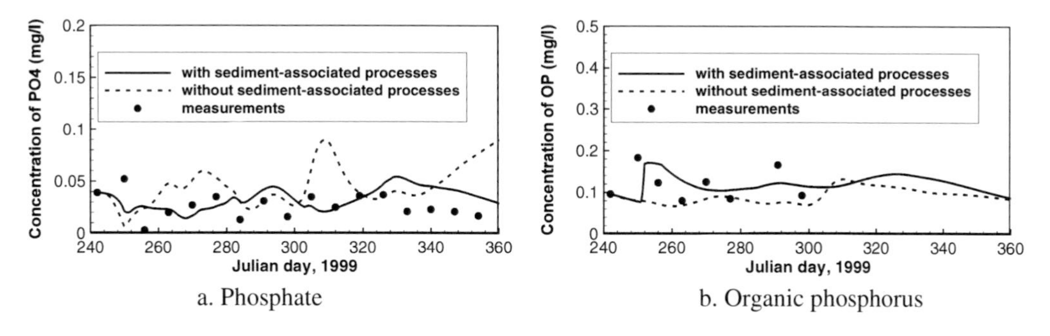

a. Phosphate

b. Organic phosphorus

Figure 23. The concentration of phosphorus at Station DH1for calibration.

CCHE3D_WQ was also applied to simulate the concentration of phosphorus in bottom sediment layer.

In this layer, the concentration is affected by the processes of biochemical kinetic, bed release, settling and re-suspension (Eq. 34).

Due to lack of field data, a hypothetical case was assumed with the initial concentrations of phosphate and organic phosphorus in sediment layer of 0.04 mg/l and 0.2 mg/l, respectively. Figure 24 shows the simulated concentrations of ortho-phosphorus and total organic phosphorus.

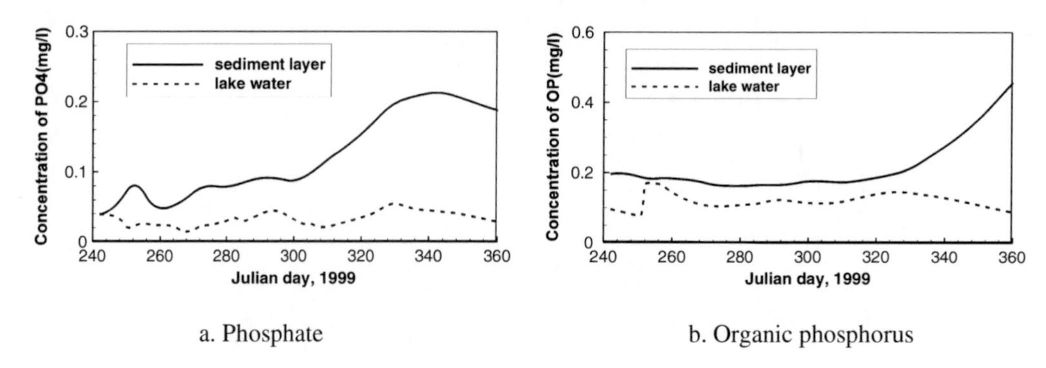

a. Phosphate b. Organic phosphorus

Figure 24. The concentration of phosphorus in bed sediment layer and lake water.

DISCUSSION

Comparison of Langmuir Equation and Linear Approach for Modeling the Adsorption-Desorption

In many water quality models, both Langmuir equation and linear approach have been applied to simulate the processes of adsorption-desorption [1, 2, 3, 37, 38]. For linear approach, the adsorption-desorption of nutrients by sediment is described by a linear isotherm, and the equilibrium adsorption concentration Q (mg /mg SS) can be expressed as

$$Q = K_p C_d \tag{52}$$

in which K_p is the partition coefficient. Based on Eqs.(17), (20) and (21), C_p and C_d can be calculated by

$$C_p = \frac{K_p s}{1 + K_p s} C_0 \tag{53}$$

$$C_d = \frac{1}{1 + K_p s} C_0 \tag{54}$$

Eqs.(53) and (54) were adopted by some water quality models for calculating the particulate and dissolved nutrient concentrations at each time step due to adsorption-desorption. In fact, this linear approach is satisfactory only for the low nutrient concentration cases [8]. If the nutrient concentration is very low, the term $(1 + KC_d)$ in Langmuir equation (Eq.19) approximately equals to 1, so Eq. (19) can be simplified as:

$$Q = Q_m K C_d \tag{55}$$

if the partition coefficient K_p is set as

$$K_p = Q_m K \tag{56}$$

then Eq.(55) is same as the linear approach equation (52). So Eq.(52) is the linear portion of the Langmuir equation.

In order to find the value of "low nutrient concentration" for adopting the linear approach, Eqs (53), (54) and (25), (26) were used to calculate the particulate and dissolved concentrations of nutrients in Deep Hollow Lake due to adsorption-desorption. For this case Langmuir adsorption constant K and the maximum adsorption capacity Q_m were taken as 0.7 *l/mg* and 0.0051 *mg /mg* SS, respectively. If the linear approach was adopted, based on Eq.(56), the partition coefficient K_p is 0.00357 *l/mg*.

Figures 25a and 25b show the concentrations of particulate and dissolved phosphate calculated based on the linear approach (Eqs. 53 and 54) and Langmuir equation (Eqs. 25 and 26) under different total initial concentrations.

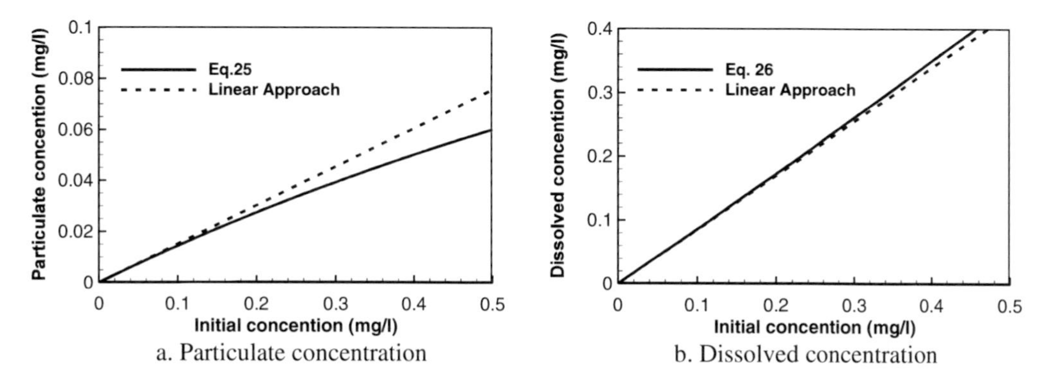

a. Particulate concentration

b. Dissolved concentration

Figure 25. Comparison of particulate and dissolved phosphate concentrations obtained based on the Linear approach and Langmuir equation.

It can be observed that the linear approach and Langmuir equation have same results if the initial concentration is less than 0.1 *mg/l*. In Deep Hollow Lake, the phosphate concentration can be greater than 0.3 *mg/l*, so Langmuir equation should give more accurate predictions.

Sensitivity of Chlorophyll Concentration to SS

Field data show that major water quality problems in Deep Hollow Lake in the late 1990s were caused by excessive sediment loads carried by runoff from surrounding cultivated lands. In order to improve the water quality of the lake, various best management practices (BMPs), such as edge-of field BMPs and agronomic BMPs were employed to reduce sediment loads on the farm lands. After the reduction of lake sediment concentration, there were significant increases in Secchi depth and chlorophyll concentration, and fish populations responded positively [32].

To show the sensitivity of chlorophyll concentration to SS, a series of hypothetical lake SS concentrations were input to the model, while the flow conditions and nutrient loadings at inlet boundaries were kept the same. The interactions between SS and nutrients in the lake were considered in the model simulation. The calibration and validation periods were selected for sensitivity study.

It was assumed the SS levels were varied from 10% to 300% of the current SS condition (base condition), and the model was applied to simulate the response of chlorophyll concentration in the lake. As expected, the concentration of chlorophyll is inversely related to the sediment concentration. Figure 26 shows the sensitivity of temporal mean chlorophyll concentration to SS for the entire simulation period at DH1 Station. In this figure, the square represents the base condition (SS=100%, Chlorophyll=100%). When lake SS was reduced by 50%, simulated mean chlorophyll concentration increased about 40%. When lake SS was doubled, the chlorophyll concentration fell to 37% of the base condition. Field observations of SS and chlorophyll in the years 1999, 2000 and 2001 were used to evaluate the model sensitivity analysis. Measured mean concentrations of SS and chlorophyll under the base condition (Year 1999) were 82 mg/l and 40 mg/l, respectively.

Mean concentrations of SS in 2000 and 2001 were reduced to 56% and 62% of base condition, and the chlorophyll concentrations increased to 142% and 137% of the base condition, respectively (Figure 26). This tendency agrees with the numerical predictions shown in Figure 26.

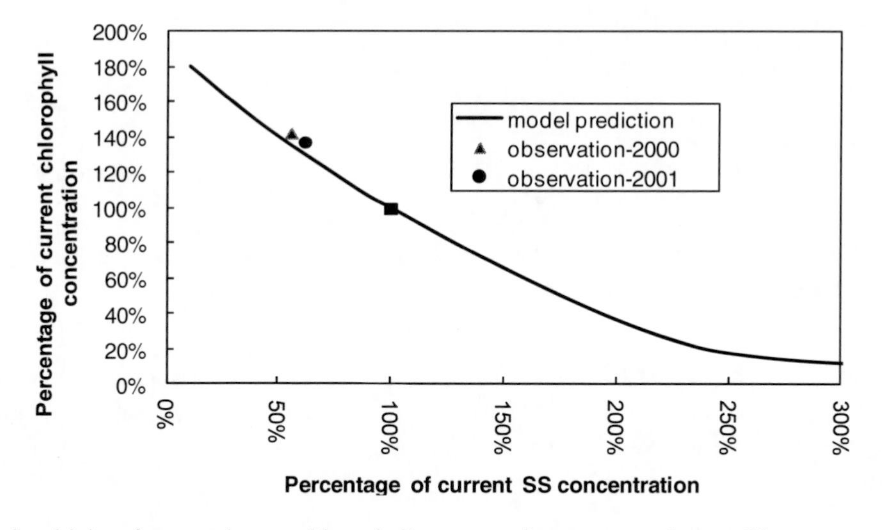

Figure 26. Sensitivity of temporal mean chlorophyll concentration to temporal mean SS.

Sensitivity of Chlorophyll Concentration to Nutrient Loadings

To examine the effects of changing nutrient loadings on chlorophyll concentration, scenarios were generated by reducing and increasing the observed concentrations of dissolved inorganic nitrogen and phosphorus. The calibration and validation periods were selected for sensitivity study. It was assumed the nutrient loadings were varied from 10% to 300% of the current nutrient conditions (base conditions), and the model was applied to simulate the chlorophyll concentration in the lake. Reducing nutrient loads produced lower chlorophyll concentrations, as expected. Figure 27a and Figure 27b show the sensitivities of temporal mean concentration of chlorophylls under the base conditions and reduction/increasing loads for inorganic nitrogen and phosphorus, respectively. In these figures, the squares represent the base condition (nutrient=100%, Chlorophyll=100%). The temporal mean chlorophyll concentration under base nutrient load was about 0.059 mg/l.

According to model simulations, reducing the lake inorganic nitrogen concentration by 50% would reduce the average chlorophyll concentration by approximately 0.0145mg/l, or about 24%. When the concentration of inorganic nitrogen was doubled, the chlorophyll concentration increased about 16.5%.

For inorganic phosphorus, reducing the concentration by 50% in the lake would reduce average chlorophyll concentration by approximately 2%. When the concentration of inorganic phosphorus was doubled, the average concentration of chlorophyll increased less than 0.5%. This analysis indicates that the concentration of chlorophyll is much more sensitive to the inorganic nitrogen than that to inorganic phosphorus in Deep Hollow Lake.

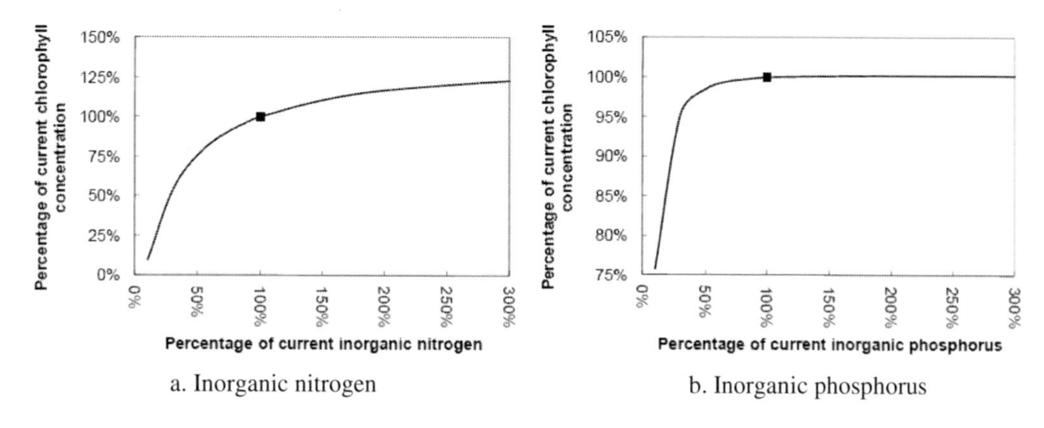

Figure 27. The effect of reduction/increasing of nutrient loadings on the chlorophyll concentration.

CONCLUSION

A three-dimensional numerical model (CCHE3D_WQ) was developed to simulate the concentration of water quality constituents in shallow natural lakes where sediment-associate processes are important. Four biochemical cycles were simulated, and eight state variables were involved in the interacting systems. Sediment-associated water quality processes were studied. A formula was obtained based on field measurements to calculate the light attenuation coefficient by considering the effects of suspended sediment and chlorophyll. The concentrations of particulate and dissolved nutrients due to adsorption-desorption were calculated using two formulas derived based on the Langmuir Equation. The release rates of nutrients from the bed sediment were calculated by considering the effects of the concentration gradient across the water-sediment interface, pH, temperature and dissolved oxygen concentration. Those formulas were tested using field measured data.

The sediment-associated processes were integrated into CCHE3D_WQ. This model was first verified using analytical solutions of pollutant transport in open channel flow, and then it was applied to the study of the concentration of water quality constituents in Deep Hollow

Lake. Realistic trends and magnitudes of nutrient and phytoplankton concentrations obtained from the numerical model generally agreed with field observations. The effects of sediment-associated processes are quite important in water quality processes. Without considering the sediment-associated processes, the model overestimated ortho-phosphorus concentration and underestimated organic phosphorus. After considering those processes, the numerical results produced better agreements with field observations.

The model was also used to conduct analyses of the sensitivities of lake chlorophyll concentration to SS concentration and nutrient loadings. Lake primary productivity is mainly limited by suspended sediment concentration, which limits light penetration. After the reduction of SS in the lake, there is significant increase in chlorophyll concentration. The numerical results also show that the concentration of chlorophyll is much more sensitive to the inorganic nitrogen than that to the inorganic phosphorus in Deep Hollow Lake.

This model provides a useful tool for predicting water quality constituents in natural lakes. It helps us understand the interaction processes between water quality and sediment in natural lakes. It is also useful for the decision makers to evaluate BMPs established in the lake watersheds.

REFERENCES

[1] T. M. Wool, et al. Water Quality Analysis Simulation Program (WASP) version 6 User's Manual, US Environmental Protection Agency, Atlanta, GA (2001).

[2] C. F. Cerco and T. Cole, User's Guide to the CE-QUAL-ICM: Three-dimensional Eutrophication Model, Technical Report EL-95-1 5, U.S. Army Corps of Engineers, Vicksburg, MS (1995).

[3] Delft Hydraulics, Delft3D-WAQ: Technical Reference Manual, Delft Hydraulics, The Netherlands (2003).

[4] Danish Hydraulic Institute, Coastal and Inland Waters in 3D, <http://www.dhi.dk>.

[5] A. R. Trancoso, S. Saraiva, L. Fernandez, P. Pina, P. Leitao and R. Neves, *Ecological Modeling,* 187, 232–246 (2005).

[6] Z.G.Ji, Hydrodynamics and Water Quality: Modeling Rivers, Lakes, and Estuaries, John Wiley, New Jersey, USA (2008).

[7] Y.Jia, S. Scott, Y. Xu, S. Huang and S.S.Y. Wang, Journal *of Hydraulic Engineering,* 131(8), 682–693(2005).

[8] S. C. Chapra, Surface Water Quality Modeling, The Mcgraw-Hill Companies, Inc, New York (1997).

[9] H. G. Stefan, et al., *Water Resources Research*, 19, No. 1, 109-120 (1983).

[10] M. Ishikawa and H. Nishmura, *Water Research*, 23, 351-359 (1989).

[11] I. Fox, M. A. Malati and R. Perry, *Water Research*, 23, 725-732 (1989).

[12] A. Appan and H. Wang, *Journal of Environmental Engineering*, 126, 993-998 (2000).

[13] M. D. Bubba, C. A. Arias and H. Brix, *Water Research*, 37, 3390–3400 (2003).

[14] L. H. Kim, E. Choi and M. K. Stenstrom, *Chemosphere*, 50 (1), 53–61 (2003).

[15] L. H. Fisher and T. M. Wood, Effect of Water-column pH on Sediment-phosphorus Release Rate in Upper Klamath Lake, Oregon, 2001, USGS Water Resources Investigation Report 03-4271 (2004).

[16] J. R. Romero, et al., Computational Aquatic Ecosystem Dynamics Model: CAEDYM v2, Science Manual, University of Western Australia (2003).
[17] M. R. Loeff, et al., *Limnology and Oceanography*, 29, 675-686 (1984).
[18] P. A. Moore, K. R. Reddy and M. M. Fisher, *Journal of Environmental Quality*, 27, 1428-1439 (1998).
[19] N. Steinberger and M. Hondzo *Journal of Environmental Engineering*, 125, 192-200 (1999).
[20] P. C. M. Boers, *Water Research*, 25, 309-311 (1991).
[21] A. Kleeberg and G. Schlungbaum, Hydrobiologia, 253, 263-274 (1993).
[22] Y. Jia, T. Kitamura and S. S. Y. Wang, *Journal of Hydraulic Engineering,* 127, 219-229 (2001).
[23] X. Chao, Y. Jia and S. S. Y. Wang, *Journal of Hydraulic Engineering*, 135, 554-563 (2009).
[24] X. Chao and Y. Jia, Three-dimensional Numerical Simulation of Cohesive Sediment Transport in Natural Lakes, in Book "Sediment Transport", INTECH Publisher, 137-160 (2011).
[25] I. K. Tsanis, *Journal of Hydraulic Engineering*, 115 (8), 1113-1134 (1989).
[26] C. Koutitas and B. O'Connor, *Journal of Hydraulic Engineering*, 106 (11), 1843-1865 (1980).
[27] K. R. Jin, J. H. Hamrick and T. Tisdale, *Journal of Hydraulic Engineering*, 126(10), 758–771 (2000).
[28] W. Huang and M. Spaulding, *Journal of Hydraulic Engineering*, 121(4), 300-311(1995).
[29] F. J. Rueda and S. G. Schladow, *Journal of Hydraulic Engineering*, 129(2), 92-101 (2003).
[30] X. Chao, Y, Jia and D. Shields, "Three Dimensional Numerical Simulation of Flow and Mass Transport in a Shallow Oxbow Lake", World Water and Environmental Resources Congress 2004, ASCE, Salt Lake City, USA, June 27-July 1 (CD-ROM).
[31] Y. Zhang, Y. Jia and S. S. Y. Wang , "Techniques for mesh density control", The 7th International Conference on Hydroscience and Engineering, Philadelphia, USA, Sep. 10 – Sep. 13.
[32] R. A. Rebich and S. S. Knight. The Mississippi Delta Management Systems Evaluation Area Project, 1995-99, Mississippi Agriculture and Forestry Experiment Station, Information Bulletin 377, Division of Agriculture, Forestry and Veterinary Medicine, Mississippi State University (2001).
[33] Y. Zhang, CCHE Mesh Generator User Manual, Technical Report, The University of Mississippi (2002).
[34] R. Portielje and L. Lijklema, *Hydrobiologia*, 253, 249-261 (1993).
[35] D. M. DiToro, Sediment Flux Modeling, Wiley-Interscience, New York (2001).
[36] M. R. Hipsey, Computational Aquatic Ecosystem Dynamics Model: CAEDYM v2, User Manual, University of Western Australia, Perth, Australia (2003).
[37] J. G. C. Smits and D. T. Molen, Hydrobiologia 253: 281-300 (1993).
[38] X. Chao, Y. Jia, C. Cooper, D. Shields and S.S.Y. Wang, *Journal of Environmental Engineering,* 132, No. 11, 1498-1507 (2006).

In: Water Quality
Editor: You-Gan Wang
ISBN: 978-1-62417-111-6

Chapter 4

INTEGRATING MAJOR ION CHEMISTRY WITH STATISTICAL ANALYSIS FOR GEOCHEMICAL ASSESSMENT OF GROUNDWATER QUALITY IN COASTAL AQUIFER OF SAIJO PLAIN, EHIME PREFECTURE, JAPAN

Pankaj Kumar[1] and Ram Avtar[2]
[1]Graduate School of Life and Environmental Sciences,
University of Tsukuba, Tsukuba, Japan
[2]Institute of Industrial Science, The University of Tokyo,
Komaba Meguro-Ku, Tokyo, Japan

ABSTRACT

A comprehensive study of major ions, silica and isotopes was carried out to understand the geochemical processes controlling groundwater quality in coastal aquifer of Saijo plain, Western Japan. Various graphs were plotted using chemical data to enable hydrochemical evaluation of the aquifer system based on the ionic constituents, water types, hydrochemical facies, and factors controlling groundwater quality.

Carbonate weathering and atmospheric precipitation are strong factors controlling the chemistry of major ions. From stable isotopic results, it was found that most of sample points plotted near the local meteoric water line (LMWL) i.e. origin of ground water is meteoric in principle; however point away from the LMWL favors exchange with rock minerals mainly salinization process. This study is crucial considering that Saijo city is known as one of the water capital of Japan and groundwater is the exclusive source of drinking water in this region.

Keywords: Groundwater, salinization, weathering

1. INTRODUCTION

Coastal aquifers are commonly stressed with enhanced pumping for water supply, ultimately results in lowering of water table, increase of land subsidence and intrusion of saline water into freshwater aquifers (Capaccioni et al. 2005; Lee et al. 2007; Trabelsi et al. 2007; Gattacceca et al. 2009). Because of dynamic nature of sea water- fresh water interface, lack of regular monitoring network for groundwater from the coastal aquifer, can't ensure its diligent sustainable management (Steyl 2010). Globally a huge number of studies have been done to understand the hydrodynamics of the coastal aquifers to assess safe yield for groundwater withdrawal in order to implement appropriate management policies (Voudouris 2006). Saltwater intrusion is the migration of saltwater into freshwater aquifers under the influence of groundwater development (Freeze and Cherry 1979). Integrated approach of geochemical and isotopic analysis proved as an efficient tool to understand interaction between groundwater and its surrounding environment which contribute to its better management (Adams et al. 2001; Schiavo et al. 2009). Combined study of major ions (HCO_3^- , Cl^- , SO_4^{2-} , NO_3^- , Ca^{2+} , Na^+, Mg^{2+} and K^+) and environmental isotopes (δD, $\delta^{18}O$ and $^{87}Sr/^{86}Sr$) is an proficient investigative means to know the aquifer matrix chemistry and water–rock interaction along the flow path in the aquifer to trace groundwater evolution and mechanism of salt water – fresh water mixing (Kim et al. 2003; Vengosh et al. 2005; Chen et al. 2007; Jorgensen et al. 2008; Schiavo et al. 2009; Langman et al. 2010). Piper and Durov diagrams (in original and modified versions) have been consistently applied to study the cation exchange and reverse cations reaction during dynamics of sea water intrusion or freshening occurring in alluvial coastal aquifer (Ray et al. 2008; Petalas et al. 2009; Forcada 2010). Kurihara (1972) described that geological structure in Saijo plain is very complex because of subsidence took place in alluvial fan of Kamo River. Nakano et al. (2008) reported that groundwater in the Saijo city shows a different quality between the eastern Saijo plain and western Shuso one, indicating that each groundwater constitutes an independent aquifer system. Saijo city office (2008) also reported that there is a decline in groundwater level in northern coastal area of the Saijo plain, which may trigger sea water intrusion. Saijo plain is known for its plenty of good quality groundwater in western Japan but despite its importance, little is known about the natural phenomena that govern the chemical composition of groundwater. The prime objective of this study is to elucidate the mechanism or chemical processes that control water chemistry with special attention on sea water- fresh water interaction pattern on this small coastal aquifer of Saijo plain.

2. STUDY AREA

Saijo plain is located in Saijo City, Ehime Prefecture, and the north-western part of the Shikoku Island in the western Japan (Figure 1). The Saijo plain is bounded on the south by the range of Ishizuchi mountains (1982 m.a.m.s.l.), the highest peak in the western Japan, on the north by the Seto Inland Sea, on the west by the Shuso plain and on the east by Toyo region. From the geological point of view, mountains mainly consist of the Sambagawa metamorphic rocks, whereas the alluvial plain is composed of the Pliocene-Quaternary sediment (with approximately 1 km thickness).

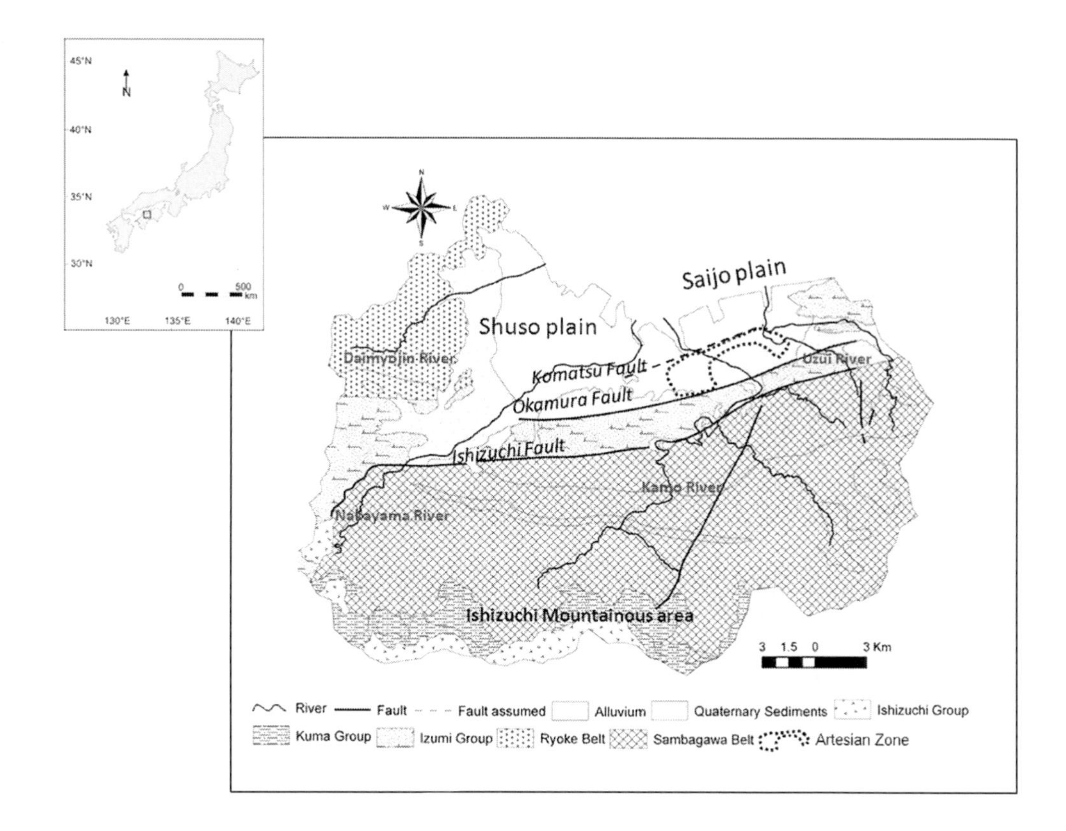

Figure 1. Geological map of Saijo pain, Western Japan.

There are several active faults with right lateral motion. The Okamura fault, an active fault segment of the Median Tectonic Line (MTL), is the boundary between the mountainous area and the plain.

The Komatsu fault is the main fault in the plain, location of which is partly unknown because of the overlying sediments. In Saijo city, drainage system mainly consists of four rivers namely Kamo, Uzui, Nakayama and Daimyojin Rivers flow from the mountain to the coast. The Kamo and the Uzui Rivers are the major component recharging the unconfined groundwater of the alluvial fan in the Saijo plain. The Saijo plain belongs to the Setouchi climate, a temperate climate with a relatively small amount of mean annual precipitation (1433 mm, Saijo city, 2008). The mean annual temperature is approximately 15°C.

The spatial distribution of groundwater under the Saijo plain is divided into three zones from southern mountains to northern coast: the spring, the artesian and the coastal groundwater zones. The spring area is mainly featured as a recharge area despite some local upwelling.

3. Methodology

Sampling wells were selected in such a way that they represent different geological formations and land use patterns at varying topography of the coastal area of Saijo plain. A total of twenty two water samples (eleven from each confined and unconfined aquifer) were collected during July, 2010. In-situ measurements for EC, pH, ORP and temperature were done using respective probes. The water samples were collected in pre-rinsed clean polyethylene bottles. The concentration of HCO_3^- was analyzed by acid titration (using Metrohm Multi-Dosimat) while other anions Cl^-, NO_3^-, SO_4^{2-}, Br^- and PO_4^{3-} were analyzed by DIONEX ICS-90 ion chromatograph. Major cations and trace elements were determined with inductively coupled plasma-mass spectrometry (ICP-MS). Oxygen and hydrogen isotopes were analyzed by mass spectrometer (MODEL MAT 252, Thermo Finnigan Inc.) in university of Tsukuba. The results for both isotopes are expressed through deviation from the VSMOW (Vienna Standard Mean Ocean Water) standard using the δ-scale according to the equation and unit is per mil.

$$\delta\ ‰ = \left[\frac{R_{Sample} - R_{VSMOW}}{R_{VSMOW}} \right] \times 1000$$

where, R is the isotopic ratio (i.e. $^2H/^1H$ and $^{18}O/^{16}O$) for the sample and standard. Analytical precisions of stable isotopes were better than 0.1‰ for $\delta^{18}O$ and 1.0‰ for δ^2H. For major ions, analytical precision was checked by normalized inorganic charge balance (NICB) (Kumar et al. 2010). This is defined as [(Tz+ - Tz-)/(Tz+ + Tz-)] and represents the fractional difference between total cations and anions. The observed charge balance supports the quality of the data points, which is better than ±5% and generally this charge imbalance came in favor of positive charge.

Factor analysis is used here for the classification and assessment of groundwater quality using the Statistical Package for Social Sciences (SPSS) software package (SPSS Inc., USA, version 12).

Factor analysis was applied on experimental data standardizing through z-scale transformation to avoid misclassification of samples due to the wide differences in data dimensionality (Singh et al., 2011). It gave information about the variables responsible for the spatial variation in groundwater quality.

4. Results and Discussions

4.1. General Water Chemistry

Statistical summary (i.e. minimum, maximum, average and the standard deviation) of the analytical results for each water-quality characteristic analyzed is given in Table 1. Average value trend for cations and anions in groundwater samples was found as $Ca^{2+} > Na^+ > Mg^{2+} > K^+$ and $HCO_3^- > Cl^- > SO_4^{2-} > NO_3^-$ respectively.

All the physico-chemical parameters within the highest desirable permissible limit recommended by WHO (World Health Organization) (2004), except Cl^- at some sampling points. High HCO_3^- represents the major source of alkalinity caused by the presence of carbonaceous sandstones in the aquifers and weathering of carbonate minerals related to the flushing of carbonate enrich water from unsaturated zone, where it is formed by decomposition of organic matter.

Table 1. Statistical summary of hydrogeochemical parameters of groundwater samples of Saijo plain, Japan

Parameters	Minimum	Maximum	Average	Std. Dev.
Temp.(℃)	14.0	23.2	17.8	2.6
Ph	5.9	7.6	6.9	0.5
EC (μs/cm)	87.0	1594.0	203.6	316.3
ORP (mv)	34.0	280.0	175.1	64.6
Na^+ (mg/L)	1.1	97.1	11.3	23.7
K^+ (mg/L)	0.7	6.0	2.2	1.3
Ca^{2+} (mg/L)	1.1	92.4	18.1	17.2
Mg^{2+} (mg/L)	0.9	76.4	6.3	15.7
Cl^- (mg/L)	1.9	502.7	29.1	106.3
SO_4^{2-} (mg/L)	6.5	56.6	14.2	11.3
HCO_3^- (mg/L)	25.3	57.1	41.0	7.6
NO_3^- (mg/L)	0.2	37.8	6.2	8.6
SiO_2 (mg/L)	5.9	33.2	11.9	5.5
$\delta^{18}O$ (‰)	-9.4	-7.8	-8.5	0.4
δD (‰)	-61.6	-53.8	-57.9	2.5

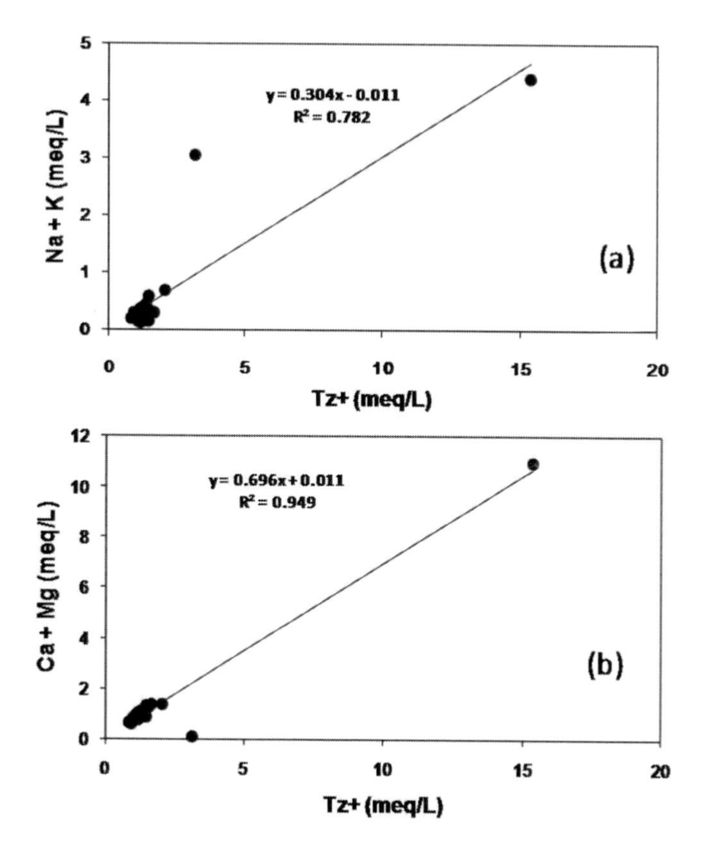

Figure 2. Scatter diagram for groundwater showing relationship between (a) $Ca^{2+}Mg^{2+}$ and Tz^+ and (b) Na^++K^+ and Tz^+ where Ca^2+Mg^{2+} accounts for most of cations in the groundwater.

Scatter plot between (Ca^{2+} + Mg^{2+}) versus Tz^+ for the groundwater yields a strong linear trend indicating that they accounts for most of the cations (Figure 2a) whereas, the trend between (Na^+ + K^+) versus Tz^+ was weak (Figure 2b).

These results suggest that the carbonate weathering is the dominant process and contribution of cations via alumino-silicate weathering is low in comparison to carbonate weathering. Low average value of silica also supports the above mentioned fact. Stiff diagram is used to make a quick visual comparison between waters from different sources and to gain better insight into the hydro-geochemical processes operating in the coastal aquifer of Saijo plain which resulted in the spatio-temporal variation in hydrochemistry (Figure 3). From this diagram, it was found that water samples were mainly dominated by Ca-HCO_3 type which is very consistent with the geological feature of the alluvial plain. On contrast, on the west side of the plain there is preliminary signature of sea water intrusion which was characterized by Ca-Cl and Na-Cl type of water samples taken from bore wells with screen depth more than 25 meter. This shows that average high value of Cl^- might be because of encounter of screen depth to the sea water – fresh water interface. On the east side of plain, water is enriched with anions SO_4^{2-} and NO_3^- indicating the dominance of anthropogenic inputs like agricultural activities and sewage effluents.

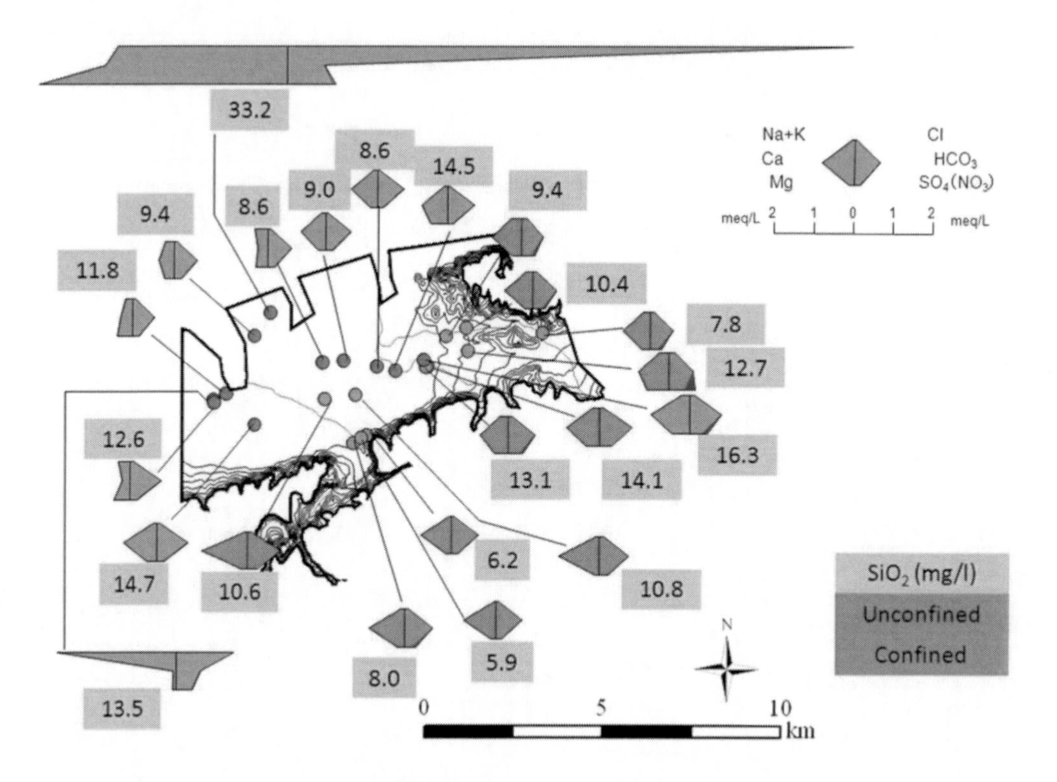

Figure 3. Hexadiagram showing groundwater qulity at spatial scale.

4.2. Isotopic Signature of Groundwater

Value for $\delta^{18}O$ ranges from (-9.4 to -7.8 ‰) while corresponding value for δ^2H ranges from (-61.6 to -53.8‰) (Table 1). The relationship between $\delta^{18}O$ and δ^2H values for confined and unconfined water is shown in Figure 4. Except few points, most of the water samples were clustered on or near the meteoric water line. Thus, these results are not affected according to the deviations isotopic compositions away from the meteoric water line, including evaporation from open surface and exchange with rock minerals. It was found that confined groundwater samples tend to have lower value than unconfined water which suggested that recharge elevation of confined groundwater is higher than one of unconfined water. Confined groundwater samples away from the LMWL might be results of evaporation and exchange with rock minerals. To confirm the process behind it, scatter plot was drawn to show relationship between $\delta^{18}O$ and chloride concentrations of the groundwater in the study region (Figure 5). The dash or broken line in this figure denotes the precipitation-seawater mixing line.

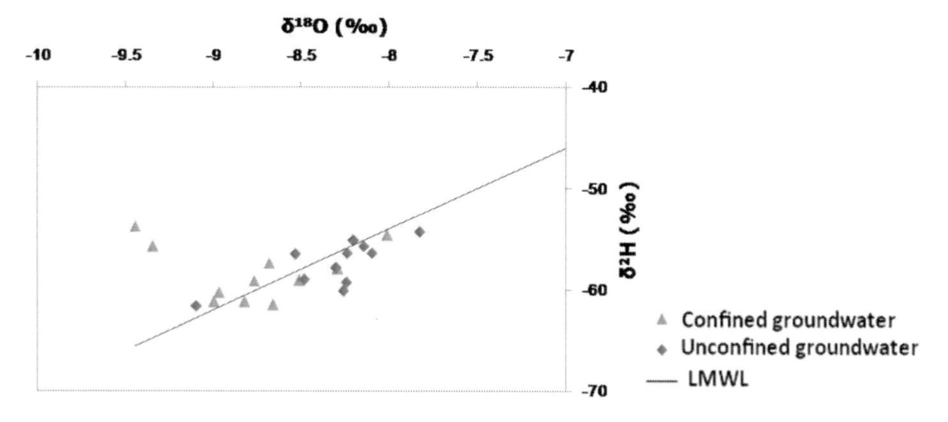

Figure 4. Scatter plot showing relation between $\delta^{18}0$ and $\delta^{2}H$ for groundwater. (Here LMWL –Local meteoric water line).

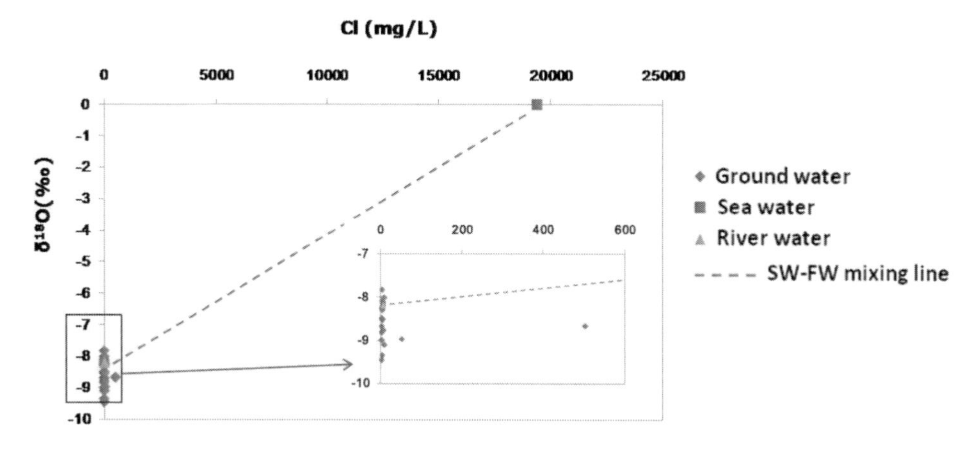

Figure 5. Scatter plot showing relation between $\delta^{18}0$ and Cl for groundwater and its trend on mixing line.

Groundwater samples were plotted along the $\delta^{18}O$ axis is due to the very low chloride concentration of groundwater samples compared to their wide range of $\delta^{18}O$ values. However, two groundwater samples are distinguishable from other samples indicating relatively high chloride concentrations with value of 52.0 and 502.7 mg/l, respectively (also shown in west side in Figure 3), and they converge toward the composition of sea water along the mixing line. This suggests that salinization is caused solely by mixing with coastal marine water.

4.3. Factor Analysis

The factor analysis (shown in Table 2) resulted in three principal components representing three main sources of variation in the hydrochemistry at the regional level. These three components accounts for 76.38 % of the variance in the hydro-chemical data. Factor 1 has higher loadings for EC, Ca^{2+}, Mg^{2+} and HCO_3^- which Indicating weathering of carbonaceous materials in the area. Factor 2 has higher loading of Cl^-, $\delta^{18}O$ and δ^2H which suggests about groundwater salinization. Factor 3 shows higher loading for SO_4^{2-} and NO_3^- indicating anthropogenic activities determining water quality at some points.

Table 2. Multivariate factor analysis score for groundwater samples of Saijo plain, Japan

Variables	F1	F2	F3
Temp.	0.35	-0.42	0.867
pH	0.22	0.37	0.13
EC	0.98	0.34	0.65
ORP	-0.49	-0.24	0.64
Na^+	0.55	0.72	-0.18
K^+	0.25	0.80	0.28
Mg^{2+}	0.73	0.13	0.16
Ca^{2+}	0.94	0.61	0.39
Cl^-	0.47	0.88	0.51
NO_3^-	0.12	0.56	0.72
SO_4^{2-}	0.38	0.27	0.86
HCO_3^-	0.92	-0.71	0.17
SiO2	0.46	-0.28	-0.02
$\delta^{18}O$	-0.03	0.77	-0.70
δ^2H	-0.32	0.51	-0.43
Eigen Value	6.86	2.67	1.92
Percentage of Variance	45.74	17.82	12.82
Cumulative percentage	45.74	63.56	76.38

Conclusion and Recommendations

Hydrochemical analysis suggests that most of the water samples in Saijo plain were showing good quality i.e. all the chemical parameters were within the highest desirable limit set by WHO, while two samples shown high chloride concentration i.e. state of salinization which needs immediate attention. From isotopic study, it was found that most of the water samples are of meteoric origin. Relation between $\delta^{18}O$ and chloride firmly supported local SW-FW mixing reason behind elevated chloride concentration in case of two water samples. Factor analysis also firmly supported the results from graphical representation. From above work it was found that for samples taken from bore well with screen depth >25 meter has a problem of salinization so it will be recommended to have an alternate tube/bore well with screen depth <15 meter in order to prevent the encounter of SW–FW interface as an alternative. To validate qualitative aspects of water quality, study for present and future status of SW intrusion through numerical simulation can be a future perspective.

ACKNOWLEDGMENTS

The authors are highly thankful to the Monbukagakusho (MEXT) Japanese Government fellowship which helped to pursue research. Authors also want to put on record contribution of Graduate School of Life and Environmental Science, University of Tsukuba for facilitating data analysis in its labs.

REFERENCES

Adams, S., Titus, R., Pietersen, K., and Tredoux, G. (2001). Hydrochemical characteristics of aquifers near Sutherland in the Western Karoo, South Africa. *Journal of Hydrology*, 241, 91-93.

Capaccioni, B., Didero, M., Paletta, C., and Didero, L., (2005). Saline intrusion and refreshening in a multilayer coastal aquifer in the Catania Plain (Sicily, Southern Italy): dynamics of degradation processes according to the hydrochemical characteristics of groundwaters. *Journal of Hydrology,* 307 (1– 4), 1–16.

Chen, K.P., and Jiao, J.J. (2007). Seawater intrusion and aquifer freshening near reclaimed coastal area of Shenzhen. *Water Science and Technology*, 7,137-145.

Forcada, E. G. (2010). Dynamics of Sea water interface using hydrochemical facies evolution diagram. *Ground Water*, 48, 2, 212-216.

Freeze, R.A., and Cherry, J.A., 1979. Groundwater, Prentice-Hall.

Gattacceca, J.C.,Coulomb, C.V., Mayer, A., Claude, C., Radakovitch, O., Conchetto, E., and Hamelin, B. (2009). Isotopic and geochemical characterization of salinization in the shallow aquifers of a reclaimed subsiding zone: *The southern Venice Lagoon coastland. Journal of Hydrology,* 378, 46–61.

Jorgensen, N.O., Andersen, M.S., and Engesgaard, P. (2008). Investigation of a dynamic seawater intrusion event using strontium isotopes (87Sr/86Sr). *Journal of Hydrology*, 348, 257– 269.

Kim,Y., Lee, K.S., Koh, D.C., Lee, D.H., Lee, S.G., Park, W.B., Koh, G.W., and Woo, N.C. (2003). Hydrogeochemical and isotopic evidence of groundwater salinization in a coastal aquifer: a case study in Jeju volcanic island, Korea. *Journal of Hydrology*, 270, 282-294.

Kumar, P., Kumar, M., Ramanathan, A.L., and Tsujimura, M. (2010). Tracing the factors responsible for arsenic enrichment in groundwater of the middle Gangetic Plain, India: a source identification perspective. *Environmental Geochemistry and Health*, 32,129–146.

Kurihara, G. (1972). Geology of the alluvial plains of the southern coastal area of Setouchi. Tohoku Univ. Inst. Geol. Pal. Center, 73, 31-65 (in Japanese with English abstract).

Langman, J. B., and Ellis, A. S. (2010). A multi-isotope (δD, δ18O, 87Sr/86Sr, and δ11B) approach for identifying saltwater intrusion and resolving groundwater evolution along the Western Caprock Escarpment of the Southern High Plains, New Mexico. *Applied Geochemistry*, 25, 159–174.

Lee, J.Y., and Song, A.S.H. (2007). Evaluation of groundwater quality in coastal areas: implications for sustainable agriculture. *Environmental Geology*, 52, 1231–1242.

Nakano, T., Saitoh, Y., and Tokumasu, M. (2008). Geological and human impacts on the aquifer system of the Saijo basin, western Japan. *Proceedings of 36th IAH Congress.*

Petalas, C., Pisinaras, V., Gemitzi, A., Tsihrintzis, V.A., and Ouzounis, K. (2009). Current conditions of saltwater intrusion in the coastal Rhodope aquifer system, northeastern Greece, *Desalination*, 237, 22-41.

Ray, R.K., and Mukherjee, R. (2008). Reproducing the Piper Trilinear diagram in Rectangular Coordinates. *Ground Water*, 46 (6), 893-896.

Saijo City Office. (2008). *Annual report of ground water for 2008. Saijo city*, life environment section, 88pp (in Japanese).

Schiavo, M.A., Hauser, S., and Povinec, P.P. (2009). Stable isotopes of water as a tool to study groundwater–seawater interactions in coastal south-eastern Sicily. *Journal of Hydrology*, 364, 40– 49.

Singh, C.K., Shashtri, S., and Mukherjee, S. (2011). Integrating multivariate statistical analysis with GIS for geochemical assessment of groundwater quality in Shiwaliks of Punjab, India. *Environmental Erath Science*, 62, 1387-1405.

Steyl, G., and Dennis, I. (2010). Review of coastal-area aquifers in Africa. *Hydrogeology Journal*, 18, 217–225.

Trabelsi, R., Zairi, M., and Dhia, H. (2007). Groundwater salinization of the Sfax superficial aquifer, Tunisia. *Hydrogeology Journal*, 15 (7), 1341–1355.

Vengosh, A., Kloppmann, W., Marei, A., Livshitz, Y., Gutierrez, A.,Banna, M., Guerrot, C., Pankratov, I., and Raanan, H. (2005). Sources of salinity and boron in the Gaza strip: natural contaminant flow in the southern Mediterranean coastal aquifer. *Water Resource Research*, 41, W01013.

Voudouris, K.S. (2006). Groundwater Balance and Safe Yield of the coastal aquifer system in NEastern Korinthia, Greece. *Applied Geography*, 26, 291–311.

WHO. (2004). Guidelines for drinking water quality-II pp 333. Geneva, *Environmental Health Criteria*, 5.

In: Water Quality
Editor: You-Gan Wang

ISBN: 978-1-62417-111-6

Chapter 5

SUITABILITY OF GROUNDWATER OF ZEUSS-KOUTINE AQUIFER (SOUTHERN OF TUNISIA) FOR DOMESTIC AND AGRICULTURAL USE

Fadoua Hamzaoui-Azaza*[*,1], *Besma Tlili-Zrelli*[1], *Rachida Bouhlila*[2] and *Moncef Gueddari*[1]

[1]Laboratory of Geochemistry and Environmental Geology, Department of Geology, Faculty of Mathematical, Physical and Natural Sciences, University Campus,Tunis, Tunisia

[2]Modeling in Hydraulic and Environment Laboratory, National Engineers School of Tunis, Tunisia

ABSTRACT

In arid and semi-arid regions from North African countries, ground water forms the major source of water supply for socio-economic growth and sustainability of the environment.

Hydrogeochemical characteristics of deep groundwater in southeast Tunisia have been assessed to identify its suitability for drinking and irrigation applications

In order to evaluate the quality of groundwater in Zeuss-Koutine, which represents the principal resource of water for Medenine, the hydrochemical data have been analyzed using geochemical methods and multivariate statistical techniques such as Principal component analyses and Cluster analyses.

In this study 14 wells were sampled in summer (July) and their water was analyzed for various variables, including temperature, pH, Total Dissolved Solids (TDS), Na^+, Cl^-, Ca^{2+}, Mg^{2+}, SO_4^{2-}, K^+, HCO_3^-, Fe^{3+}, Mn^{2+}, Zn^{2+}, Al^{3+}, Pb^{2+}, Cr^{3+}, Cu^{2+} and F^-

The results showed that the majority of ions are above the maximum desirable limit although trace metals are within the maximum permissible limit for drinking water. Besides and on the basis on the water quality index WQI, groundwater has been

* Corresponding autors: Fadoua Hamzaoui-Azaza, 1Laboratory of Geochemistry and Environmental Geology, Department of Geology, Faculty of Mathematical, Physical and Natural Sciences, University Campus.Tunis. Tunisia; fadoua_fst@yahoo.fr.

classified into “unsuitable for drinking purposes”, “Very poor water “and “Poor water” and can’t be used for drinking purposes without any treatment. The calculated values for sodium adsorption ratio (SAR) indicate well to permissible use of water for irrigation signifying that sodicity is very low. According to the USSL diagram, the most dominant classes of water sampling are C4a-S1 and C4b-S1 indicating very high and extremely high salinity waters which are unsuitable for irrigation with a restrict drainage.

In the whole domain, a significant increase in the degree of water mineralization was observed from southwest to northeast, following the regional flow direction.

Interpretation of hydrochemical data reveals that Water quality is mainly dominated by dissolution of evaporates minerals. Results obtained from principal component analyses (PCA) showed that 3 components explaine more than 75 % of the total variance in the groundwater quality and demonstrated that the variable responsible for water quality are largely related to soluble salts species (Na^+, Cl^-, Ca^{2+}, Mg^{2+}, SO_4^{2-} and K^+). Cluster analysis showed that sites sampling can be grouped in tow clusters.

INTRODUCTION

Groundwater is an important source of water supply throughout the world. It is estimated that approximately one-third of the world’s population use groundwater for drinking (Nickson et al. 2005). In arid zones, water is a rare and precious resource. The exploitation of water resources is a complex problem in the framework of sustainable agricultural development in these regions (Hamzaoui-Azaza et al., 2009).

The current situation of the water resources and their uses in Tunisia, which is the most drought-stressed countries in the middle East and North Africa region, present a common stakes with many areas of the Mediterranean basin: limited and already largely exploited resources to answer the growth of the needs, the increasing use of so-called non conventional resources, the overdependence on groundwater to meet increasing demands of domestic, agriculture, and industry sectors, an increasing merchandising of the resources, constraining climatic conditions which come to reinforce the tensions around water. Under such conditions, Tunisia as all other countries subject to these same constraints, have no choice but to increase their vigilance around the issues related to rational management and conservation of water resources. While the demography is expanding in South Tunisia, the water needs of the three main sectors, irrigation, drinking water supply and industry, are expected to increase in order to provide the population with employment and life conditions enabling people to settle in their homeland. In fact in this region of Tunisia, the groundwater is currently the only available water resources upon which depend all the socioeconomic sectors. In the medium term, desalination of sea water, preferably using solar energy available in these latitudes, could be one of the best alternatives to the supply of freshwater.

In terms of quality, different classifications for water use are commonly cited and geochemical and statistical methods enable us to deduce the sources of water mineralization. The chemical composition of groundwater is controlled by many factors that include the climate conditions, the geological structure and mineralogy of the watershed and the aquifer, and the geological processes within the aquifer (Rosen and Jones 1998). The interaction of all these factors leads, for a specific aquifer, to a particular type of water whose chemical composition determines its uses.

Geochemical methods and statistical techniques including Hierarchical cluster analysis (HCA) and Principal component analysis (PCA) are respectively used to identify factors and the dominant mechanisms controlling groundwater chemistry.

A water quality index (WQI), which can serve as a useful tool for correct evaluating the quality of groundwater and surface water (Abassi 1999, Adak et al., 2001), was generated for the entire study area to define the suitability of water for drinking purpose and to classify groundwater in the area into spatial water quality types. Generally, indices have been developed to summarize water quality data in an easily and understandable format (Saeedi 2010).

The high salinities of groundwater from some of the wells in the area and the large dependence of the communities on groundwater for irrigation require a global evaluation of the quality of the resource in the area for irrigation purposes.

Numerous parameters such as Sodium adsorption ratio (SAR), Percent Sodium (% Na), Residual Sodium Carbonate (RSC) and Percent Magnesium (% Mg) have been used to evaluate the suitability of groundwater for irrigation purposes (Paliwal 1972, Domenico and Schwartz 1990, Tijani 1994, Shaki 2006).

In the current study, a first attempt was made to assess the groundwater quality of Zeuss Koutine aquifer and its suitability for drinking and agricultural uses, thus helping to the effective and sustainable management of groundwater resources in this area. Sustainable development of water sources must go hand in hand with improved sanitation and hygiene practice.

STUDY AREA

The study site, Zeuss-Koutine aquifer, is described in details by Hamzaoui Azaza et al., 2011. Tt belongs to the region of south eastern Tunisia. It is situated in the north of the city of Médenine between the Jeffara plain and Matmata mountains, actually ending in the saline depression (Sebkha) of Oum Zessar before ending into the Gulf of Gabès in the Mediterranean sea. It is bordered, at the north, by the Segai plain and the town of Mareth, in the south by the Trias Sandstone aquifer and in the south east by the Jorf Miocene aquifer (Figure 1). The study site is characterized by steppe vegetation. It is characterized by an arid to semi-arid climate with a low and highly irregular total rainfall (150-240 mm). Temperature differences are extreme between the seasons ranging from-3°C (winter) to as high as 48°C (summer) (Ouassar 2007).The hydrogeographic network is quite dense.

It is constituted primarily by three big Wadiis named Zeuss Wadi, Koutine-Om Ezzassar Wadi and Zigzaou Wadi. The geological formations consist of alternating continental and marine origin. The oldest submerging layers are represented by a marine, superior Permian, and the most recent ones are of the recent Quaternary (Morhange and Pirazzoli 2005).

Between the Permian and Quaternary formations we find confined strata of different ages, which are generally declining in northward direction (Gaubbi 1988). Zeuss-Koutine region is bordered by three main structures that define the South-Eastern Tunisia: the Dahar monocline, West and North-West, the Medenine Tebaga monocline, South, and the plain of the Djeffara, East and North (Hamzaoui Azaza et al., 2011).

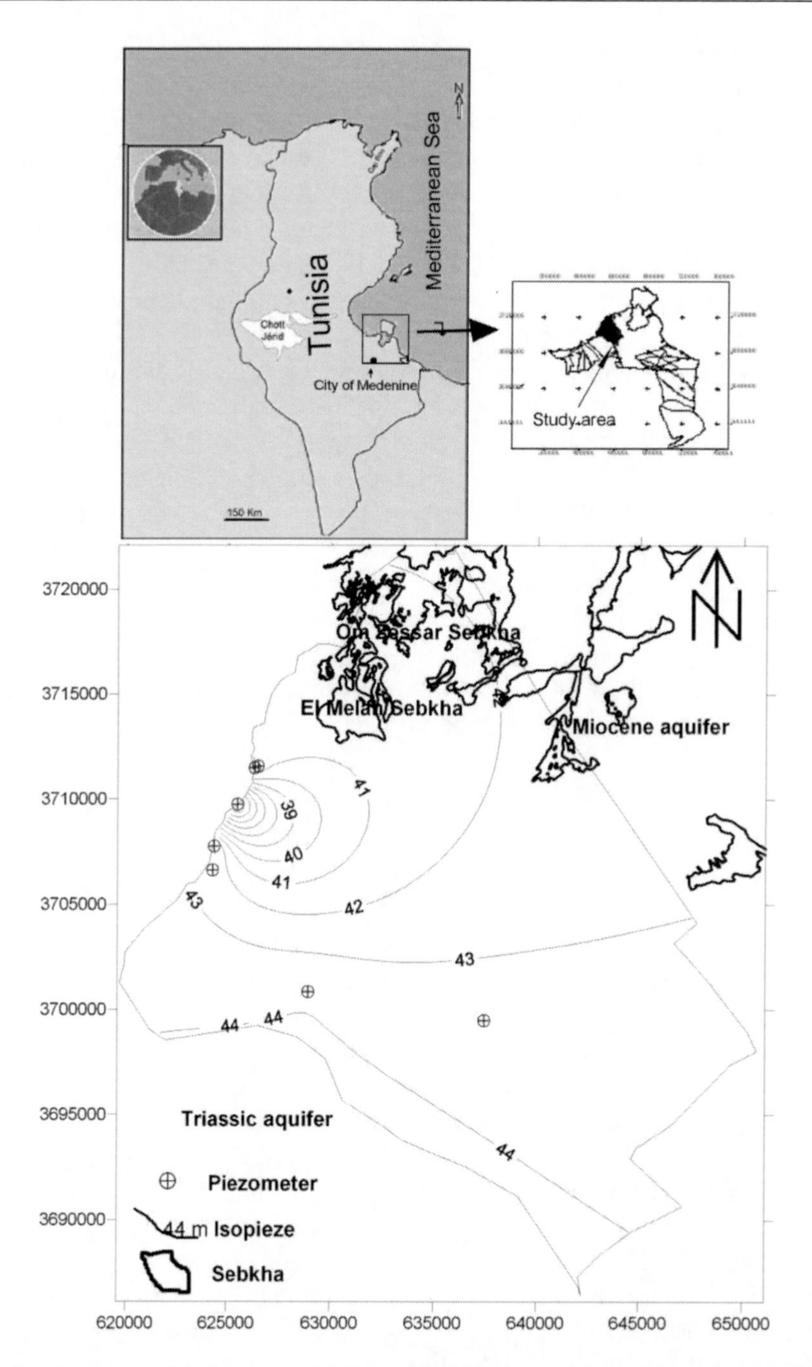

Figure 1. Location map of the study area and Spatial distributions of groundwater level in 2004.

The Zeuss-Koutine aquifer unit constitutes an important water source used for various purposes at an average abstraction estimated at a rate of 350 l/s. The extraction of water from this reservoir started in 1962. The continuous monitoring has shown that the withdrawal rate increased from 102 l/s in 1974 to 440 l/s in 2005 resulting in a decline of the mean piezometric level of 0.33 m in 1982 and 1.02 m in 2004 (DGRE, 2004).

Zeuss-Koutine aquifer is constituted by the following litho-stratigraphic units: Jurassic limestone and dolomite; Albo-Aptian calcitic dolomites; Turonian dolomites and dolomitic limestone and Lower Senonian limestone. However, in Medenine region, the Jurassic limestone and dolomite constitute the main aquifer material because of its considerable thickness reaching up to 120 m. The thickness of this aquifer is between 30 and 200 m; and its depth varies between 70 and 250 m (Hamzaoui-Azaza 2011).

Results of aquifer tests indicate that transmissivities of the aquifer varies between 0.055 and 0.2 $m^2 s^{-1}$ (OSS 2005).

This aquifer unit is recharged by water flowing from Matmata Mountains, where important outcrops are made up of Jurassic limestone, as well as by local infiltration from several rivers. Besides, The Zeuss-Koutine aquifer unit is fed by the underlying Triassic Sandstone aquifer. The general groundwater flow in the horizontal plane s takes part from South-West to North-East (OSS 2005).

Water of the Zeuss-Koutine aquifer is used unevenly by different economic sectors. However, drinking water supply remains the primary use. Anthropogenic activities in the study area rely mainly on agricultural.

SAMPLE COLLECTION AND ANALYTICAL TECHNIQUES

In order to study the groundwater quality in the Zeuss-Koutine aquifer 14 groundwater samples, taken from wells used for drinking, industrial, irrigation, and other domestic purposes, were collected in July 2005 from boreholes ranging in depth from 91 to 577 m after 10 min of pumping. The geographical location of the sampling sites is shown in Figure 4. Groundwater samples were collected in sterilized polythene bottles through 0.45-μm in-line filters. All samples were transported to the laboratory to analysis and kept in a refrigerator below 4° C. Each of the groundwater samples was analyzed for 17 parameters such as physico-chemical parameters, majors and trace elements.

Unstable parameters temperature, pH and salinity were measured in the field using well-calibrated digital sensors. The pH electrode was calibrated against pH 4, 7, and/or 10 buffers. Samples for the analysis of cations were acidified to pH 2 by adding several drops of ultra-pure nitric acid. The analyses for various chemical parameters to assess the groundwater quality were carried out using standard procedures (Rodier, 1996). The water samples were analyzed at an approved chemical laboratory in Tunisia. Calcium and magnesium concentration were determined by complexometric titration method using Ethylene Diamine Tetra-acetic Acid (EDTA). Sodium and potassium concentrations were determined using Flame Photometer. Bicarbonate concentration was determined by acidimetric titration method using methyl orange as indicator. Chloride was determined by using 0.1N $AgNO_3$ solution. Sulphate was determined by gravimetric method.

Metal ion concentrations were determined by atomic absorption spectrometry. Quantification of metals by means of a calibration curves of aqueous standard solutions of respective metals. These calibration curves were determined several times during the period of analysis. The quality of the chemical analyses was carefully inspected by checking ion balances. The ion balance errors for the analyses were within ± 5%.

Visually communicating iso-concentration/contour maps were constructed using Surfur 7.0 software to delineate spatial variation of physico-chemical characteristics of groundwater samples.

RESULTS AND INTERPRETATION

Understanding the groundwater chemistry is important as it is the main factor determining its suitability for drinking, agricultural and industrial purposes (Subramani et al. 2005). The chemical composition of groundwater is generally controlled by several factors that include Climate, soil characteristics, geological structure and mineralogy of the watersheds and aquifers and geochemical processes within the aquifer (Jallali, 2010). The mixing/non-mixing of different types of groundwater may also play important roles in determining the quality of the groundwater (Reghunath et al., 2002). The interaction and the combination of all factors leads to different water types. Physical and chemical parameters including statistical measures such as minimum, maximum, average, and median are given in Table 1. The physical and chemical parameters of the analytical results of groundwater were compared with the standard guideline values recommended by the World Health Organisation (WHO 2004) for drinking water.

Table 1. Summary statistics of the analytical data such as minimum, maximum, average, and median

Parameters	Average	Mean	Min	Max	Variance	Standard deviation	WHO 2004
T°C	27.80	27.10	22.00	36.10	11.29	3.36	
pH	7.56	7.60	7.00	7.90	0.05	0.22	Optimum 6.5-9.5
Na(mg/l)	641.66	740.00	270.00	1327.56	95845.22	309.59	200 (mg/l)
K (mg/l)	18.02	20.00	8.60	33.24	47.53	6.89	30 (mg/l)
Ca (mg/l)	273.77	302.00	107.00	445.00	11319.35	106.39	200 (mg/l)
Mg (mg/l)	121.71	133.00	73.00	194.08	1208.14	34.76	200 (mg/l)
SO4 (mg/l)	945.62	930.00	534.05	1484.64	95057.76	308.31	400 (mg/l)
HCO3 (mg/l)	184.08	191.64	106.00	240.34	1533.39	39.16	380 (mg/l)
Cl (mg/l)	1001.65	1135.97	338.00	2159.11	293109.19	541.40	250 (mg/l)
TDS (mg/l)	3199.93	3480.00	1520.00	5400.00	1486496.07	1219.22	1000 (mg/l)
Fe	0.18	0.01	0.00	1.67	0.22	0.47	
Mn (ug/l)	0.01	0.00	0.00	0.06	0.00	0.02	0.4 (mg/l)
Cu (ug/l)	0.02	0.00	0.00	0.05	0.00	0.02	2 (mg/l)
Al (ug/l)	0.20	0.06	0.00	0.79	0.06	0.25	0.2 (mg/l)
Si (ug/l)	7.54	5.67	2.81	13.64	17.66	4.20	
Zn (ug/l)	0.06	0.03	0.00	0.20	0.00	0.06	3 (mg/l)
Pb (ug/l)	2.38	0.00	0.00	6.30	4.50	2.12	0.01 (mg/l)
Cr (ug/l)	0.37	0.03	0.00	1.40	0.24	0.49	0.05 (mg/l)
F (mg/l)	2.18	2.20	1.64	2.80	0.10	0.31	1.5 (mg/l)

Water chemical data has been first approached by a description of the spatial variation of some ions, than has been used for computation irrigation quality parameters Sodium adsorption ratio (SAR), Percent Sodium (% Na), Residual Sodium Carbonate (RSC) and Percent Magnesium (% Mg), besides, tow statistical analyses methods were applied : Principal Component Analysis (PCA) and Cluster Analyses.

Physico-Chemical Parameters

Temperature and pH

The temperature variation of the groundwater in the study area ranges from 22◦ C to 30.6◦C. The pH of the groundwater in the study area ranges from 7.0 to 7.9 indicating that the waters are generally neutral to slightly alkaline. The pH values of groundwater samples are within the permissible limit suggested by WHO (WHO, 2004).

Salinity

In natural waters, dissolved solids consists mainly of inorganic salts such as bicarbonates, sulfates, chlorides, calcium, phosphates, and magnesium, potassium, sodium, etc., and a small amount of organic matter and dissolved gases. The salinity value ranging from 1520 mg/l to 5400 mg/l with an average value of 3199 mg/l. The highest values of salinity are generally registered when the movement of groundwater is at its least, hence the salinity is influenced with depth/time and recharge/discharge area relationships. Water from recharge areas is usually diluted in contrast in the discharge areas, it is often relatively saline (Chilton 1992).

Consequently, the increase in the groundwater salinity of the aquifer from southwest towards the north and northeast may be attributed to the farthest distance from the naturally recharged area and the increased saturated thickness of the aquifer in this direction.

Spatial distribution maps, showing an increase towards groundwater flow (Figure 2). The map shows also that P9, P10, P11 and P12 wells that are located in the upstream section of the aquifer (recharge zone) have the lowest salinity levels. Relatively low TDS values characterize wells located near the River and reveal the dilution of groundwater by the water from Triassic aquifer. About 100% of the samples analyzed were found above the desirable limit of 1000 mg/L (WHO 2004) This spatial variations reflects the variation of Na, Cl, Ca, Mg, K and SO4 concentrations, with the linear correlations (R^2= 0.98, 0.96, 0.97, 0.79, 0.91 and 0.77) between salinity and the concentrations of these ions, suggesting that the increase of these elements et especially Na and Cl concentrations contributes to the salinity increases.

Major Ions

Chlorides and Sodium

Chloride is a usually distributed element in nature in one or in combination with other elements. It has a high affinity towards sodium. Therefore, its concentration is high in ground waters, where the temperature is high and rainfall is less (Geetha et al, 2008; Ramakrishnaiah et al., 2009). Chloride Concentrations vary between 338 and 2159 mg/l with an average value of 1002 mg/l. Those of the sodium are ranging between 270 and 1328 mg/l with an average value of 740 mg/l. Sodium is the dominant cation in Zeuss-Koutine aquifer, whereas Chloride is the dominant anion. Excess in Chloride and sodium concentration is found in wells.

The possible source of sodium and chloride concentration in groundwater would be due to the dissolution of rock salts and weathering of sodium-bearing minerals (Krishna Kumar et al., 2009).

If the halite dissolution process is responsible for the sodium, Na/Cl ratio should be approximately 1, while the Na/Cl ratio grather than 1 reflect generally a release of sodium from silicate weathering (Meyback, 1987), if Na/Cl ratio is less than 1, the reduction of Na concentration may be due to ion exchange process (Jeevanandam et al., 2007). In this present study, Na/Cl ratio is greater than 1 in the predominant groundwater samples (9 samples).

The chloride and sodium ions concentration in groundwater of the study area exceeds the maximum allowable limit of 600 mg/l in respectively nine and ten locations (WHO 2004).

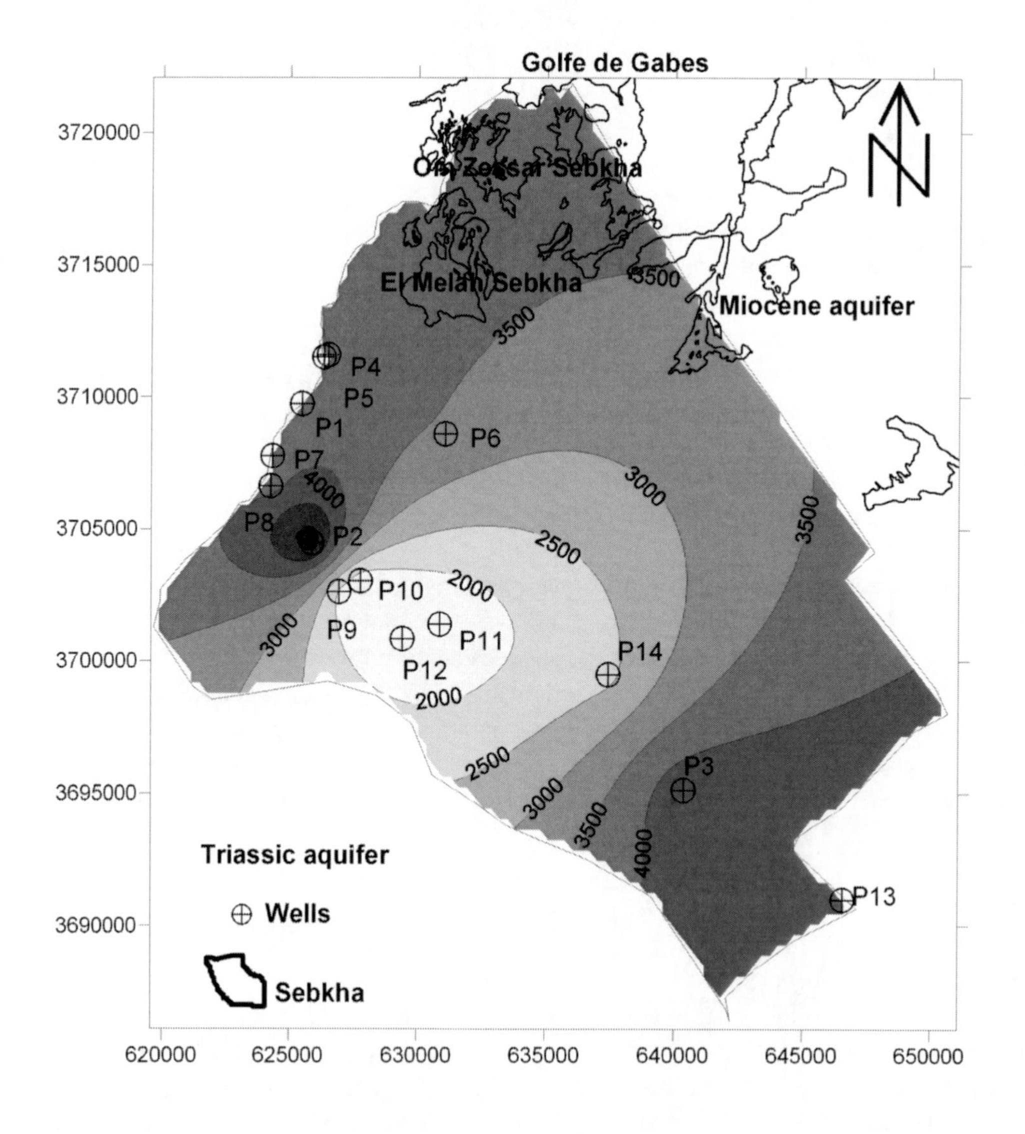

Figure 2. Spatial distribution of Salinity and location of the sampling wells.

Calcium and Magnesium

The origin of Calcium and magnesium ions present in groundwater is generally the weathering of dolomites, limestone, gypsum and anhydrites, whereas the cation exchange process may explain the contribution of calcium in aquifer (Subramani et al., 2010).

The calcium concentrations oscillate between 107 and 445 mg/l, 64 % of the samples exceed the desirable limit of 200 mg/l as per WHO standard (WHO 2004). Those of Magnesium range between 73 and 194 mg/l. The recommended limit for Magnesium in natural water is 150 mg/l (WHO 2004). 80% of the groundwater samples fall below this limit.

Sulfates

Natural *concentrations* of sulphates may be enriched by weathering of *sulfate* minerals, such as *gypsum* and anhydrite. Sulfate concentrations varied from 534 to 1485 mg/l with an average value of 945 mg/l. The spatial distribution maps of sulfate concentration show that high concentrations are located in North-Est, in P4 and P5 well and in P13 and P3 well at the South-East. All groundwater samples exceed the desirable limit of 250 mg/l as per WHO standard (WHO 2004).

Potassium

Potassium concentrations in the Zeuss-Koutine aquifer were nearly homogeneous within each sample site. The values range from 9 to 33 mg/L, Potassium concentration in groundwater of the study area is within the maximum allowable limit in all the sample locations (WHO 2004). This low level of potassium is explained by the tendency of this element to be fixed by clay minerals and to participate in the formation of secondary minerals (Matheis 1982, Nwankwoala and Udom 2011).

Alkalinity

Water alkalinity of the Zeuss-Koutine aquifer is the only product of bicarbonate ions, given that pH values are almost near to neutral. The bicarbonate ion concentrations ranged from 106 to 240 mg/l with an average value of 184 mg/l. The bicarbonate concentration in groundwater is derived from carbonate weathering:

$$CaCO_3 + CO_2 + H_2O \rightarrow Ca^{2+} + 2\,HCO_3^- \text{ and } CO_2 + H_2O \rightarrow H^+ + HCO^-_3$$

The spatial distribution of alkalinity is rather different from the maps of the main ions. In fact, wells where water shows the lowest concentration major ion are characterized by the highest bicarbonate concentration levels. The increase of bicarbonate concentration in the groundwater may be due to the availability of carbonate minerals in the recharge areas and silicate weathering (Elango et al., 2003, Krishna Kumar et al., 2009). The alkalinity values also exceed the desirable limit of 200 mg/l in about 30% of the samples.

Trace Elements

Minor elements are less abundant and are often concentrated under certain conditions. In fact, many physical and chemical processes may control amounts of these elements in groundwater such as dispersion, complexation, acid-base reactions, oxidation-reduction, precipitation-dissolution…(Fetter 2001).

The trace elements concentrations have a few extreme values. The concentrations of Fe^{3+}, Mn^{2+}, Zn^{2+}, Al^{3+}, Pb^{2+}, Cu^{2+} and Cr^{3+} were lower than the maximum permissible level prescribed by WHO standards set for drinking water. The observed variations are not explained by concurrent variations in TDS. The comparison of the hydrochemical data with the WHO standards shows that 100 % of the samples exceeded the guide value for fluoride ion (1.5 mg/l), which explains the existence of many cases of dental fluorosis in the south of Tunisia. In fact, research on the relationship between fluorine concentration in drinking water and endemic fluorosis has been conducted in many parts of the world (Guo et al., 2007).

In general, the main fluoride source in groundwater is related to the mineralogy of the bedrock and their concentration dispersion values are due to different bedrock types. Dissolution of Fluorite (CaF_2) is a plausible source of fluoride ion in groundwater (Abu Rukaha and Alsokhny 2004).

Hydrochemical Facies

Based on the major cation and anion, The chemical groundwater character of the study area were represented by drawing piper trilinear diagram to identify and compare water types (Piper, 1944). Calcium and sodium are mainly the dominant cations in Zeuss-Koutine groundwater and among the anions, chloride and sulfate are dominant.

Two major groundwater groups can be distinguished in the piper diagram (Figure 3): the Na–Cl group, which is characterising discharge area and the Ca–SO4–Cl group characterising recharge area.

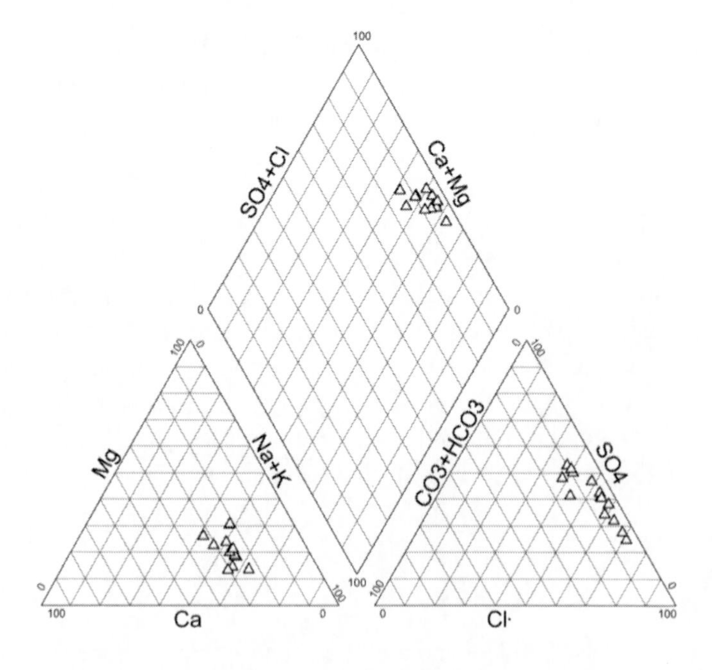

Figure 3. Piper trilinear diagram for the various cations and anions composition of the water samples.

Suitability for Drinking Purposes Using Water Quality Index (WQI)

Water Quality Index is a useful and efficient method for assessing the quality of water and its suitability for drinking purposes. WQI is a mathematical instrument used to transform large quantities for water quality data into a single number which represents the water quality level (Saedi 2010, Chander 2011). The adopted methodology to develop a WQI involves three steps. The first one is to assign weight (w_i) to the sampling points based on the concentration of the physico-chemical parameters and/or biological constituents of the water and their relative importance in the quality of water for drinking purposes (Table 2) (Yiadana 2010).

Table 2. Weight and Relative weight of chemical parameters used for WQI computation

Chemical parameter (mg/l)	WHO (2004)	Weight (w_i)	Relative weight W_i
TDS	1000	5	0.172
Na^+	200	3	0.103
Ca^{2+}	200	3	0.103
Mg^{2+}	200	3	0.103
K^+	30	3	0.103
F^-	1.5	5	0.172
Cl^-	250	3	0.103
SO_4^{2-}	400	3	0.103
HCO_3^-	380	1	0.034

The second step consists on the computation of the relative weight (W_i) which is calculated from (Eq.1)

$$W_i = w_i / \sum^{n}_{i=1} w_i \qquad \text{(Eq.1)}$$

where w_i is the weight of each parameter,
n is the number of parameters

In the third step, a quality rating scale (q_i) from each parameter is computed from (Eq.2)

$$q_i = (Ci/Si) * 100 \qquad \text{(Eq.2)}$$

where Ci is the concentration of each chemical parameter in each water sample in mg/l S_i is a world health organization's standard for each of the major parameters in drinking water (WHO), 2004.

The sub index SI_i of ith parameter and the water quality index WQI were respectively computed from Eqs. (3) and (4).

$$SI_i = W_i * q_i \qquad \text{(Eq.3)}$$

$$WQI = \sum SI_i \text{(Eq.4)}$$

Depending on WQI values, water quality was than categorized into five consecutive classifications (Table 3) (Sahu and Sikdar, 2008). In the current study, a total of nine chemical parameters (Total Dissolved Solids (TDS), Na^+, Cl^-, Ca^{2+}, Mg^{2+}, SO_4^{2-}, K^+, HCO_3^- and F^-) of 14 water samples were used to calculate WQI, in order to assess a groundwater suitability for domestic (drinking) use of the Zeuss Koutine aquifer.

Table 3. Water quality index (WQI) ranges

WQI range	Water quality description
WQI<50	Excellent water
50 < WQI < 100.1	Good water
100 <WQI< 200.1	Poor water
100 <WQI< 200.1	Very poor water
100 <WQI<2 00.1	Water unsuitable for drinking purposes

Table 4. WQI and water description for location samples

	WQI	Water quality description
P1	303.41	Water unsuitable for drinking purposes
P2	446.52	Water unsuitable for drinking purposes
P3	358.80	Water unsuitable for drinking purposes
P4	298.82	Very poor water
P5	347.70	Water unsuitable for drinking purposes
P6	287.27	Very poor water
P7	287.54	Very poor water
P8	262.26	Very poor water
P9	148.61	Poor water
P10	139.21	Poor water
P11	139.24	Poor water
P12	146.08	Poor water
P13	324.63	Water unsuitable for drinking purposes
P14	206.73	Very poor water

The weights assigned to the parameters ranged between 1 and 5 (Table4), and were based on the extent of health effects of those parameters. The highest weight (5) were assigned to TDS and F- due to their major importance in water quality assessment and their health implications when they have high concentration in water (Serinivasamoorthy et al., 2008; Yiadana et al., 2010). Furthermore, Fluoride and TDS are the most sensitive water quality parameters in the study area, and hence they need to be monitored regularly with higher accuracy.

Based on the water quality index, in the analyzed samples of the study area, none fall within the “Excellent” and “good” categories. 35.7%, 35.7% and 28.6% fall in “unsuitable for drinking purposes”, “Very poor water “and “Poor water” respectively. Poor and very poor water can’t be used for drinking water without any treatment and conventional disinfection, whereas water “unsuitable for drinking purposes” could only be used for aquaculture, irrigation and industrial purposes (Jindal and Sharma 2011).

Suitability for Irrigation

The water quality for irrigation purposes is assessed on the basis of its salinity level and its sodicity (Kelly 1951). The longtime effects of irrigation water on soil physical properties and crop productivity depend on the total salt, sodium, bicarbonate and carbonate concentrations of the irrigation water, and also the soil's initial physical properties (Yidana 2010).

In the study area, irrigation suitability of groundwater was evaluated based on sodium percentage % Na, Sodium adsorption ratio SAR, Residual sodium carbonate RSC, Magnesium percentage % Mg and permeability index PI.

SAR Sodium Adsorption Ratio

The sodicity hazard of water is generally described by the sodium adsorption ratio. Indeed, there's a significant relationship between SAR values of irrigation water and the extent to which sodium is adsorbed by the soils (Subba 2006). If water used for irrigation has high sodium concentrations and low calcium concentrations, the cation exchange complexe may become saturated with sodium which can destroy the soil structure due to the dispersion of clay particles (Todd, 1980).

SAR (Sodium Adsorption Ratio) is computed from Eq. (5) where concentrations are reported in meq/l.

$$SAR = Na^{+}/\sqrt{((Ca^{2+} + Mg^{2+})/2)} \quad \text{Eq. (5)}$$

Excessive sodium concentrations in irrigation water can also result in sodium hazard, particularly in dry climates such as south of Tunisia, causing reduced permeability. Indeed, in arid climates, where plants tend to uptake more water than in cooler climates, more dissolved solids are concentrated in the root zone as the soils dry up, resulting in salinity hazard (Shaki 2006). The SAR values in the study area range between 4.08 and 13.9 with mean of 7.87. 78.57% of groundwater samples have SAR<10 and 21.42% have SAR >10.

To ameliorate a water classification for irrigation suitability, the SAR has been plotted with the salinity measurement on USSL diagram (Richards, 1954) (Figure 4).

As seen in the figure, 42.85% of water samples analyzed fall in C4-bS1 field indicating extremely high salinity with low SAR, 35.71% in C4a-S1 (very high salinity with low SAR) and 21.43% in C4b-S2 field (extremely high salinity with medium SAR).

Very high and extremely high salinity waters are unsuitable for irrigation with a restrict drainage. An adequate drainage with low salinity waters and plants having good salt tolerance should be selected.

Percent Sodium % Na

Sodium concentration plays an important role in evaluating the groundwater quality for irrigation because sodium causes an increase in the hardness of soil as well as a reduction in its permeability (Tijani 1994).

Excessive Na+ causes a dispersion of soil mineral particles and a decrease of water penetration (Jalali 2007). Sodium percentage is calculated using the formula given below (Eq.6), where all the ionic concentrations are expressed in milliequivalents per litre (meq/l).

$$\% \text{ Na} = (Na^+ / Na^+ + Ca^{++} + Mg^{++} + K^+) * 100 \qquad \text{Eq. (6)}$$

% Na of all samples ranged between 40 and 60 indicates that the groundwater is permissible for irrigation purposes.

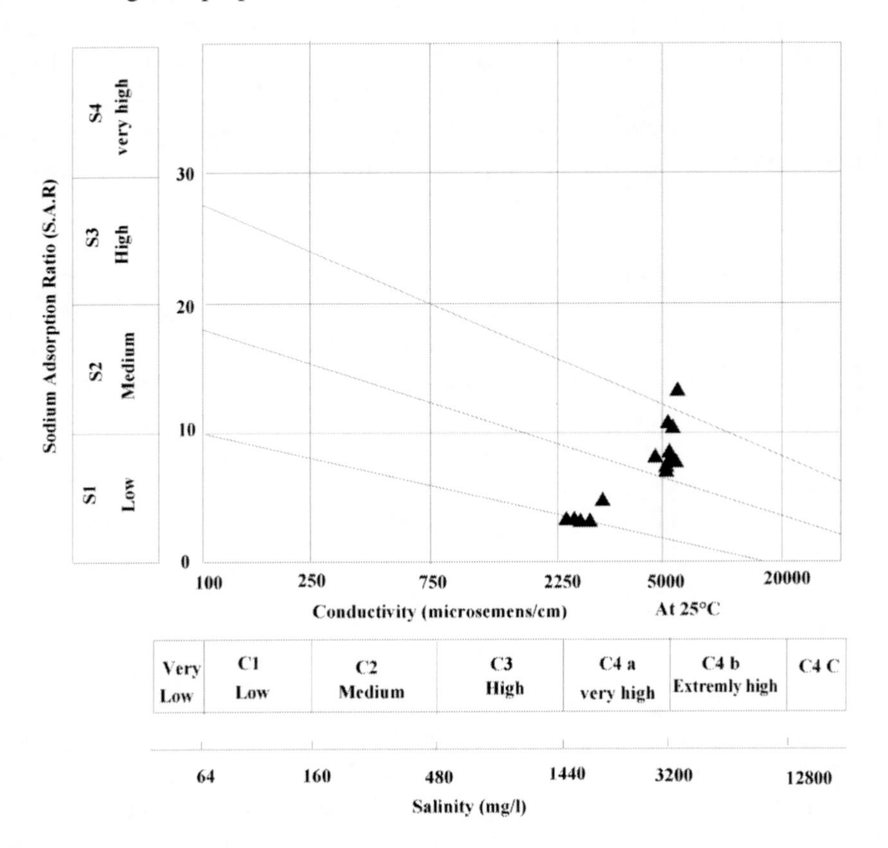

Figure 4. USSL diagram for the study area.

Residual Sodium Carbonate RSC

In water with high bicarbonates concentration, there is tendency of calcium and magnesium to precipitate as carbonates and, consequently the water of the soils becomes more concentrated. As a result, the cumulative concentration of carbonates and bicarbonates react with sodium as sodium bicarbonate which influences the suitability of groundwater for irrigation (Shaki 2006, Kumar et al., 2007).This is denoted as residual sodium carbonate RSC, which is suggested by Eaton, 1950 and calculated as

$$RSC = (CO_3^{--} + HCO_3^{-}) - (Ca^{++} + Mg^{++}) \qquad \text{Eq. (7)}$$

Thus, estimation of residual sodium carbonate RSC is another way to examine irrigation water. Computed RSC indicate that all samples have negative RSC showing that the cumulative concentration of CO_3^{--} and HCO_3^{-} is lower than the combined Ca^{++} and Mg^{++} concentrations which involve that there is no residual carbonate to react with sodium in the soil. Based on RSC, all groundwater of the study area are suitable for irrigation.

Percent Magnesium % Mg

The cation exchange behavior of magnesium is similar to that of calcium. Both ions are strongly adsorbed by clay minerals and other surfaces having exchange site (Hem, 1985). Excess of magnesium in water will adversely affect the soil quality rendering it alkaline, resulting in decreased and adversely affected crop yields (Ravikumar et al., 2011). An index for calculating the magnesium ratio was developed by Paliwal (1972) and it's computed by the following equation (Eq.8).

$$\% \, Mg = (Mg / (Ca + Mg)) * 100 \qquad \text{Eq. (8)}$$

where all the ionic concentrations are expressed in milliequivalents per litre (meq/l).

Percent of magnesium % Mg of samples analyzed range between 29.8% and 58.43%. In 85.7% of samples, % Mg is less than 50% suggesting their suitability for irrigation, while only 14.3% fall in the unsuitable class with % Mg more than 50% indicating their adverse effect on crop yield.

Permeability Index

The permeability index is an important factor which influences quality of irrigation water in relation to soil for development in agriculture (Srinivasamoorthy et al., 2011). The permeability index PI of a water samples is computed from Eq. (9) where the concentration of all ions is in meq/l.

$$PI = ((Na^{+} \sqrt{HCO_3^{-}}) / (Na^{+} + Ca^{++} + Mg^{++})) * 100 \qquad \text{Eq. (9)}$$

Permeability indices were plotted with the total ionic content of the groundwater samples on a Doneen's chart (Domenico and Schwartz 1990), which represent three different classes: CI with best water type for irrigation, CII water generally acceptable and CIII waters unacceptable. In the study area, the PI ranged from 52.44 to 64.09 with mean of 56.13 (Figure 5) indicating that all the samples are within CI best water type for irrigation purposes.

Finally; table 5 has been established to summarize several parameters throughout the study period to evaluate suitability of Groundwater for irrigation purposes.

Multivariate Data Analysis

To carry out multivariate statistical techniques such as cluster analysis and principal components analysis, the computer program ANDAD 6.00, performed by the Geo-Systems Center of Instituto Superior Tecnico Portugal (CVRM, 2000), was used. Many hydrogeochemical studies applied these techniques in an attempt to analyse and understand chemical quality of groundwater data set. The multivariate statistical analysis is a quantitative and independent technique of groundwater classification permitting the ranging of groundwater samples and examining relationships between chemical parameters and groundwater at the same time (Cloutier et al., 2008). For the multivariate analysis the data for each parameter were standardized to a range of –1 to 1.

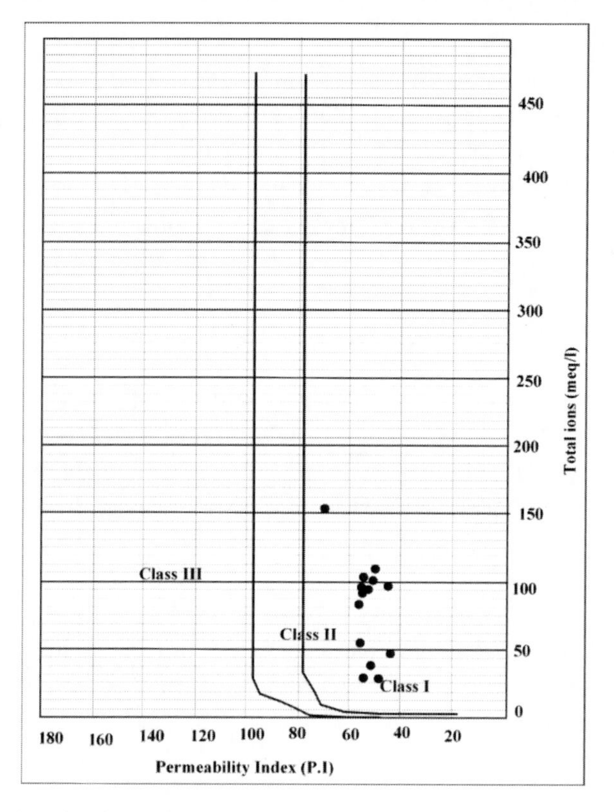

Figure 5. A Doneen's chart for the study area.

Table 5. Suitability of Groundwater for irrigation based on various classifications

Parameters	Range	Class	N° of samples	Percentage of samples
SAR	<10	Low	11	78.57%
	10-18	Medium	3	21.42%
	18-26	High	-	-
	> 26	Very high	-	-
% Na	<20	Excellent		
	20-40	Good		
	40-60	Permissible	1-2-3-4-5-6-7-8-9-10-11-12-13-14	100%
	60 -80	Doubtful		
	> 80	Unsuitable		
RSC	< 1.25	Good	1-2-3-4-5-6-7-8-9-10-11-12-13-14	100%
	1.25 – 2.5	Doubtful		
	> 2.5	Unsuitable		
% Mg	<50	Suitable	1-2-3-4-5-6-7-8-9	85.7%
	>50	Unsuitable	10-11	14.3%
	<75	Soft		
PI (%)	0 - 75	CI best water type for irrigation	1-2-3-4-5-6-7-8-9-10-11-12-13-14	100%
	75 - 80	CII water generally acceptable		
	80 - 120	CIII waters unacceptable.		

Principal Component Analysis (PCA)

The PCA was carried out for data reduction in order to systematize the interpretation of large sets of data and to characterize the linear correlations and loadings of the water quality parameters. For application of PCA, we considered the data of the 14 water points. PCA was applied to a matrix of 14 rows (wells) by 8 columns (variables). Table 6 presents the eigenvalues and the cumulative eigenvalue of variance.

PCA results reveals show that the 2 first axes explain approximately more than ¾ of the total variance in the data set, the first component accounted for about 66.55 % and the second component accounted for about 14.70 %. These tow components together accounted for about 81.26 % of the total variance and the rest of the components only accounted for about 18.74 %. Thus, only the 2 factorial plans were retained for interpretation. Most significant variables in the components, represented by loadings higher than 0.6, are taken into consideration for interpretation (Figure 6). The 1st factorial axis (F1) is interpreted by the relative projection of the major ions in a specific group on the positive side, as a mineralization axis. The second factor is best represented by HCO_3^-. We can note that this variable is not associated with any other elements, showing an independent behavior regarding the other groups of variables.

Table 6. Results from the principal component analysis: eigenvalues, %total variance and % cumulative variance (principal vectors are in bold)

Factors	Eigenvalues	%total variance	%cumulative variance
1	**6.655649**	**66.556488**	**66.556488**
2	**1.470906**	**14.70905**	**81.265541**
3	1.050427	10.50427	91.769814
4	0.493581	4.935808	96.70562
5	0.18061	1.806096	98.511719
6	0.099812	0.998121	99.509842
7	0.032324	0.323242	99.833076

Cluster Analysis (CA)

Cluster analysis is a commonly used method for identifying and selecting the homogeneous groups from the hydrogeochemical datasets. There are two types of cluster analysis: *R* and *Q*-modes (Cruz and França 2006). The approach used in this study is based on Q-mode so that similarities between different parameters could be revealed rather than similarities between variables.

In our study, and in order to perform CA, an agglomerative hierarchical clustering was developed using a combination of the Ward's linkage method as a clustering algorithm and Euclidean distances as a measure of similarity. The monitoring result of such analyses is a graph, called dendrogram showing the degree of similarity between parameters (Hamzaoui et al., 2011).

On the basis of chemico-physicals parameters and major ions concentrations, all 14 sampling sites were clustered into two homogenized water quality groups (Figure 7):

Cluster 1 (wells P1, P3, P 4, P13, P2, P7, P5 and P8) located in the downstream region.

Cluster 2 (Wells P6, P7, P9, P10, P11, P12, P13 and P14), these wells are nearer to the upstream zone.

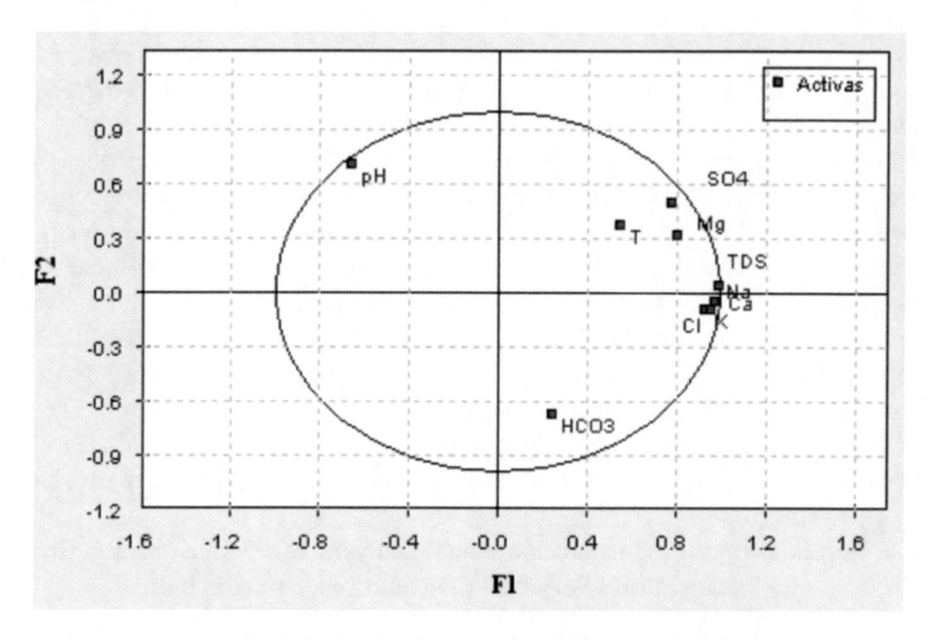

Figure 6. Representation of the parameters in the first factorial plane (axis 1 and 2).

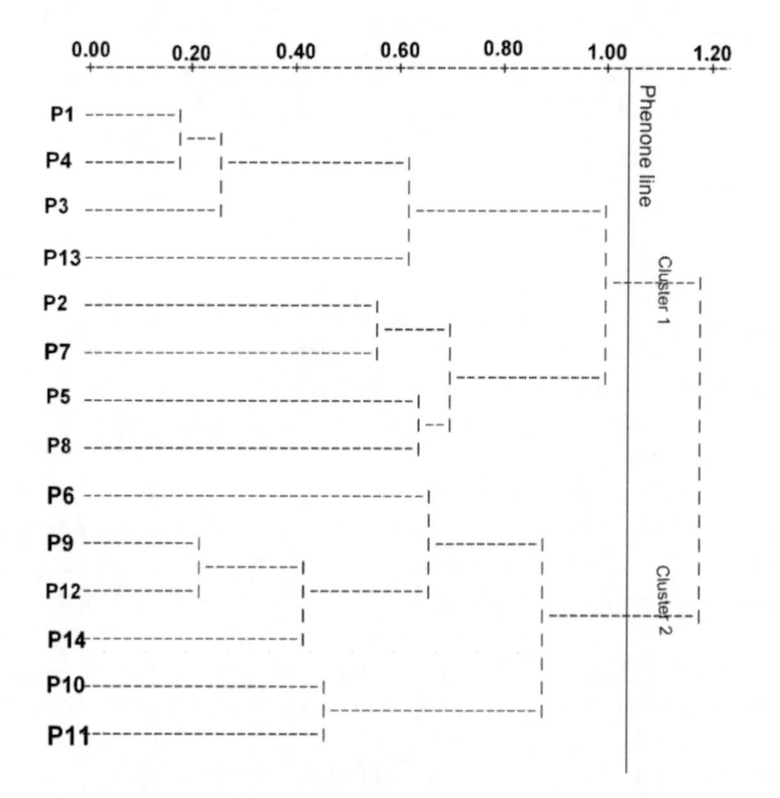

Figure 7. Dendrogram of the Q-mode cluster analysis.

CONCLUSION

Zeuss–Koutine aquifer, located in southeastern Tunisia, is used intensively as a unique water resource to meet the growing needs of the various sectors. The groundwater quality monitoring in this region is significant. In this work and thanks to collected and measured data, the groundwater quality in Zeuss–Koutine aquifer has been assessed for its use for domestic and irrigation purposes. The salinity value ranges from 1520 mg/l to 5400 mg/l with an average value of 3199 mg/l and obviously, the fresh water is rather encountered near the recharge zones while downstream, water in saline. .

The assessment of water samples according to the permissible limits prescribed by WHO for drinking purposes indicated that the values of the majority of parameters (Salinity, major elements except potassium concentrations, Temperature, pH and Fluorine concentration) exceed the WHO limits. Furthermore, the suitability of water for drinking purposes was examined using WQI indicating that groundwater in this region is classified into"unsuitable for drinking purposes", "Very poor water "and "Poor water" and can't be used for drinking purposes without any treatment .

Due to very high to extremely high salinity hazard, all water samples of Zeuss Koutine aquifer are unsuitable for irrigation with a restricted drainage and need rather an adequate drainage, with low salinity waters and selected plants having a good salt tolerance.. Nevertheless, according to Doneen's chart, all samples fall in within CI of best water type for irrigation purposes.

According to the order of the cations and anions dominance, Zeuss-Koutine groundwater is divided into two facies: a Na–Cl facies, which characterises discharge area and Ca–SO4–Cl facies in recharge area.

Application of multivariate statistical analysis including Cluster Analysis and Principal Components Analysis, on the 14 samples collected from Zeuss-Koutine aquifer confirms results obtained by conventional geochemical methods.

ACKNOWLEDGMENTS

The authors gratefully thank the National Society of Drinking Water in Tunisia (SONEDE), the Resources Water Direction of Tunis (DGRE) and the Regional Direction of Agriculture and Water Resources of Medenine (Southeaster Tunisia). We would like to acknowledge Mr Taher Atoui for his gently help during the field visits.

REFERENCES

Abbasi, SA. *Environmental pollution and its control*. Cogent. International. Philadelphia and Pondicherry, 1999, 442 p.

Abu Rukaha, Y.; Alsokhny, K. Geochemical assessment of groundwater contamination with special emphasis on fluoride concentration, North Jordan. *Chemie der Erde*. 2004. 64, 171-181.

Chilton ,J. Water Quality Assessments - *A Guide to Use of Biota, Sediments and Water in Environmental Monitoring* - Second Edition. Edited by Deborah Chapman. Great Britain. 1992, 88 p.

Cloutier, V; Lefebvre, R; Therrien, R; and Savard, MM. Multivariate statistical analysisof geochemicaldata as indicative of the hydro-geochemical evolution of groundwater in a sedimentary rock aquifer system. *Journal of Hydrology*. 2008, 353, 294–313.

Cruz, JV.; França, Z. Hydrogeochemistry of thermal and mineral water springs of the Azores archipelago (Portugal). *Journal of Volcanology and Geothermal Research*. 2006, 151(4), 382-398.

CVRM. Programa ANDAD. Manual do Utilizador: CVRM-Centro de Geosistemas, Instituto Superior Tecnico, Lisbon, Portugal. 2000.

DGRE., a. Annuaires de l exploitation des nappes profondes en Tunisie. 2004, 200 p. DGRE., b. Annuaires piézométriques en Tunisie. 2004, 150 p.

Domenico, PA; and Schwartz FW. Physical and chemical hydrogeology.Wiley, New York, 1990, 410–420.

Eaton, AD.; Clesceri, LS.; and Greenberg, AE. *Standard methods for the examination of water and wastewater* (19th ed.). Washington DC: American Public Health Association. 1995. 1325 p.

Elango, L; Kannan, R. and Kumar, S. Major ion chemistry and identification of hydrogeochemical processes of groundwater in a part of Kancheepuram District, Tamil Nadu, India. *Journal Environmental Geosciences*, 2003, 10(4), 157-166.

Fetter, CW. *Applied Hydrogeology* (4th ed.), Prentice-Hall, Upper Saddle River, New Jersey. 2001, 598p.

Gaubbi, E. Evolution de la piézométrie et de lagéochimie de la nappe de Zeuss–Koutine. Master thesis. Tunis, Tunisia: University El Manar. 1988, p 63.

Geetha, A; Palanisamy PN; SIVAKUMAR, P; Ganesh kumar, P. and Sujatha M. Assessment of Underground Water Contamination and Effect of Textile Effluents on Noyyal River Basin In and Around Tiruppur Town, Tamilnadu. *E-Journal of Chemistry*, 2008, 5, 4, 696-705.

Guo, H., Wang, Y. Geochemical characteristics of shallow groundwater in Datong basin, northwestern China. *Journal of Geochemical Exploration*. 2005, 87, 109–120.

Hamzaoui-Azaza, F.; Bouhlila, R.; and Gueddari, M. Geochemistry of fluoride and major ion in the groundwater samples of Triassic aquifer (south eastern Tunisia), through multivariate and hydrochemical techniques: *Journal Applied Sciences Research*. 2009, 5, 1941–1951.

Hamzaoui-Azaza, F., 2011, Géochimie et Modélisation des Nappes de Zeuss-Koutine, des Gre `s du Trias et du Mioce `ne de Jorf-Jerba-Zarzis: Unpublished Thesis, Department of Geology, University of Tunis El-Manar, Faculty of Science of Tunis; 2011262 p.

Hamzaoui Azaza, F ; Ketata, M; Bouhlila, R. ; Gueddari,M.; and Riberio, L. Hydrogeochemical characteristics and evaluation of drinking water quality in Zeuss-Koutine aquifer, south-eastern Tunisia. *Environmental Monitoring and Assessment*, 2011, 174, -283-298.

Hem, JD. Study and interpretation of the chemical characteristics of natural water: United States; *Geological Survery Supply paper*. 1985, 263 p.

Jalali, M. Assessment of the chemical components of Famenin groundwater, western Iran: *Environmental Geochemal Health*. 2007, 29, 357–374. no access.

Jalali, M. Groundwater geochemistry in the Alisadr, Hamadan, western Iran. *Environmental Monitoring and Assessment*. 2010, 166, (1-4), 359-369.

Jeevanandam, M.; Kannan, R.; Srinivasalu, S; and Rammohan, V. Hydrogeochemistry and Groundwater Quality Assessment of Lower Part of the Ponnaiyar River Basin, Cuddalore District, South India. *Environmental Monitoring and Assessment*. 2007, 132 (1-3), 263-274.

Krishna Kumar, S.; Rammohan, V.; Dajkumar Sahayam J.; and Jeevanandam M. Assessment of groundwater quality and hydrogeochemistry of Manimuktha River basin, Tamil Nadu, India. *Environmental Monitoring and Assessment*. 2009,159 (1-4), 341-351.

Kumar, M.; Kumari, K.; Ramanathan, AL.; and Saxena R. A comparative evaluation of groundwater suitability for irrigation and drinking purposes in two intensively cultivated districts of Punjab, India. *Environmental Geology*. 2007, 53,553–574.

Mathess, G. *The properties of groundwater*. Wiley, New York; 1982.

Meybeck, M. Global chemical weathering of surficial rocks estimated from river dissolved loads. *American Journal of Science*. 1987, 287, 401–428.

Morhange, C.; and Pirazzoli, PA. Mid-Holocene emergence of Southern Tunisian coasts. *Marine Geology*. 2005, 220, 205–213.

Nwankwoala, HO.;and Udom, GJ. Hydrochemical Facies and Ionic Ratios of Groundwater in Port Harcourt, Southern Nigeria. *Research Journal of Chemical Sciences*. 2011, 1(3), 87-101.

OSS : Observatoire du Sahara et du Sahel.Etude Hydrogéologique du Système Aquifère de la Djeffara Tuniso-Libyenne: Rapport de synthèse, Tunisia. 2005, 209 p.

Ouessar, M. *Hydrological Impacts of Rainwater Harvesting in Wadi Oum Zessar Watershed* (Southern Tunisia): Unpublished Thesis, Faculty of Bioscience Engineering, Ghent University, Ghent, Belgium. 2007, 154.

Paliwal, KV. *Irrigation with saline water*. Monogram, New Delhi: IARI. 1972, 198p.

Piper, AM. *A graphic procedure in the geochemical interpretation of water analysis*. American Geophysical Union, Tranc. 1944, 914-923.

Ramakrishnaiah C R; Sadashivaiah C. and Ranganna G. Assessment of Water Quality Index for the Groundwater in Tumkur Taluk, Karnataka State, India. *E-Journal of Chemistry*, 2009, 6(2), 523-530.

Ravikumar, P.; Somashekar, RK.; and Angami, M. Hydrochemistry and evaluation of groundwater suitability for irrigation and drinking purposes in the Markandeya River basin, Belgaum District, Karnataka State, India. *Environmental Monitoring Assessement* . 2011, 173, 459–487.

Reghunath, R; Sreedhara Murthy, TR. and Raghavan, BR. The utility of multivariate statistical techniques in hydrogeochemical studies:an example from Karnataka, India. *Water Research*. 2002, 36, 2437–2442.

Rodier, J. L'Analyse de l'Eau: Eaux Naturelles, Eaux Résiduaires, Eau de Mer, 8th ed.: DUNOD, Paris. 1996,1384 p.

Rosen, MR. and Jones, S. Controls on the groundwater composition of the Wanaka and Wakatipu basins, Central Otago, New Zealand. *Hydrogeology Journal*. 1998, 6, 264-281.

Saeedi, M; Sharifi, OA; and Meraji, H. Development of groundwater quality index. *Environmental Monitoring Assessement*. 2010, 163, 327–335.

Sahu, P.; and Sikdar, PK. Hydrochemical framework of the aquifer in and around East Kolkata wetlands, West Bengal India. *Environmenetal Geology*. 2008, 55, 823–835.

Service Quality. 2005, 15(2), 195-208.

Shaki, AA; and Adeloye, A.J. Evaluation of quantity and quality of irrigation water at Gadowa irrigation project in Murzuq basin, southwest Libya. *Agricultural water management*. 2006, 84, 193 – 201.

Srinivasamoorthy, K.; Chidambaram, M.; Prasanna, M.V.; Vasanthavigar, M.; John Peter, A.; and Anandhan, P. Identification of major sources controlling Groundwater Chemistry from a hard rock terrain A case study from Mettur taluk, Salem district, Tamilnadu, India. *Journal of Earth System Sciences*. 2008, 117(1), 49–58.

Subramani, T.; Elango, L.; and Damodarasamy, SR. Groundwater quality and its suitability for drinking and agricultural use in Chithar River Basin, Tamil Nadu, India. *Environmental Geology*. 2005, 47, 1099-1110.

Tijani, MN. Hydrochemical assessment of groundwater in Moro area, Kwara State, Nigeria. *Environmental Geology*. 1994, 124, 194–202.

Venugopal, T; Giridharan, L. and Jayaprakash, M. Groundwater Quality Assessment Using Chemometric Analysis in the Adyar River, South India. *Arch Environ. Contam Toxicol.*2008, 55, 180–190.

World Health Organization (WHO). *World Health Organization Guidelines for Drinking Water Quality; Volume 1: Recommendations*, 3rd ed.: World Health Organization, Geneva, Switzerland. 2004, 188 p.

Yidana, SM; Bruce Banoeng, Y.; and Akabzaa, TM. Analysis of groundwater quality using multivariate and spatial analyses in the Keta basin, Ghana. *Journal of African Earth Sciences*. 2010, 58, 220–234.

In: Water Quality
Editor: You-Gan Wang
ISBN: 978-1-62417-111-6

Chapter 6

APPLICATION OF WATER QUALITY INDICES (WQI) AND STABLE ISOTOPES ($\delta^{18}O$ AND $\delta^{2}H$) FOR GROUNDWATER QUALITY ASSESSMENT OF THE DENSU RIVER BASIN OF GHANA

Abass Gibrilla[1], Edward Bam[1], Dickson Adomako[1], Samuel Ganyaglo[1] and Hadisu Alhassan[2]

[1]Nuclear chemistry and Environmental Research Centre, National Nuclear Research Institute, Ghana Atomic Energy Commission, Kwabenya-Accra, Ghana
[2]Ghana Urban Water Company Ltd, Weija

ABSTRACT

Groundwater and surface water (Densu River) were collected for physical, chemical and stable isotope analysis to determine their suitability for drinking and agricultural purposes. Rain water was also sampled on event basis at Koforidua for stable isotope analysis. The results showed that, groundwater in the study area are generally fresh and slightly acidic to neutral while the surface water is slightly alkaline. The WQI values were found ranging from 0-50 belonging to "excellent" and "good" water quality. An integrated approach of heavy metal evaluation indices using Contamination index (C_d), heavy metal pollution index (HPI) and heavy metal evaluation index (HEI) were used to evaluate the extent of pollution and suitability of the samples for drinking with respect to heavy metals. The three indices showed similar trends with strong correlations but with different water quality classifications. Whereas the C_d showed that 95% of groundwater and 100% of surface water were highly polluted, HPI and HEI indicated that, 4.76% and 0% of groundwater and 25% and 0% of the surface water were polluted. A modification of C_d and HPI using multiple of mean criteria showed a comparable classification with HEI. C_d, HPI and HEI showed that 4.8%, 0% and 0% of the groundwater were highly polluted while 25%, 0% and 0% of the surface water were respectively classified as highly polluted. Chemical indices like percentage of sodium, Sodium adsorption ratio, residual sodium carbonate, and permeability index (PI) indicate that, the groundwaters in the study area are suitable for irrigation. A comparison of the isotopic data of the rain water, Local Meteoric Water Line (LMWL) and Global Meteoric Water Line (GMWL) indicates that the groundwater in the study area is mainly meteoric with few groundwater and all the surface water showing an evidence of evaporation. The d-excess values show that the groundwater has undergone dilution with the rainfall and this is observed from the decrease of the d-excess of the groundwater with increase in Oxygen-18. This observation also suggests a modern day recharge to the groundwater.

INTRODUCTION

In Ghana, groundwater is the major source of potable water for most rural communities. Groundwater is a valuable natural resource; it occurs almost in all geological formations beneath the earth surface not in a single widespread aquifer but in thousands of local aquifer systems with similar characteristics [1]. The presence of dissolved minerals coupled with some special characteristics of groundwater as compared to surface water makes it a preferred choice for many purposes [2], [3]. In many rural communities in the Densu River basin, groundwater is the major source of water for domestic and other uses. This is partly due to pollution of the Densu River and its tributaries. In recent years, these communities are experiencing rapid growth due to urbanization. The potential threat to groundwater resources in the area due to over exploitation, agriculture and improper waste disposal practices are envisaged [4]. Knowledge of the occurrence, quality and recovery of the groundwater resources is, therefore, essential for proper implementation of integrated water resources management programme [5]. The quality and suitability of groundwater for domestic, industrial and agricultural purposes depends on the quality of recharge water, atmospheric precipitation, in-land surface water, and on sub surface geochemistry. Because of the potential of absorption and transportation of waste materials (domestic, industrial and agriculture), river basins are highly vulnerable to pollution, hence the need for regular monitoring and control of water quality in these areas [6].

Interest in chemical and trace metal pollution has generated a desire both on the national and international scales, for integrating numerous parameters associated with water quality in a specific index, hence, the development of several water quality indices and metal pollution indices [7], [8], [9]. Some of these pollution indices had been successfully used by [5], [10], [11], [12] to assess the quality of surface and groundwater with respect to chemical and heavy metal. These pollution indices are intended to provide a useful and comprehensible guiding tool for water quality executives, environmentalist, decision makers and potential users of a given water system [11]. Stable isotopes of ^{18}O and ^{2}H have also been widely used in water resources management. This is because, they allow conclusion to be drawn as regards the recharge process, the location of recharge and discharge areas, aquifer continuity, sources of ions in water and turnover time [13], [14]. Groundwater hydrochemistry in the Densu river basin has been fairly studied by various authors [15], [16], [17], [18] and their properties are well known to the extent that groundwater is tapped in commercial quantities to meet both domestic and industrial needs of the people. All these authors generally appear to attribute the groundwater hydrochemistry to rock-water interaction.

Later studies by [16],[19] using stable isotopes of ^{18}O and ^{2}H showed an evidence of evaporated waters recharging the groundwater system in some areas; this implied that anthropogenic activities on the surface and the unsaturated zone [4] may pose a serious challenge to the groundwater quality in the near future. In this chapter, Water Quality Index (WQI), percentage of sodium (Na%), Sodium Adsorption Ratio (SAR), Residual Sodium Carbonate (RSC), permeability index (PI), contamination index (C_d), heavy metal pollution index (HPI) and heavy metal evaluation index (HEW) which are regarded as one of the most effective ways to communicate water quality [20] will be used. The objective of this chapter, therefore, is to study the suitability of the groundwater and the surface water for drinking and irrigation purposes using the data obtained through quantitative analysis and [21] water quality standards. The study will also employ stable isotopes as complementary tool to study the origin of groundwater in the study area.

METHOLOGY

Study Area

The Densu river basin lies between latitude 5° 30’ N to 6° 20’ N and longitude 0° 10’ W to 0° 35’ W (Figure 1). The river shares its catchment boundary with the Odaw and Volta basins to the east and north respectively, the Birim basin in the northwest and the Ayensu and Okrudu in the west.

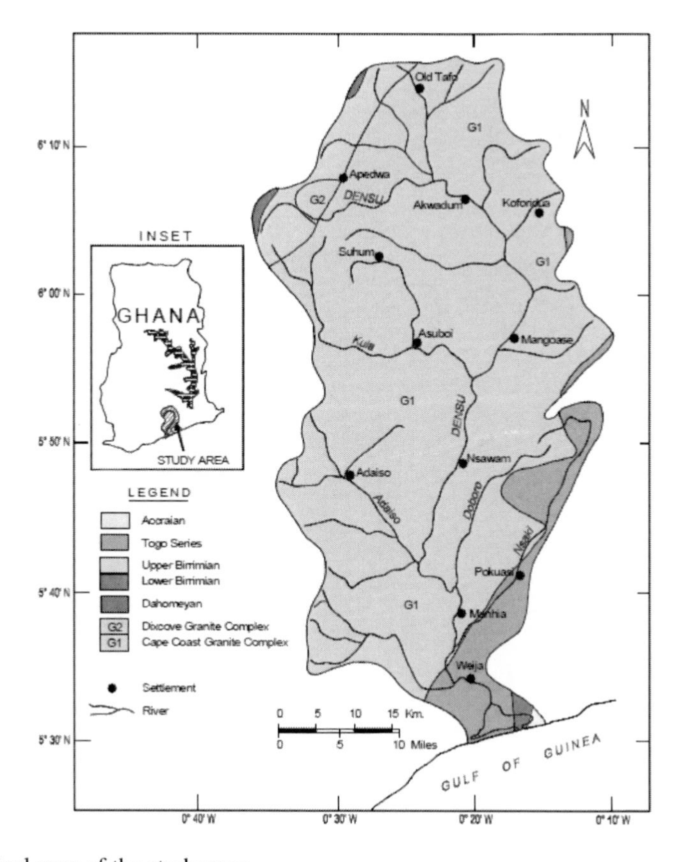

Figure 1. Geological map of the study area.

The Densu River takes its source from Atewa-Atwiredu mountain range near Kibi (East Akyem District) in the Eastern Region of Ghana. The river is about 116 km long with a catchment area of 2564 km^2 covering nine administrative districts.

The main tributaries include rivers Adeiso, Nsakyi, Dobro, and Kuia (Figure 1). The Densu River enters the Weija reservoir, one of the two main sources of water supply for the city of Accra and finally discharges into Sakumo lagoon and Gulf of Guinea near Bortianor west of Accra. Most communities upstream and midstream of the river depend on groundwater and to a lesser extent the raw water without any form of treatment.

The year 2000 population and housing census estimated the total population of the Densu River basin to be about 1.2 million people [22]. The economic activities in the catchment are mainly cultivation of crops such as cocoa, maize, cassava, vegetables, pineapples and cocoyam with few livestock and fish farming. Artisanal mining popularly called "galamsey" in some of the major rivers is fast becoming a lucrative business in the study area.

Climate and Geology

The basin falls under two distinct climatic zones characterized by two rainfall regimes with different intensities [23]. The major rainy season extends from April/May to July. The minor season occurs between September and November.

The mean annual rainfall recorded for 10 years during the period 1993 to 2003 obtained from the Ghana Meteorological Agency (GMA) varies from about 1200 mm at Nsawam to about 1487 mm in the river source area at Kibi. The mean annual temperature is about 27 °C, with March/April being the hottest (32°C) and August being the coldest month (23°C). Maximum and Minimum monthly temperature, normal rainfall distributions and relative humidity are shown in Figure 2.

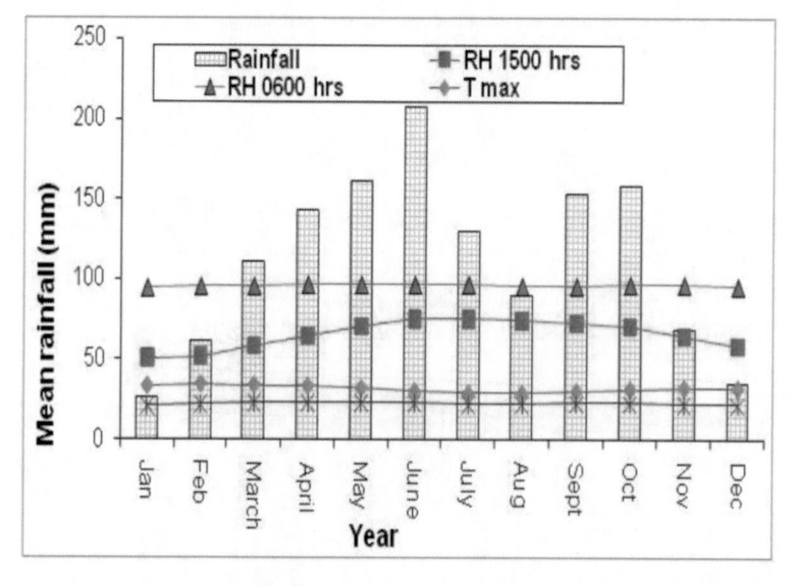

Figure 2. Distribution of rainfall, temperature and relative humidity in the study area, adapted from [25].

The middle to northern portion under study is mainly underlain by Precambrian crystalline rocks, comprising of Birimian (upper and lower) and Cape Coast Granites while the southern portion is underlain by the Togo series. Detailed description of the geology and hydrogeology can be found in our previous work [19]. The dominant soils are ochrosols, with patches of gleisols and lithosols.

Data and Field Work

The principal dataset presented in paper includes the physical, chemical, isotopic and trace metals analyses collected by [24] and also from [15] and [25].

Rainwater samples were collected from the study area on an event basis (2006–2008) at Kibi meteorological stations which is about 320 m a.s.l. for isotope analysis. The rain water samples were collected by a 500ml vial through a 200mm diameter funnel with a pingpong ball to avoid evaporation. The rain water samples were then collected into 60ml air-tight polyethylene vials soon after the rain event for the analysis.

A total of twenty one boreholes, one hand-dug well and four Surface water points (River Densu) were sampled at various locations. Figure 1. All the water samples were collected in 500ml pre-conditioned high density polyethylene bottles.

They were first conditioned by washing with five (5%) percent nitric acid, and then rinsing several times with distilled water. This was carried out to ensure that the sampling bottles were free from contaminants. Samples for isotopes analysis were collected in 60ml glass bottle filled to the brim and securely capped.

At the sampling points, the boreholes were pumped to purge the aquifer of stagnant water to acquire fresh samples for analysis. Most of the wells were being used for domestic water supply during the sampling period; therefore, purging lasted for 5-10 minutes.

pH, temperature, electrical conductivity (EC) and total dissolve solids (TDS) measurements were conducted in situ in the field using a pre calibrated HACH pH meter and HACH conductivity meter. Alkalinity titration was done at the wellhead using a HACH digital titrator. For chemical and trace metal analysis, the samples were filtered on site through 0.45μm cellulose filters with the aid of a hand operated vacuum pump and collected in the 500ml bottle. The sample for metal analysis was preserved by adjusting to $pH<2$ with 6N ultrapure nitric acid [26]. The bottles and caps meant for collecting the samples were rinsed three times with the filtered water after which they were filled to the brim and caped. All the samples were then kept in an ice chest containing ice bricks and transported to the laboratory.

Laboratory Analysis

Na^+ and K^+ were analyzed using flame photometer, magnesium (Mg^{2+}) and calcium (Ca^{2+}) using Fast Sequential Atomic Absorption Spectroscopy (Varian AA240FS). Chloride (Cl^-), Sulphate (SO_4^{2-}) and Nitrate (NO_3^-) were analyzed using ion chromatography system (Dionex ICS-90) at Nuclear Chemistry and Environmental Research Centre, NNRI, GAEC. The ionic-balance-error was computed, taking the relationship between the total cation (Ca^{2+}, Mg^{2+}, Na^+ and K^+) and the total anions (HCO_3^-, Cl^-, SO_4^{2-}) for each set of complete analysis of water samples. Only samples which fall within ±5% were reported in this work. The acidified samples were used to determine the trace metals concentrations using Instrumental Neutron Activation Analysis (INAA) using Ghana research reactor 1 (GHARR 1) at Ghana atomic Energy Commission (GAEC). The stable isotope analysis of the samples was carried out at HMGU, Institute of Groundwater Ecology (Neuherberg/Germany) using Isotope mass spectrometry.

The variation in isotope ratio D/H and $^{18}O/^{16}O$ in water samples are expressed in terms of per mille deviation (‰) relative to internal standards that were calibrated using the Vienna-Standard Mean Ocean Water (V-SMOW). The data was then normalized following [26] as follows

$$\delta\text{‰} = \left[\frac{R_{sample}}{R_{V-SMOW}} - 1\right] * 1000$$

where

R_{sample} represent either the $^{18}O/^{16}O$ or the D/H ratio of the samples.
$R_{V\text{-}SMOW}$ represents either the $^{18}O/^{16}O$ or the D/H ratio of the V-SMOW.

The analytical reproducibility was 0.1‰ for oxygen and 1‰ for deuterium.

Sample Preparation

0.50ml of each water sample (weighing 0.50g) was pipette using a calibrated Eppendorf tip ejector pipette (Brinkmann Ins., Inc., Westbury, NY) into clean pre-weighed 1.5ml polyethylene vials, and fitted with polyethylene snap caps and heat-sealed. Four of these sample vials were placed into a 7.0ml volume polyethylene vial and heat-sealed (for medium lived radionuclide) for irradiations. Two replicates were prepared for each sample. However, for short lived radionuclide, only one sample was put into the 7.0ml vial. The elemental comparator standard used in this work was synthetic standards prepared by pipeting aliquots of multielement and single elements NIST standard solution and validated against IAEA Standard Reference Material (SRM) 1547 Peach leaves prepared as the samples.

Sample Irradiation, Counting and Analysis

All the samples, synthetic standards prepared by pipeting aliquots of multi-element and single elements standard solution and standard reference materials were irradiated in the inner pneumatic irradiation sites of the Ghana Research Reactor-1 (GHARR-1) facility operating at half full power of 15 kW with corresponding thermal neutron flux of 5.0 x 10^{11} n/cm^{2}s^1. The scheme for irradiation and counting was chosen according to the half lives of the elements of interest Table 1. The detector used in this work was an n-type high purity germanium (HPGe) detector Model GR 2518 (Canberra Industries Inc.) with a resolution of 0.85 keV (FWHM) and 1.8 keV (FWHM) for ^{60}Co gamma-ray energies of 1332 keV.

Table 1. Nuclear data and irradiation scheme of elements determined in this work, adapted from [25]

Element	Target isotopes	Formed isotope	Half-life	Gamma ray energy Kev	Irradiation time (ti)	Cooling time (td)	Counting time (tc)
Cu	^{65}Cu	^{66}Cu	5.10 min	1039.2	120 sec	60-300 sec	600 sec
Al	^{27}Al	^{28}Al	2.24 min	1778.9	120 sec	60-300 sec	600 sec
Mn	^{55}Mn	^{56}Mn	2.58 h	846.8	3600 sec	60-300 sec	600 sec
As	^{75}As	^{76}As	26.32 h	559.1	14400 sec	2 days	600 sec
Zn	^{64}Zn	^{65}Zn	243.9 days	1115.6	14400 sec	3-30 days	12-24 h
Fe	^{58}Fe	^{59}Fe	44.49 days	1099.2	14400 sec	3-30 days	12-24 h
Cr	^{50}Cr	^{51}Cr	27.7 days	320.0	14400 sec	3-30 days	12-24 h

Table 2. Validation result for elemental concentration of NIST-1547 Peach Leaves Certified Reference Material (CRM: mg/kg), adapted from [25]

Element	NIST 1547 Peach leaves			
	This work	Recommended value	%RSD	Z-Score
Al	252±6	249±8	2.38	0.38
Cu	3.9±0.2	3.7±0.4	5.13	0.50
Cr	1.01±0.01	1.02±0.03	0.99	-0.03
Zn	17.7±0.3	17.9±0.4	1.69	-0.50
Fe	210±10	218±14	4.76	0.57
Mn	97.6±7	98±8	7.17	-0.05
As	0.06±0.01	0.05±0.01	16.67	1.00

Concentrations are reported as mean ± SD.

$\%RSD = (\sigma / x) * 100$

$Z - score = \frac{(x-\mu)}{\sigma}$,

where

x= measured mean value.

μ= recommended value.

σ= standard deviation of the recommended value.

The detector operated on a bias voltage of (-ve) 3000 V with relative efficiency of 25% to NaI detector. A Microsoft Soft window based software MAESTRO was used for the spectra analysis, employing the relative standardizing (comparator) method.

The analytical results of the present studies were validated using NIST-1547 Peach Leaves Certified Reference Material as shown in Table 2 where mean elemental concentrations and standard deviation, certified/ recommended value and standard deviation, percentage RSD and Z-score were tabulated. The high precision in the data were suggested by the low RSDs (in %) which were < 10% for all the elements except As where RSDs is 16.67%. All the Z-score values were below 3, suggesting that the data are within 95% confidence limit. Hence, the elemental concentration data for samples analyzed in this study are reliable within ±10%. The detection limits of the technique for the determined elements were calculated under identical experimental conditions and are: Al=1.5mg/l, Cr=0.001mg/l, Cu=0.5mg/l and Zn=2mg/l for groundwater and surface water.

Estimation of the Water Quality Index (WQI)

Water Quality Index (WQI) is a very reliable, useful and efficient method for assessing and communicating the information on the overall quality of water [28], [29]. The determination of WQI helps in deciding the suitability of groundwater sources for its intended purpose. From the early 1960s, different WQI have been developed [30], [31]. This work will employ the use of WQI proposed by [9] in assessing the suitability of the water in the study area for drinking.

$$WQI = Anti\log\left[\sum W_{n=1}^{n} \log_{10} q_n\right]$$

where

Wn = Weightage factor and calculated from the following equations

$$W_n = K(S_i)^{-1}$$

K proportionality constant derived from

$$K = \left[\sum_{n=1}^{n} (S_i)^{-1}\right]^{-1}$$

S_i are the [21], [32] standards values of the water quality parameter. The calculated Weightage factors of each parameter are given in Table 3.

Table 3. Water Quality Parameters, their standard values, their ideal values and the assigned weightage factors, adapted from [25]

Parameter	Standard Value, S_i	Ideal value, C_{id}	$1/S_i$	Assigned Weightage factor, Wi
pH	8.5	7	0.1176	0.00396
Total dissolve solids	500	0	0.0020	6.73E-05
Alkalinity	120	0	0.0083	0.000281
Electrical conductivity	1400	0	0.0007	2.4E-05
Chloride	250	0	0.0040	0.000135
Sulphate	250	0	0.0040	0.000135
Sodium	200	0	0.0050	0.000169
Nitrate	50	0	0.0200	0.000674
Calcium	75	0	0.0133	0.000448
Magnesium	30	0	0.0333	0.001122
Mn	0.5	0	2.0000	0.067400
Fe	0.3	0	3.3333	0.112202
Cu	2.0	0	0.5000	0.01683
Cr	0.05	0	20.000	0.673211
Zn	3.0	0	0.3333	0.01122
Al	0.3	0	3.3333	0.112202
			$\sum$=29.71	$\sum$= 1

Table 4. Water Quality Index Scale, adapted from [25]

Water quality	Description
0-25	Excellent
26-50	Good
51-75	Poor
76-100	Very poor
>100	Unfit for drinking(UFD)

Quality rating (q_n) was calculated using the formula

$$q_n = \left[\frac{(V_{actual} - V_{ideal})}{V_{standard} - V_{ideal}} \right] * 100$$

where,

q_n = Quality rating of ith parameter for a total of n water quality parameters

V_{actual} = Value of the water quality parameter obtained from laboratory analysis

V_{ideal} = Value of that water quality parameter can be obtained from the standard tables, V_{ideal} for pH = 7 and for other parameters it is equivalent to zero.

$V_{standard}$ = WHO, 2004 / ISI, 1993 standard of the water quality parameter

The calculated WQI values are then used rate the ground water quality as excellent, good, poor, very poor and unfit for human consumption (Table 4).

Heavy Metals Indexing Approach

Three standard methods for heavy metal contamination evaluated in this study are Contamination index (C_d) developed by [33], Heavy Metal pollution Index (HPI) proposed by [34] and Heavy Metal evaluation Index (HEI) proposed by [35].

Contamination Index (C_d)

This method evaluates the quality of water by determining the degree of contamination. The C_d is the sum of the contamination factors of individual components exceeding the upper permissible limit for each water sample analysed. For this reason, the C_d can be used to give a summary of the combined effects of heavy metals in drinking water. The C_d is calculated using the relation

$$C_d = \sum_{i=1}^{n} C_{fi}$$

where

$$C_{fi} = \left[\frac{C_{Ai}}{C_{Ni}}\right] - 1$$

C_{fi} = contamination factor for the i-th components
C_{Ai} = measured value for the i-th components
C_{Ni} = upper permissible limit of the i-th component (N denoting the 'normative value').

Even though, the original authors did not consider the measured values below the upper permissible concentration value, this study will use all the measured parameters irrespective of their value for the sake comparison. The upper permissible concentration values C_{Ni} was taken as the maximum admissible concentration (MAC) and was obtained from [21] guidelines for drinking water quality.

The waters were then classified based on their C_d values into three groups as follows: C_d <1 as low, C_d=1-3 as medium and C_d>3 as high.

Heavy Metal Pollution Index (HPI)

HPI signifies the overall quality of water with respect to heavy metals. The HPI is based on weighted arithmetic mean quality method following two steps. The first step involves establishing a rating scale to weigh each selected parameter. The Second step involves the selection of the pollution parameters on which the index is to be based. The rating system is an arbitrarily value between zero to one. This is defined by the perceived importance of the individual quality parameter under considerations in a comparative way or it can be assessed by making values inversely proportional to the recommended standard for the corresponding parameter [31], [34].

In this study, the weightage factor (W_i) for each parameter was defined as the inverse of the recommended standards denoted as (S_i) as suggested by [36]. The concentration limits (highest permissible value for drinking water S_i and maximum desirable value I_i) for each parameter were taken from [21] and [32]. The S_i refers to maximum allowable concentration in groundwater in the absence of alternative source. The HPI was calculated using the model [34] and is given by

$$HPI = \frac{\sum_{i=1}^{n} W_i Q_i}{\sum_{i=1}^{n} W_i}$$

where

Q_i is the sub-index of the i-th parameter.

W_i is the unit weightage of the i-th parameter

n is the number of parameters measured

The sub-index (Q_i) of the parameters was calculated by

$$Q_i = \sum_{i=1}^{n} \frac{|M_i - I_i|}{S_i - I_i} * 100$$

where

M_i is the measured value of the heavy metal of the i-th parameter

I_i is the ideal value of the i-th parameter

S_i is the standard value of the ith parameter

Generally, pollution indices are determined to assess the use of water for specific use; the calculated indices will be used for the purpose of drinking water. The critical HPI value for drinking water is 100, above which the water is unsuitable for drinking.

Heavy Metal Evaluation Index (HEI)

The HEI is also a useful method that gives an overall quality of water with respect to heavy metal.

$$HEI = \sum_{i=1}^{n} \frac{H_c}{H_{mac}}$$

where

H_c is the measured value of the i-th parameter

H_{mac} is the maximum admissible concentration of the i-th parameter.

RESULTS AND DISCUSSIONS

The statistical summary of the physico-chemical parameters and trace metals measured in the groundwater and surface waters are presented in Table 5. The pH of the groundwater ranges from 5.58-7.09. The lowest pH occurred at ANY while the highest pH was found at DPOT. The mean pH was 6.47. The very acidic groundwater were found in the Birimian ANY, AK, NK, POT (5.58-6.66) while in the granite where most of the samples were located, the pH are slightly acidic to neutral (5.98-7.09). This deviation can be attributed to the activities occurring in the unsaturated zone which might have effect in the groundwater before recharge since these areas are characterized by intensive agricultural activities.

Table 5. Statistical summaries of the physical parameters and trace metal in the study area

	Groundwater				Surface water			
Parameter	Min	Max	Mean	SD	Min	Max	Mean	SD
pH	5.5800	7.0900	6.4710	0.3920	7.3800	7.4800	7.4450	0.0451
Temp	26.1000	28.6000	26.9333	0.6843	26.1000	26.7000	26.3750	0.2500
TDS	49.5000	361.0000	168.9905	85.3246	50.9000	96.0000	68.7750	19.9075
COND	98.8000	722.0000	337.9619	171.2090	101.9000	191.2000	137.3500	39.4394
Sal	0.0000	0.4000	0.1524	0.0981	0.0000	0.1000	0.0500	0.0577
Alkalinity	11.2600	193.2100	85.2390	53.6163	19.4400	85.2900	53.1150	31.2376
Na^+	30.8000	226.7000	100.4238	65.8449	24.9000	66.5000	39.4800	19.0500
K^+	0.8000	52.9000	10.1429	11.4894	1.9000	9.9000	5.6300	4.1700
Cl^-	3.4000	124.5000	36.9619	31.1747	4.9000	11.3000	9.1500	2.9500
HCO_3^-	12.4000	233.6000	103.6862	65.4912	23.4600	102.8000	63.9700	37.2100
Mg^{2+}	0.3700	68.0000	7.7814	14.1066	2.0500	3.7600	2.8900	0.7000
Ca^{2+}	2.8400	99.6000	19.9014	31.6991	2.1800	4.5400	3.4400	0.9700
SO_4^{2-}	4.9500	26.6800	17.8181	5.4075	9.3800	14.3400	11.9300	2.0400
NO_3^-	0.1790	64.4600	23.3362	19.1288	2.8900	7.1600	4.4800	1.9100
PO_4^{3-}	0.0200	26.1300	1.3871	5.6700	0.0200	0.1700	0.1000	0.0600
Cu	0.0100	9.5700	4.1340	3.2579	3.6200	9.7800	7.0150	2.6406
Al	0.9320	2.6680	1.5854	0.4840	1.1680	3.2040	2.0115	0.9893
Mn	0.0025	2.2775	1.1496	0.7039	0.1950	0.9850	0.4463	0.3650
Cr	0.0010	0.0030	0.0013	0.0006	0.0010	0.0020	0.0014	0.0005
As	0.0010	0.0200	0.0099	0.0076	0.0010	0.0200	0.0103	0.0078
Fe	0.1870	0.6860	0.3206	0.1230	0.2890	0.4480	0.3403	0.0729
Zn	2.0000	3.2000	2.1619	0.3584	2.0000	9.1000	3.9000	3.4708

All units are in mg/l except pH in pH units, COND in µs/cm, Temp in oC, Sal in ppt.

All the surface waters D POT, D AKD, D MAN and NS were found to have pH values very close (7.38-7.48). The WHO recommended limit for potable water is 6.5-8.5. This implies that about 42.30% of the samples fall outside the recommended range while 58.7% fall within the range.

Electrical conductivity (EC) values are generally low. Minimum and maximum values are 98.8µs/cm^{-1} and 722µs/cm^{-1} respectively, with the mean value of 337.96. Total dissolve solid (TDS) ranged from 49.5-361mg/l. According to TDS classification by [37], all the groundwater's are fresh (TDS<1000). The surface water EC and TDS range from 101.9 to 191.2 and 50.9 to 96 mg/l with mean values of 137.35 and 68.78 respectively. Major cations (Ca^{2+}, Mg^{2+}, Na^+ and K^+) in both groundwater and surface water were also generally low with Na^+ being the most dominant cation. HCO_3^- is also the most dominant anion with values ranging from 12.4 to 233.6 mg/l for groundwater and 23.46 to 102.8 for the surface water.

The nitrate in the groundwater varied from 0.18 - 64.46mg/l with an average of 23.34 mg/l. Eventhough it has been observed that igneous rocks contain small amounts of nitrate [38], most nitrate in water comes from fertilizers, nitrification by leguminous plants and animal excreta. Industrial and domestic sources can also contribute to higher elevation of nitrate in water. Nitrogen is an essential component of protein hence occurs in all living organisms. When these materials decay through microbial activities, the complex protein changes through amino acid to ammonia, nitrite and finally nitrate. The nitrate produce may then leach to the groundwater.

Fears have been expressed that nitrate contaminated water supplies carries the risk of methaemoglobinaemia (blue-baby syndrome) and stomach cancer. The main pollution risk for the aquifers is vertical infiltration of precipitation and flushing of pollutants from the soil. About 57.7% of the samples have nitrate concentrations above the recommended value of 10mg/l by WHO.

There was no general trend in the different geological formations; nitrate from the rocks is therefore not likely to be the source of the nitrate, but since the area is a forest zone dominated by agriculture activity and also heavily populated, higher nitrate concentrations can be attributed to decay of organic matter, nitrogen fixation, fertilizer applications and sewage.The surface water nitrate was also observed to be almost uniform with little variation from the upper to the lower portion of the study area. This can also be attributed to run-offs, sewage and industrial effluents being discharge into the river.

SO_4^{2-} and PO_4^{3-} were observed to have generally low concentration in both the surface and the groundwater in the study area.

Water Quality Index (WQI)

The chemistry of groundwater has been utilized as a measure to outlook the quality of water for drinking and other purposes [5], [12]. Tiwari and Mishra [9] specifically used WQI to determine the suitability of groundwater for drinking purpose. A location wise calculated WQI values for the different geological formations and the surface water were presented in Table 6.

Table 6. Results of the calculated WQI of the sampling points, adapted from [25]

Location	CODE	Geological type	WQI	Quality
Nkroso	NK	Birimian	27.94	Good
Potroase1	POT 1	Birimian	11.14	Excellent
Potroase Yawofori	POT YOF	Birimian	17.48	Excellent
Anyinase	ANY	Birimian	15.00	Excellent
Akooko	AK	Birimian	19.20	Excellent
Potroase2	POT 2	Birimian	16.83	Excellent
Tinkong1	TK 1	Cape Coast Granitoid	10.90	Excellent
Tinkong2	TK 2	Cape Coast Granitoid	10.26	Excellent
Crig	CR	Cape Coast Granitoid	10.83	Excellent
Maase1	MS 1	Cape Coast Granitoid	11.77	Excellent
Maase2	MS 2	Cape Coast Granitoid	16.24	Excellent
Maase T 1	MST 1	Cape Coast Granitoid	13.89	Excellent
Maase T 2	MST 2	Cape Coast Granitoid	14.08	Excellent
Adowkwanta	ADK	Cape Coast Granitoid	15.95	Excellent
Omenako	OM	Cape Coast Granitoid	7.46	Excellent
Kukua	KK	Cape Coast Granitoid	8.76	Excellent
Metemano	MT	Cape Coast Granitoid	6.43	Excellent
Teacher Mante	TM	Cape Coast Granitoid	14.27	Excellent
Asosotwene	AST	Cape Coast Granitoid	11.20	Excellent
Afabeng Borehole	AF B	Cape Coast Granitoid	25.09	Good
Afabeng Hand dug well	AF H	Cape Coast Granitoid	28.83	Good
Densu Potroase	D POT	Surface water	15.25	Excellent
Densu mangoase	D MAN	Surface water	11.93	Excellent
Densu Akyem Odumase	D AKD	Surface water	15.84	Excellent
Densu Nsawam	D NS	Surface water	27.92	Good

The results of the computed WQI values of the Birimian ranges from 11.14 to 27.94, while the Cape Coast granitoid and the surface water ranged from 7.46 to 28.83 and 11.93 to 27.92 respectively. All the two geological formations showed 'excellent water' quality in most locations with the Birimian having 83.33%, Cape Coast granitoid 81.25% and the surface water 75%. Few locations NK in the Birimian, AF B and AF H in the Cape-Coast granitoid and D NS for the surface water, however, showed 'good' water quality.

It is evident that, though the geological materials contribute to the presence of dissolved ions in the water, these areas of 'good' water quality might be affected by leaching from point source pollutants from nearby effluents, domestic disposal site or agricultural wastes (agro-chemicals, fertilizers etc.) A comparison of the WQI values of the samples reveals that, groundwater in all the geological formations and the surface water based on the parameters measured are suitable for drinking. However, Nkroso in the Birimian, Afabeng borehole and hand dug wells showed evidence of gradual contamination.

Groundwater and Surface Water Classification

The groundwater and the surface water were classified using [39] modification of [40] method. When pyrites and other sulphide minerals in the aquifer material undergo oxidation, acid solutions are produced in which heavy metals become highly mobile.

The relationship between pH and total metal content (mg/l) for the analysed samples are presented in Figure 3. The total metal content was calculated as sum of all the measured metal (Zn, Fe, Mn, As, Cr, Al and Cu). All the surface water and about 85.71% of the groundwater samples plot in the field of near neutral-high metal while 14.29% of the groundwater plot in the region of acid-high metal. In all the cases, the samples plotted in the region of high metal.

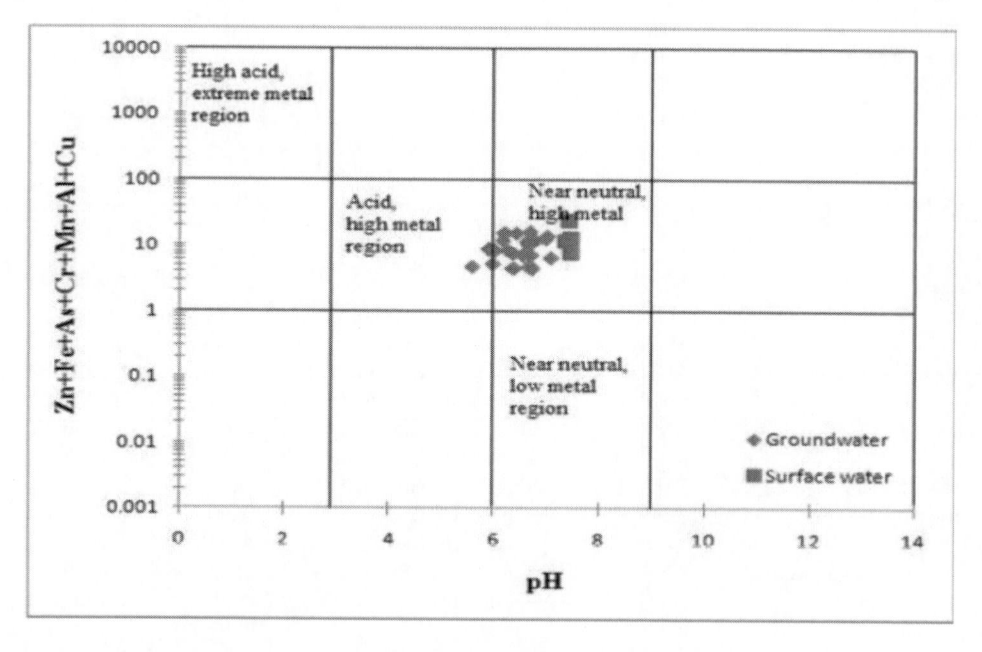

Figure 3. Classification of water based metal load and pH.

Even though, the metal levels were below the WHO guidelines for drinking water, this is a source of worry because all the groundwater and the entire stretch of the surface water (Densu River) are used for drinking and irrigation purposes [11]. The high metal concentration in most of the groundwater may originate from the geology and to a lesser extent corrosion of pipes or pipe fittings. Most of these pumps have not seen any major rehabilitation with some functioning with difficulty due to corrosion.

Heavy Metal Pollution Indices

The results of the statistical and location wise C_d, HPI and HEI of the groundwater and surface water are presented in Table 7 and Table 8 respectively.

Contamination Index (C_d)

C_d values ranged from -0.64 to 20.19 with mean value of 8.44. The computed C_d values showed that, with the exception of MT (C_d = -0.64) which belongs to the low category, all the groundwater belong to high contamination level (C_d >3). The surface water C_d ranged from 3.35 to 20.08 with mean value of 10.16 representing 100% contamination.

Table 7. Heavy metal pollution indices, mean deviation (MD) and percentage deviation (%D) for groundwater in the study area

Location	Code	C_d	MD	% D	HPI	MD	% D	HEI	MD	% D
Tinkong1	TK1	3.74	-4.70	-55.73	53.63	-10.61	-16.52	10.74	-4.70	-30.46
Tinkong2	TK2	3.63	-4.81	-57.02	53.25	-10.99	-17.11	10.63	-4.81	-31.17
Crig	CR	3.66	-4.78	-56.64	51.81	-12.43	-19.35	10.66	-4.78	-30.96
Maase1	MS1	4.80	-3.63	-43.07	55.36	-8.88	-13.83	11.80	-3.63	-23.54
Maase2	MS2	10.06	1.62	19.25	72.88	8.64	13.45	17.06	1.62	10.52
Maase T 1	MST1	9.21	0.77	9.14	69.84	5.60	8.71	16.21	0.77	4.99
Maase T 2	MST2	9.61	1.17	13.86	62.87	-1.37	-2.14	16.61	1.17	7.57
Adowkwanta	ADK	9.74	1.31	15.47	77.11	12.87	20.03	16.74	1.31	8.46
Nkroso	NK	16.96	8.52	101.00	87.03	22.79	35.47	23.96	8.52	55.20
Potroase1	POT1	4.42	-4.02	-47.62	47.16	-17.08	-26.59	11.42	-4.02	-26.03
Potroase2	POT2	11.31	2.87	34.05	72.82	8.58	13.35	18.31	2.87	18.61
Potroase Yawofori	POT YOF	10.72	2.28	27.07	78.28	14.04	21.85	17.72	2.28	14.79
Omenako	OM	2.50	-5.93	-70.33	40.32	-23.92	-37.24	9.50	-5.94	-38.44
Kukua	KK	2.06	-6.38	-75.58	41.04	-23.20	-36.12	9.06	-6.38	-41.31
Metemano	MT	-0.64	-9.08	-107.55	29.34	-34.90	-54.33	6.36	-9.08	-58.79
Teacher Mante	TM	10.46	2.02	23.98	62.68	-1.56	-2.43	17.46	2.02	13.11
Asosotwene	AST	6.95	-1.49	-17.64	51.54	-12.70	-19.77	13.95	-1.49	-9.64
Afabeng Borehole	AF B	20.19	11.75	139.24	109.22	44.98	70.01	27.19	11.75	76.10
Afabeng Hand dug well	AF H	15.01	6.57	77.86	92.23	27.99	43.57	22.01	6.57	42.55
Anyinase	ANY	3.81	-4.63	-54.85	51.12	-13.12	-20.43	10.81	-4.63	-29.98
Akooko	AK	16.32	7.88	93.43	79.47	15.23	23.70	23.32	7.88	51.06
	Min	-0.64			29.34			6.36		
	Max	20.19			109.22			27.19		
	Mean	8.44			64.24			15.44		

Table 8. Heavy metal pollution indices for surface water in the study area

Location	C_d	Mean Deviation	% Deviation	HPI	Mean Deviation	% Deviation	HEI	Mean Deviation	% Deviation
D POT	3.35	-6.81	-67.06	44.34	-21.12	-32.26	10.35	-6.81	-39.71
D MAN	6.5	-3.67	-36.07	50.62	-14.84	-22.67	13.5	-3.67	-21.36
D AKD	10.73	0.57	5.58	66.28	0.82	1.26	17.73	0.57	3.3
D NS	20.08	9.91	97.55	100.59	35.13	53.67	27.08	9.91	57.76
Min	3.35			44.34			10.35		
Max	20.08			100.59			27.08		
Mean	10.16			65.45			17.16		

Heavy Metal Pollution Index (HPI)

The groundwater showed a wide variation in HPI ranging from 29.34 to 109.22 with mean value of 64.24. The HPI showed that 95.24% of the groundwater samples were below the critical value of 100 with only AF B slightly exceeding the limit representing 4.76% of the samples in the study area.

Table 9. PHI calculation for groundwater in the Densu River Basin based on WHO, 2004 and Indian Standard 1991, 10500

Metals	Mean concentration (M_i) (ppb)	Highest permitted value for drinking water (S_i) (ppb)	Desirable maximum value (I_i)(ppb)	Unit weighting factor (W_i)	Sub-index (Q_i)	W_i*Q_i	HPI
Cu	4134	1500	50	0.00067	281.655	0.18777	64
Al	1585.38	200	30	0.00500	914.93	4.57465	
Mn	1149.64	300	100	0.00333	524.821	1.749405	
Cr	1.34286	10	0	0.10000	13.4286	1.342857	
As	9.85714	50	0	0.02000	19.7143	0.394286	
Fe	320.571	1000	100	0.00100	24.5079	0.024508	
Zn	2161.9	15000	5000	0.00007	28.381	0.001892	

$\sum W_i = 0.13007$, $\sum W_i*Q_i = 8.2754$

Table 10. HPI calculation for surface water in the Densu River Basin based on WHO, 2004 and Indian Standard 1991, 10500

Metals	Mean concentration (M_i) (ppb)	Highest permitted value for drinking water (S_i) (ppb)	Desirable maximum value (I_i)(ppb)	Unit weighting factor (W_i)	Sub-index (Q_i)	W_i*Q_i	HPI
Cu	7015	1500	50	0.00067	480.345	0.32023	65
Al	2011.5	200	30	0.00500	1165.59	5.827941	
Mn	446.25	300	100	0.00333	173.125	0.577083	
Cr	1.35	10	0	0.10000	13.5	1.35	
As	10.25	50	0	0.02000	20.5	0.41	
Fe	340.25	1000	100	0.00100	26.6944	0.026694	
Zn	3900	15000	5000	0.00007	11	0.000733	

$\sum W_i = 0.13007$, $\sum W_i*Q_i = 8.5126$.

The surface water HPI ranged from 44.34 to 100.59 with a mean value of 65.46. This represent 75% low contamination. The detailed average calculation of the HPI with unit weightage (W_i) and standard permissible values (S_i) are presented for groundwater and surface water in Table 9 and Table 10 respectively.

Heavy Metal Evaluation Index (HEI)

The HEI developed by [35] also gave a meaningful insight into the extent of the heavy metal pollution in the study area. The groundwater HEI ranged from 6.36 to 27.19 with a mean value of 15.44.

The surface water values also ranged from 10.34 to 27.08 with a mean value of 17.16. The HEI values were grouped based on the mean values using multiple of mean into three signifying different levels of contaminations.

The proposed grouping criteria are low (HEI<15), medium (HEI=15-30) and high (HEI>30). Using this criteria, 47.6%, 52.4% and 0% of the groundwater samples show low, medium and high contamination, respectively, whereas, 50%, 50% and 0% of the surface water low, medium and highly contaminated.

Comparison of the Three Indices

A comparative study of the three indices (C_d, HPI and HEI) showed a similar trend in all the sampling points Figure 4. A steady rise in all the indices as the river flows from upstream to downstream is an indication of effluent, domestic or industrial waste entering the river.

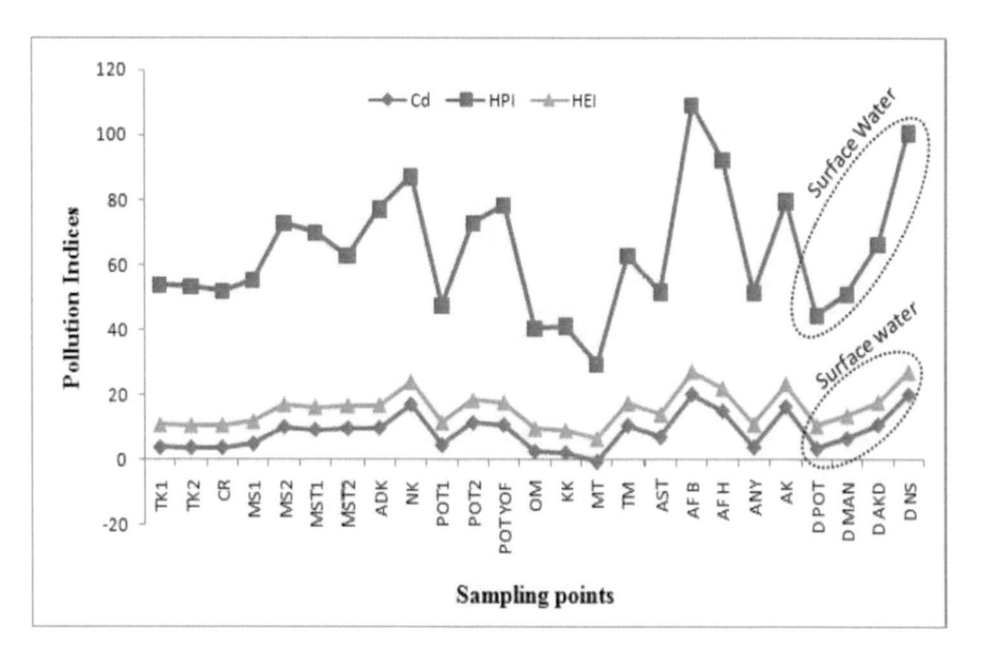

Figure 4. Comparison and spatial distribution of the three pollution indices.

Table 11. Classification of groundwater and surface water quality based on modified categories of the pollution indices in the study area

Index method	Class	Degree of pollution	No. of location	Percentage (%)
Groundwater				
	<10	Low	13	61.9
C_d	10-20	Medium	7	33.3
	>20	High	1	4.8
HPI	<60	Low	10	47.6
	60-120	Medium	11	52.4
	>120	High	0	0
HEI	<15	Low	10	47.6
	15-30	Medium	11	52.4
	>30	High	0	0
Surface water				
	<10	Low	2	50.0
C_d	20-40	Medium	1	25.0
	>20	High	1	25.0
HPI	<60	Low	2	50.0
	60-120	Medium	2	50.0
	>120	High	0	0
HEI	<15	Low	2	50.0
	15-30	Medium	2	50.0
	>30	High	0	0

It is interesting to also note that, despite, the similar trends in the indices; they presented somehow different water quality information. For instant, whereas C_d suggest 95% of groundwater and 100% of surface water as highly polluted, HPI and HEI indicated that, 4.76% and 0% of groundwater and 25% and 0% of the surface water were polluted.

For the purpose of comparison with HEI, the classification C_d and HPI were slightly modified using the multiple of mean criteria as HEI. The reclassification scheme showed comparable results and water quality information, thus for C_d, 61.9% of groundwater and 50% of surface water were classified as low, 33.3% and 25% of groundwater and surface water were classified as medium, while 4.8% and 25% of the groundwater and surface were highly contaminated respectively.

The HPI also classified 47.6% of groundwater and 50% of surface water as low, 52.6% of groundwater and 50% surface water classified as moderately polluted with 0% high pollution in both cases Table 11.

The relationship between the indices and the metals responsible for the calculated indices were also examined using Pearson correlation matrix (Table 12). C_d , HPI and HEI showed strong correlation (C_d vrs HPI =0.751, C_d vrs HEI =0.820 and HPI vrs HEI =0.993). A similar strong correlation was observed for C_d, HPI and HEI with Cu, Al and Mn suggesting that, these metals are the dominant or controlling parameters of these indices.

In summary, it can be said that, despite the comparable trends shown by the three indices, the HPI and HEI are more suitable for water quality index determination in the study area. Furthermore, the simplicity of HEI as compared to HPI and C_d methods makes it a more preferable choice. This study corroborates well with the findings of [11] and [35].

Table 12. Correlation matrix among metals, pollution indices and the physical parameters in the study area

	pH	T °C	TDS	EC	Sal	Alk	Cu	Al	Mn	Cr	As	Fe	Zn	Cd	HPI
pH															
T °C	-0.343														
TDS	**0.548***	-0.351													
EC	**0.549****	-0.350	**1.000****												
Sal	**0.511***	-0.444*	**0.880****	**0.880****											
Alk	**0.813****	-0.392	**0.633****	**0.632****	0.472*										
Cu	0.287	-0.011	0.275	0.278	0.303	0.132									
Al	0.176	-0.113	0.263	0.263	0.366	0.101	**0.503***								
Mn	-0.273	0.235	-0.074	-0.077	-0.010	-0.167	0.328	**0.574****							
Cr	0.042	0.097	.0469*	0.468*	0.404	-0.068	0.327	0.181	0.293						
As	0.172	-0.105	-0.084	-0.083	-0.156	0.192	0.032	-0.298	-0.161	-0.363					
Fe	0.277	-0.346	0.233	0.234	0.225	0.345	-0.105	0.000	-0.432	-0.256	-0.284				
Zn	0.385	0.193	0.329	0.332	0.259	0.402	0.350	0.302	0.086	0.139	0.250	-0.164			
Cd	0.252	0.025	0.281	0.283	0.321	0.139	**0.969****	**0.663****	**0.510***	0.346	-0.023	-0.147	0.425		
HPI	0.020	0.047	0.236	0.235	0.302	-0.011	**0.582****	**0.863****	**0.850****	0.466*	-0.208	-0.301	0.292	**0.751****	
HEI	0.059	0.052	0.249	0.249	0.312	0.018	**0.668****	**0.858****	**0.827****	0.458*	-0.180	-0.287	0.335	**0.820****	**0.993****

* Correlation is significant at the 0.05 level (2-tailed).

** Correlation is significant at the 0.01 level (2-tailed).

Water for Irrigation Purpose

The quality of water used for irrigation is vital for crop yield, maintenance of soil productivity and protection of the environment [41]. At the same time, the quality of irrigation water is very much influenced by the land constituents of the water source. For the purpose of this work, Sodium Adsorption Ratio (SAR), Sodium Percentage (Na%), Residual Sodium Carbonate (RSC) and Permeability Index (PI) were used to determine the suitability of the groundwaters for irrigation purposes.

Sodium Absorption Ratio (SAR)

High Sodium concentration leads to development of an alkaline soil, these results in an excess Na+ in water producing the undesirable effects of changing the soil properties (formation of crust, water-logging, reduced soil aeration, reduced infiltration rate and reduced soil permeability [42]. Therefore, in assessing the suitability of groundwater for irrigation, Na+ concentration is essential. The degree to which irrigation water enters into cation exchange reactions in soil can be indicated by SAR [43]. The Na^+ replacing adsorbed Ca^{2+} and Mg^{2+} is a hazard as it causes damage to the soil structure, making it compact and impervious. SAR is defined as

$$SAR = \frac{Na^+}{\sqrt{\frac{Ca^{2+} + Mg^{2+}}{2}}}$$

where, the concentrations are reported in meq/L.

Applying this index to the samples indicates that, 100% of the Birimian and surface water belong to excellent category with SAR ranging from 2.62 to 6.7 and 2.62 to 5.56 respectively. The Cape Coast granitoid however, yielded only 73% in the excellent category with 20% and 7% falling under Good and doubtful category.

Sodium Percentage (%Na)

Soils containing a large proportion of sodium with carbonate as the predominant anion are termed alkali soils; those with chloride or sulphate as the predominant anion are known as saline soils [44]. Na+ concentration in water is widely use in assessing the suitability of water for irrigation purposes [45] and plays a vital role in the classification of river water for irrigation. This is due to the fact that, sodium reacts with soil resulting in clogging of particles, thereby reducing the permeability [44]; [46] and [47]. It is usually expressed in terms of percent sodium (Na%) can be calculated by the following relation

$$\%Na = \left(\frac{Na^+ + K^+}{Ca^{2+} + Mg^{2+} + K^+ + Na^+} \right) * 100$$

where all ionic concentrations are in meq/l

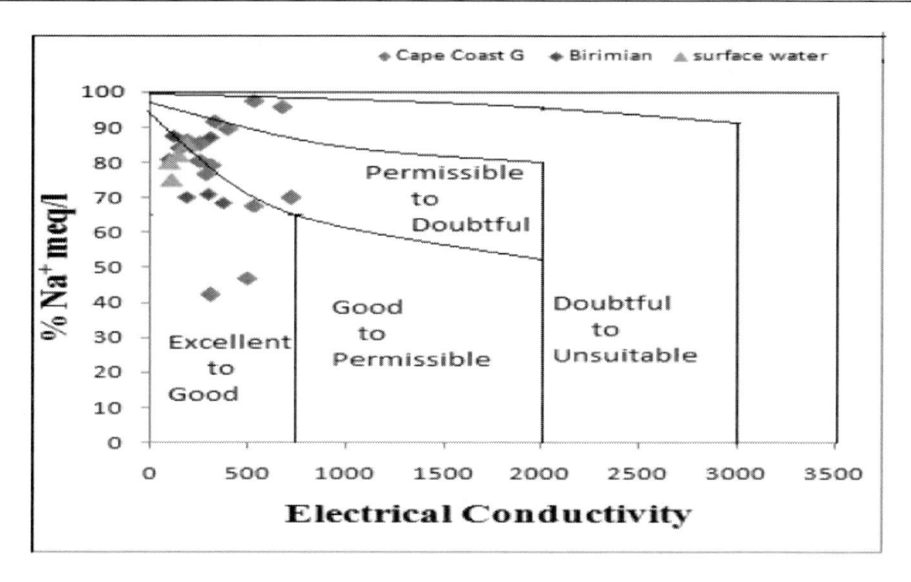

Figure 5. Rating of groundwater samples on the basis of electrical conductivity and percent sodium (after Wilcox, 1955), adapted from [25].

The calculated Na% (meq/l) in the samples were plotted against Electrical conductivity in µs/cm [45] diagram in Figure 5. The results revealed that 100% of the surface water and 83% of the Birimian fell in the field of excellent to good while 16% of the Birimian samples fell under permissible to doubtful.

The Cape-Coast granitoid however, showed a wide variation in the Wilcox diagram with 53% of the samples belonging to field of excellent to good, 33% belong to Permissible to doubtful and 13% being doubtful to unsuitable.

Residual Sodium Carbonate

In addition to the SAR and Na%, the difference between the excess sum of carbonate and bicarbonate in groundwater against the sum of calcium and magnesium also has a significant effect on groundwater suitability for irrigation. When total carbonate levels exceed the total amount of calcium and magnesium, the water quality may be deteriorated. This is because, high excess carbonate concentration often known as "residual" combines with calcium and magnesium to form a solid scale like material which settles out of the water [48]. This excess is denoted by 'residual sodium carbonate' (RSC) and is determined as suggested by [49].

$$RSC = \left(HCO_3^- + CO_3^{2-}\right) - \left(Ca^{2+} + Mg^{2+}\right)$$

where all ionic concentrations are expressed in meq/l.

Irrigations water with high RSC have high pH values; hence, soils or land irrigated by such waters becomes infertile and lowers crop yield owing to the deposition of sodium carbonate responsible for the black colour of the soil [46] and [50].

Residual sodium carbonate (RSC) has been calculated to determine the suitability or otherwise of the groundwater in the study area for agricultural purpose. The classification of irrigation water was done according to the [51]. RSC < 1.25 meq/l is safe for irrigation, values between 1.25 to 2.5 meq/l is of marginal quality and a value > 2.5 meq/l is unsuitable for irrigation [46]. In this study, the surface water and the Birimian samples show RSC values ranging from −0.03 to −1.14 meq/l and -0.38 to 1.13 respectively representing 100% safe for irrigation.

The Cape Coast granitoid samples ranges from -0.01 to 3.37 representing 73% safe for irrigation and 20% marginal quality for irrigation and 7% unsuitable for irrigation. Furthermore, the negative RSC at all sampling sites indicates that there is no complete precipitation of calcium and magnesium [52].

Permeability Index (PI)

The PI is also a useful tool which indicates whether water samples are suitable for irrigation. Irrigated waters influenced by sodium, calcium, magnesium and bicarbonate contents affect the permeability of soil after a long term use. [46]. Doneen, [53] classified water based on PI which is given as

$$PI = \frac{\left(Na^{+} + \sqrt{HCO_3^{-}}\right)*100}{Na^{+} + Ca^{2+} + Mg^{2+}}$$

According to the criteria water can be classified as Class I, II and III. Class I and II are categorized as good for irrigation with 75% or more maximum permeability. Class III water is unsuitable with 25% maximum permeability.

All the Birimian and surface waters Figure 6 fall in class I. 87.5% of the Cape Coast granitoid belongs to Class I while the rest of the 12.5% belong to Class II group, hence, based on Doneen's chart all the samples are suitable for irrigation.

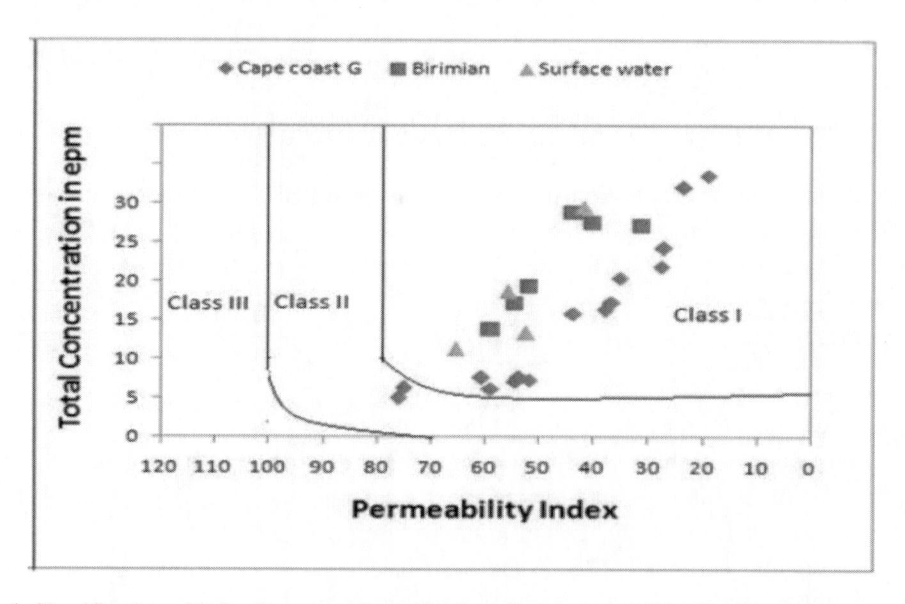

Figure 6. Classification of irrigation water for soils of medium permeability (Doneen chart).

Stable Isotope Analysis

The statistical summary of the isotopic ($\delta^{18}O$ and $\delta^{2}H$) composition in rain water, groundwater and surface water are shown in Table 13.

Rain Water Isotopic Composition

The majority of the rain water stable isotopes (40) were sampled at Koforidua covering a two year period (2007-2008) with few samples from Accra (7) from March to July 2007.

The relationship between δ D and $\delta^{18}O$ based on the 47 samples of the rain water are shown in Figure 7. According to the measured δ D and $\delta^{18}O$ in the 47 sets of rainfall, the data define a linear trend that represents the local meteoric water line for the Densu river basin. The slope of 7.3 and intercept of 8.6 are both slightly below the global meteoric water line [54]. The $\delta^{18}O$ values of the rainfall in the Densu river basin ranged from -6.01 to -1.80 with a mean value of -4.25, the δ D ranged from -35.70 to -2.80 with a mean value of -22.59. The seasonal variation of rainfall amount and δ D are shown in Figure 8. The lighter isotopic values are observed in May to September in both years. These months represents the period of heavy rainfall events, where the humidity is high in both years, hence, much of the groundwater recharge may occur during these months.

Table 13. Statistical summary of the $\delta^{18}O$, $\delta\,^{2}H$ and d-excess of the rainwater, groundwater and surface water in the study area

	$\delta^{18}O$				$\delta\,^{2}H$				d-excess			
	Min	Max	Mean	SD	Min	Max	Mean	SD	Min	Max	Mean	SD
Groundwater	-3.63	-1.43	-2.82	0.49	-16.46	-3.48	-11.24	3.04	7.01	14.08	11.35	1.45
Surface water	-3.34	-2.21	-2.76	0.35	-16.1	-10.3	-12.54	1.88	5.58	11.22	9.50	1.88
Rain water	-6.35	-1.92	-4.33	1.35	-37.8	-3.7	-23.75	9.81	4.28	14.12	10.90	2.49

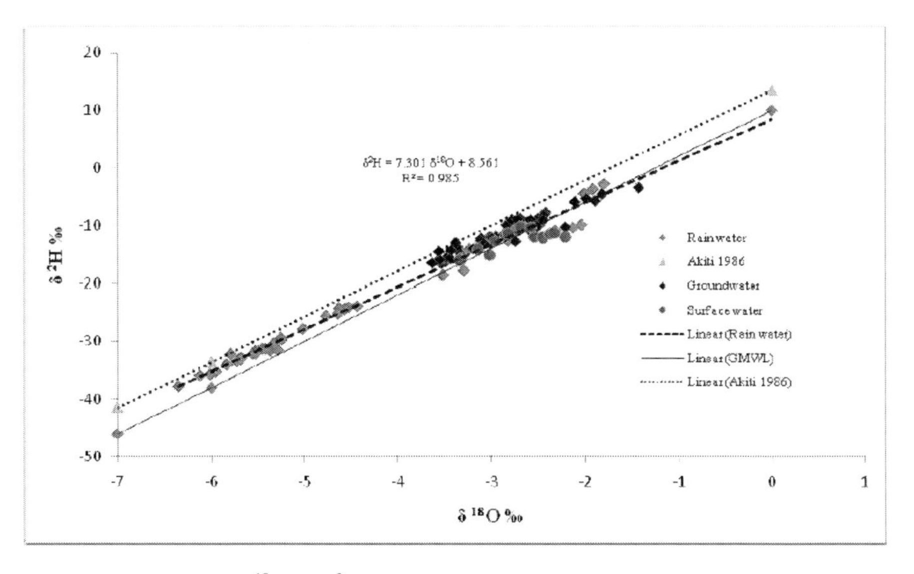

Figure 7. Relationship between ^{18}O and ^{2}H in the rain water, groundwater and surface water.

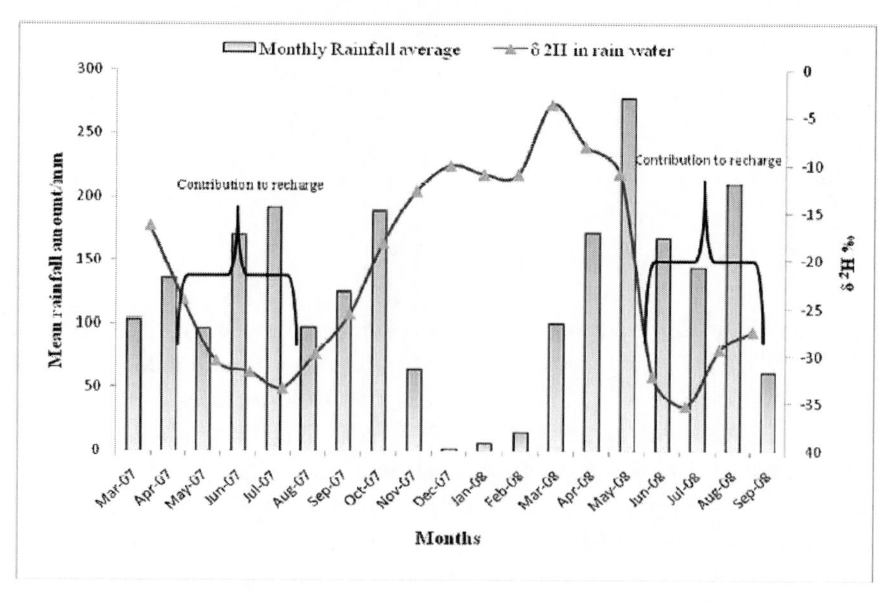

Figure 8. Seasonal variation of rainfall amount and δ ^{2}H in the study area.

Groundwater Isotopic Composition

Deuterium values of twenty one (21) groundwater ranges from -16.46 to -3.48‰ while oxygen-18 values range from -3.63 to -1.43 ‰ VSMOW. The average values of δ ^{2}H and δ ^{18}O were -11.24±1.62‰ and -2.82±0.37‰ respectively. The best fit regression line of the groundwater was $\partial^2 H = 5.83\partial^{18}O + 5.2$ with a correlation coefficient of r^2 =0.896. The surface water had δ ^{2}H and δ ^{18}O ranging from -6.1 to -10.3‰ and -3.34 to -2.21‰ respectively with an average of -12.54±1.64 for δ ^{2}H and -2.76±0.09 for δ^{18}O. The best fit regression line was $\partial^2 H = 4.08\partial^{18}O - 1.3$ with a correlation coefficient of r^2=0.578. It was observed that both the groundwater and the surface water have slopes lower than the rainfall. The lower slope implies primary evaporation during precipitation as the moisture moves allowing time for more contact and exchange with the atmosphere. The observed isotopic variation in the rainfall might be due to different rainfall events.

Origin of Groundwater

The isotopes of oxygen ^{18}O and hydrogen δ ^{2}H are a sensitive tracer and widely used in studying the natural water circulation and groundwater movements. Differences in the content of δ ^{2}H and δ ^{18}O in groundwater, surface water and rainfall collected at Koforidua were exploited on a similar graph to show the extent of variation in the study area Figure 7. A local meteoric water line (LMWL) obtained by [55] for the Accra plains (which forms part of the study area) defined by the equation

$$\partial^2 H = 7.86\partial^{18}O + 13.61$$

and the Global meteoric water line [54] defined by the equation

$$\partial^2 H = 8\partial^{18} O + 10$$

were also inserted in the graph.

The stable isotope composition relative to Global Meteoric Water Line (GMWL) reveals important information on the groundwater recharge patterns, relationship between ground and surface water. As evident from Figure 7, most groundwater samples plot in between the GMWL and LMWL with few samples falling on the lines. This suggests a meteoric origin for the groundwater. The groundwater in the study area appears to group in a narrow range, signifying a well-mixed system with relatively constant isotopic composition. A few (15%) of the groundwater samples plot slightly away from the meteoric water line showing an evidence of small isotopic enrichment by evaporation on the surface or in the unsaturated zone before recharge. This shows that, generally, the meteoric water recharging the groundwater system in the area is homogeneous with evaporation playing an insignificant role on the infiltrating water. Similar observations were made in the southern portion of the basin [56]. All the surface waters plot relatively below the global meteoric water line indicating a degree of isotopic enrichment. This observation can be attributed partly to the open flow of the river and to some extent the isotopic enrichment could be reflecting the integration of isotopic composition of the Densu river tributaries. The narrow range in the stable isotopic composition of the surface water shows a homogeneous and well mixed system between the Densu River and its tributaries.

It is interesting to note that few of the groundwater samples exhibit similar isotopic composing to that that of the surface water (River Densu), this suggest a possible hydraulic connection between the aquifers and the river water, some degree of fractionation both on land surface and in the unsaturated zone and most probably mixing mechanisms by anthropogenic activities such as irrigation, which might result in the groundwater being recharge by enriched waters.

Deuterium Excess (D-excess)

The d-excess reflects the conditions that lead to kinetic isotope fractionation between water and vapour during primary evaporation in the oceans [57]. This number also shows the extent of deviation of a given sample from the meteoric water line. The calculated deuterium excess of the rainfall was found to range from 7.01 to 14.08‰ with most samples plotting together confirming a common moisture source for the rainfalls. The groundwater d-excess values ranges from 5.58 to 11.22‰, while that of the surface water ranges between 4.28 to 14.12‰. As the $\delta^{18}O$ increases (more enriched) the deuterium excess in all the samples decreases gradually Figure 9.

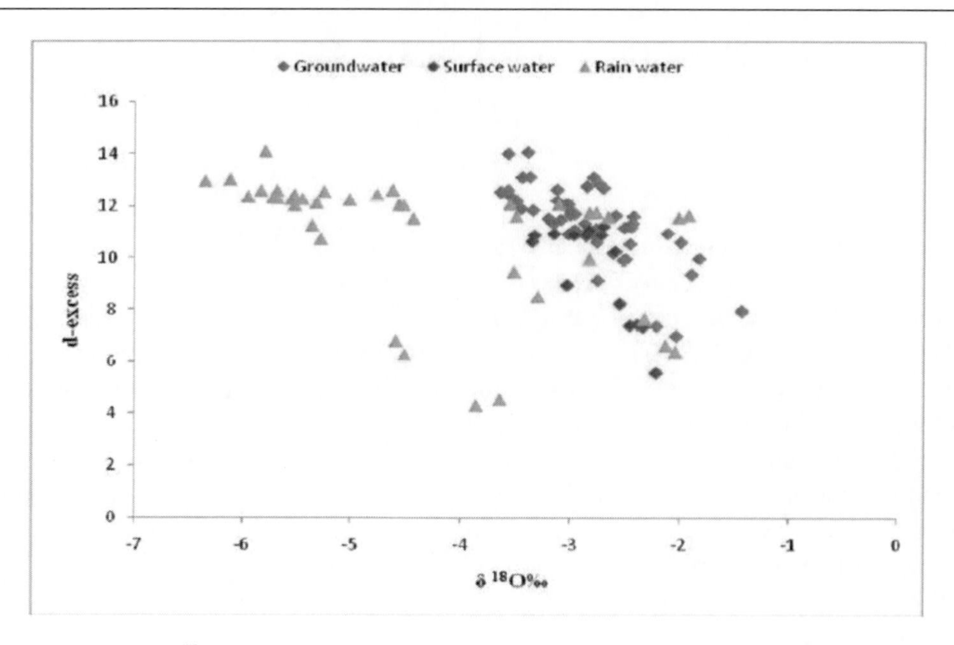

Figure 9. d-excess vs $\delta^{18}O$ plot in the study area.

It is also noticed that the groundwater samples plot together with the rainfall and the surface water. This observation can be partly attributed to the dilution of the groundwater with the rainfall which could increase the d-excess [58]. Furthermore, the tendency towards the rainfall may imply a modern recharge to the groundwater systems.

Conclusion

The study investigated the suitability of groundwater and surface water in the Densu river basin for drinking and agricultural purposes using a wide range of water quality indices. The results showed that, the groundwater and surface in the basin are generally suitable for drinking and irrigation. The stable isotopes showed that, the groundwater are generally of meteoric origin and of modern day recharge.

Acknowledgment

The authors wish to thank the technicians of the isotope hydrology lab of Ghana Atomic Energy Commission for their support in sample analysis. We also wish to thank Prof Amuesi and Mr. Oware Kesse for providing transport to the field.

References

[1] M. Vasanthavigar, K. Srinivasamoorthy, R. Rajiv Ganthi, K. Vijayaraghavan, V. S. Sarma, *Arabian Journal of Geosciences*. 5, 245, (2012).

[2] P. N. Rajankar, S. R. Gulhane, D. H. Tambekar, D.S. Ramteke, S.R. Wate, *E-Journal of Chemistry*. 6, 905 (2009).
[3] P. K. Goel, *Water Pollution – Causes, Effects and Control*, New age Int. (P) Ltd, New Delhi (2000).
[4] K. P. Bam, Major and trace elements in soil profile of the unsaturated zone of the Densu river basin, Ghana, University of Ghana, *Unpublish Mphil thesis.* (2009).
[5] M. Vasanthavigar, K. Srinivasamoorthy, K. Vijayaragavan, R.R. Ganthi, S. Chidambaram, P. Anandhan, R. Manivannan, S. Vasudevan, *Environmental Monitoring and Assessment*. 171, 595, (2010).
[6] P. Simeonova, V. Simeonov, G. Andreev, *Central European of Chemistry*. 1, 136, (2003).
[7] S.V. Mohan, P. Nithila and S.J. Reddy, *J. Environ. Sci. Health A*. 31, 283, (1996).
[8] N. Nishidia, M. Miyai, F. Tada, S. Suzuki, *Environ. Pollution*. 4, 241, (1982).
[9] T. N. Tiwari and M. Mishra, *Indian J. Environmental Protection,* 5, 276, (1985).
[10] B. Prasad and K.K. Mondal, *Mine Water Environ.*, 27, 40, (2008).
[11] M.A.H. Bhuiyan, M.A. Islam., S.B. Dampare, L. Parvez., S. Suzuki, *Journal of Hazardous Materials*. 179, 1065, (2010).
[12] W. M. Edmunds, J. J. Carrillo-Rivera and A, *Journal of Hydrology,* 258, 1, (2002).
[13] J.Ch Fontes, Environmental Isotopes in groundwater hydrology In *handbook of Environmental Isotopes Geochemistry*, (1980).
[14] IAEA, Guidebook on Nuclear Techniques in Hydrology, *Technical Report series No. 91, IAEA*, Vienna, (1983).
[15] A. Gibrilla, T.T. Akiti, S. Osae, D. Adomako, S.Y. Ganyaglo, E.P.K. Bam, A. Hadisu, *Journal of Water Resource and Protection,* 2, 1071, (2010a).
[16] D. Adomako, S. Osae, T.T. Akiti, S. Faye, P. Maloszewski, *Environ. Earth Sci.*, DOI 10.1007/s12665-010-0595-2, (2010).
[17] J. R. Fianko, S. Osae, D. Adomako, D.G. Achel, *Environmental Monitoring Assessment*, 30, 145, (2008).
[18] WRC, Groundwater assessment: an element of integrated water resources management: the case of Densu River Basin, *Technical report,* Water Resources Commission, Accra. (2006).
[19] A. Gibrilla, T.T. Akiti, S. Osae, D. Adomako, S.Y. Ganyaglo, E.P.K. Bam, A. Hadisu, *Journal Water Resource and Protection*, 2, 1010, (2010b).
[20] D. K. Sinha, A. K. Srivastava, *Indian Journal of Environmental Protection,* 14, 340, (1994).
[21] WHO, *World Health Organisation guidelines for drinking water quality*, Third edition. Geneva, ISBN 92 45 154638 7, (2004).
[22] Ghana Statistical Service, *2000 population and housing census*, (2000).
[23] K. B. Dickson and G. Benneh, *A New Geography of Ghana*. Longman, London. (1980).
[24] A. Gibrilla, pollution, hydrochemical and isotopic studies of groundwater in the northern Densu river of basin, Ghana. University of Ghana. *Unpublish Mphil thesis*.
[25] A. Gibrilla, E.K.P. Bam, D. Adomako , S. Ganyaglo, S. Osae, T.T. Akiti, S. Kebede, E. Achoribo, E. Ahialey, G. Ayanu, E.K. Agyeman, *Water Qual Expo Health. DOI 10.1007/s12403-011-0044-9.* 3, *63,* (2011).
[26] M. Radojevic,. and V.N. Bashkin, , Organic matter. In: *Practical Environmental Analysis*. The Royal Society of Chemistry, Cambridge. 325-329. 1999.

[27] T. B. Coplen, *Chemical Geology*, 72, 293, (1988).
[28] S. S. Asadi, P. Vuppala, M.A. Reddy, *International Journal of Environmental Research and Public Health,* 4, 45, (2007).
[29] S. K. Pradhan, D. Patnaik, S.P. Rout, *Indian. Journal of Environmental Protection, 21,* 355 (2001).
[30] R. D. Harkins, *Journal Water Pollution Control Federation*, 46, 588, (1974).
[31] R. K. Horton, *Journal Water Pollution Control Federation*, 37, 300, (1965).
[32] ISI, *Indian standard specification for drinking water.* New Delhi. ISI, 10500, (1993).
[33] B. Backman, D. Bodis, P. Lahermo, S. Rapant, and S Tarvainen, *Environmental Geology* 36, 55, (1997).
[34] S.V. Mohan, P. Nithida and S.J. Reddy, *Journal Environmental Science and Health,* A31, 238, (1996).
[35] A.E. Edet,. and O.E. Offiong, *GeoJournal,* 57; 295, (2002).
[36] S.J. Reddy, Encyclopedia of environmental pollution and control. *Environmental Media,* Karlla, India, 1, 342, (1995).
[37] R. A. Freeze and J.A Cherry, *Groundwater.* New Jersey: Prentice Hall. (1979).
[38] S. N. Daviest and R.J.M. Dewiest, *Hydrogeology*, John Willey and sons, New York, (1966).
[39] R. Caboi, R. Cidu, L. Fanfani, P. Lattanzi and P. Zuddas, *Chron. Rech. Miniere* 534, 21, (1999).
[40] W.H. Ficklin, G.S. Plumee, K.S. Smith and J.B. McHugh, Geochemical classification of mine drainges and natural drainages in mineralised areas. In: Kharaka Y.K. and Maest A.S. (eds), *Water-rock interaction.* Vol 7. Balkema, Rotterdam, pp 381 (1992).
[41] V. Singh, U.C. Singh, *Indian Journal of Science and Technology*, 1, 1 (2008).
[42] W.P. Kelly, *Alkali soils – Their formation, Properties and Reclamation*, Reinhold Publication, New York (1951).
[43] T. M. Manjusree, S. Joseph, J. Thomas, *Journal of the Geological Society of India*, 74, 459 (2009).
[44] T. Suresh, N.M. Kottureshwara, *Rasayan J. Chem.*, 2, 221 (2009).
[45] L.V. Wilcox, *Classification and use of the irrigation waters*, U.S. Department of Agriculture Circular No. 969, Washington, District of Columbia, (1955).
[46] A. Nagaraju, S. Suresh, K. Killham, K, Hudson-edward, *Turkish J. Eng. Env. Sci.* 30,203 (2006).
[47] D. K. Todd, *Ground water hydrology*, New York: Wiley (1980).
[48] S. K. Sundaray, B. B. Nayak, D. Bhatta, *Environmental Monitoring and Assessment,* 155, 227 (2009).
[49] L.A. Richard, Diagnosis and improvement of saline and alkali soils, *Agricultural handbook* 60, Washington, DC: USDA (1954).
[50] F.M. Eaton, *Soil Sci*, 69, 123 (1950).
[51] U S Salinity Laboratory staff et al, Diagnosis and improvement of saline and alkali soils, *U.S. Deptt. Agr. Handbook*. (1954).
[52] T.N. Tiwari and A. Manzoor, *Indian Journal of Environmental Protection*, 8, 494 (1988).
[53] L.D. Doneen, *Notes on water quality in Agriculture*, Published in water Science and Engineering, Paper 4001, Department of water Science and Engineering, University of California. (1965).

[54] H. Craig, Isotopic variation in meteoric water, *Science*, 133,1702 (1961).
[55] T. T. Akiti, Environmental isotope study of groundwater in crystalline rocks of the Accra Plains, 4th Working Meeting on Isotopes in Nature, proceedings of an advisory group meeting, *IAEA*, Vienna. (1986).
[56] Goucey et al. Application of Isotope Techniques for the Assessment of Groundwater Resources: *Densu river basin report*. (2008).
[57] W. Dansgaard, *Tellus*, 16, 436 (1964).
[58] F. Yuan and S. Miyamoto, Characteristics of oxygen-18 and deuterium composition in waters from Pecos River in American Southwest *J. Chemical Geology*. 255. pp 220-230 (2008).

In: Water Quality
Editor: You-Gan Wang

ISBN: 978-1-62417-111-6

Chapter 7

EVALUATION OF COMMUNITY WATER QUALITY MONITORING AND MANAGEMENT PRACTICES, AND CONCEPTUALIZATION OF A COMMUNITY EMPOWERMENT MODEL: A CASE STUDY OF LUVUVHU CATCHMENT, SOUTH AFRICA

L. Nare*[*] *and J. O. Odiyo
Department of Hydrology and Water Resources, University of Venda, Thohoyandou, South Africa

ABSTRACT

South Africa has projected herself as a democratic society and one of the cornerstones of democracy is community involvement and participation in development. Therefore, a study to evaluate and conceptualise a model for community empowerment in water quality monitoring and management was carried out in Luvuvhu Catchment of South Africa. The first task was to prove that the communities in the catchment were vulnerable to water quality problems. Potential sources of pollution and types of pollutants were identified. Community vulnerability and risk caused by the pollutants was confirmed by analysing results obtained from DWA surface and groundwater quality monitoring stations dotted around the catchment. Water treatment efficiencies at three water treatment plants supplying water to communities in the catchment were calculated to confirm community vulnerability due to inadequate supplies and poor water quality. The contemporary water quality monitoring and management practice was evaluated to identify any gaps relating to community involvement and participation. The willingness of the community to be involved and participate in water quality monitoring was evaluated. Indigenous knowledge systems related to water quality management were investigated to identify ways in which these could be incorporated into the national water quality monitoring framework as a way of strengthening the community's voice in water quality monitoring and management. The results showed that communities in Luvuvhu Catchment are vulnerable to water pollution problems. The contemporary water quality

[*] E-mail: leratonare@yahoo.com.

monitoring and management practice does not promote community involvement and participation despite the fact that the communities were willing to participate in water quality monitoring and were even prepared to pay for the services. The study found a wealth of indigenous knowledge related to water quality monitoring and management. Based on the above results, a model for empowering communities to participate in water quality monitoring and management was proposed and its acceptability and reasonableness determined. The model is anchored on three frameworks; technological, empowerment and communication frameworks. The model envisages a situation where simple technologies based on indigenous knowledge are developed to monitor water quality at the community level as a way of empowering communities to take responsibility for managing their own resources including water.

Keywords: Community, participation, empowerment, model, Luvuvhu Catchment

1. INTRODUCTION

There are a number of factors that compel South Africa to adopt effective community participation in water quality monitoring and management. The first factor relates to water scarcity in the country. Although Luvuvhu catchment has abundant water resources for now, the catchment will face problems of water availability in the near future. The surplus water is provided by Nandoni Dam, but the dam was built to supply the growth in urban and industrial demands in Thoyandou and Louis Trichardt (DWAF, 2006). According to the Thulamela Municipality Integrated Development Plan (IDP) (2008), there was a backlog of 93 121 households without access to reticulated water in 2008. When these households are connected to the municipal water supply system, the amount of water available in the catchment is likely to be reduced. Luvuvhu Catchment also supplies other catchments with water. About 2.4 million m^3/year of water is transferred from the Albasini dam on the Luvuvhu River to Makhado Municipality which is in the Sand River Catchment. Vondo Regional Water Scheme on Mutshundudi tributary of the Luvuvhu River supplies water to some villages which fall under Letaba River Catchment (DWAF, 2005). The high population growth and economic development in Makhado Municipality and other areas which obtain water from the Luvuvhu Catchment will also worsen the situation as more water will be required to meet the demand. The impacts of climate change are also expected to reduce the available water resources (DWAF, 2005).

Pollution of water is becoming a significant problem in South Africa as the country continues to industrialize. In many reservoirs, water quality has deteriorated as a result of increasing salinity, nutrient enrichment and bacteriological pollution upstream (Silberbauer et al., 2001). Major sources of pollution for surface waters are agricultural drainage and wash-off (irrigation return flows, fertilizers, pesticides, and runoff from feedlots), urban wash-off and effluent return flows (bacteriological contamination, salts and nutrients), industries (chemical substances), mining (acids and salts) and areas with insufficient sanitation services (microbial contamination). The worst examples of pollution in South Africa include the Klipspruit stream case in Soweto where the stream was extremely polluted with acid and toxic metals from old mine dumps and tailings (Moyo and Mtetwa, 2002). Sulphate concentrations were more than 2000 times over the WHO standards. Van Zyl (1999) gives an example of the Juskei River in Alexandria township of Johannesburg where sewage from an over populated settlement, heavily contaminated the river, with *E. coli* counts averaging 3 million per 100ml. SANS 241 require that 95% of all samples taken from a point should show no viable growth of *E. coli* for the water to be declared safe for human consumption.

In 2005 basic sanitation services were available to only 67% of the population in South Africa (Anderson et al., 2008). This means that there were still over 30% of the people without access to basic sanitation and thereby posing a threat to the quality of water resources in the country.

The perception of water pollution as a community problem in South Africa is very low. A study conducted by Marie Wentzel of Statistics South Africa in 2008 found that slightly less than 11% of all households who participated viewed water pollution as a community problem. This poses a problem because the same author argues that *"awareness of environmental problems and the willingness to deal with them are more likely to be present among populations with more enlightened perception of environmental issues".*

In South Africa, 3.6 million people do not have access to safe drinking water, while 26 600 deaths annually are related to diarrhoea, in most cases preventable (Barwell, 2006). Investigations show that an unacceptably high incidence of poor drinking water quality occurs in non-metro South Africa (Hodgson and Manus, 2006).

In 2000, South Africa experienced one of the worst cholera epidemics in the country's recent history (Genthe and Steyn, 2006). By the end of the year, the cholera outbreak had spread to eight of South Africa's nine provinces, with a total of 1 0638 99 reported cases and 229 reported deaths. On 5 September 2005, a typhoid outbreak was confirmed in Delmas resulting in five reported deaths and the hospitalisation of 17 people (Department of Health, 2005b). Delmas is in Mpumalanga, a province which is adjacent to Limpopo province where Luvuvhu catchment lies. Thus the population in the study area is vulnerable to waterborne diseases which could be prevented with good water quality monitoring and management measures.

Therefore, management of South Africa's water quality and availability is essential since it is predicted that the demand for water will outstrip its supply by 2025 (Hirji et al., 2002). South African water resources are expected to decline markedly in the years to come for the reason that the ratio of runoff to rainfall is amongst the lowest of any populated region of the world (Oberholster et al., 2007). Involvement and participation of communities in water quality monitoring and management will increase the capacity of the nation to manage properly the quantities and quality of the scarce water resources.

It is apparent that there is need to go beyond the usual framework for monitoring and managing water quality. Involvement of indigenous communities in this case could be a viable option. Tsiho (2007) states that communities all over the world have developed their own knowledge and practices for observing, measuring, and predicting environmental quality change, which are embedded in their indigenous languages and cultural beliefs. He argues that *"there is little doubt that people at the grassroots have knowledge of their environment that transcends conventional social, economic and biological indicators."* Therefore there is need to create space for this indigenous knowledge to be incorporated into water quality management strategies currently being used. Indigenous knowledge is not only at its best when it is matched with contemporary science (Hens, 2006). Bell and Davies (2001) argued that, community involvement in Integrated Water Resources Management (IWRM) or in other environmental issues is based on "*the need to use indigenous knowledge (IK) as well as opinions that are vital for environmental protection, including proper water resource use and management".*

The use of participatory approaches in the management of water resources is one of the Principles of the Dublin Convention. Although many researchers agree on the importance of local community involvement in IWRM, the level of involvement is still low in most developing countries (GWP, 2000). Public participation in decision-making, especially focusing on historically disadvantaged and marginalised communities, concerning water resource protection is one of the basic principles of IWRM in South Africa (Department of Health, 2005a). The participation of stakeholders in decision making processes and therefore governance of water resources is a critical strategy in ensuring the sustainability of watersheds in the provision of resources (Ong'or, 2005). The National Water Act of 1998 (Chapter 7) advocates for the promotion of community participation in protecting, using, developing, conserving, managing and controlling of water resources (Karar, 2003). Communities are the primary stakeholders in the watersheds where they live and therefore they need to be empowered with knowledge and tools that will enable them to effectively participate in the governance of water resources (DWAF, 2005).

Community participation involves holding discussions and open forums between community members themselves and with government authorities or non governmental organizations involved in advocacy so as to contribute ideas for inclusion in policy development and change in operation strategy (DWAF, 2005). If given a chance, communities can participate effectively in matters relating to water resources management. In Kalomo (Zambia), the local community was mobilized to manage provision of water services, whereby villagers protected a catchment area by building a fence around a borehole and regularly cleaned the water point (Dungumaro and Madulu, 2002). Evidence from Gujarat (India) demonstrates the linkages between local community involvement in water project management and empowerment of stakeholders, especially imparting them with the capacity to negotiate with other stakeholders at higher levels concerning issues that affect their livelihood and lifestyle (Dungumaro and Madulu, 2002).

The WHO Protocol on Water and Health of 2006 recognizes that water "has social, economic and environmental values, and should therefore be managed so as to realize the most acceptable and sustainable combination of these values". The Constitution of South Africa (Chapter 2, Section 24), gives the citizens the right "to an environment that is not harmful to their health and well-being" (Anderson et al., 2008). Despite improved access to safe water and sanitation services and the creation of a legal framework to protect the country's natural resource base, inequities and inequalities in both these areas remain (Hemson and O'Donovan, 2005).

All the above arguments may not be realized unless the communities are adequately empowered to deal effectively with issues that relate to the management of their natural resources including water. They need to be armed with knowledge, skills and tools that will enable them to sit and discuss with technocrats on a relatively equal footing. Kauzeni and Madulu (2000) found that, though community participation is emphasized in developing land use plans, in many cases local communities and their local knowledge are ignored by planners in developing and managing land and water resources. This study therefore sought to identify indigenous knowledge, attitudes and practices in Luvuvhu Catchment relating to water quality monitoring and management that could be incorporated into the formal water quality monitoring framework and enable communities to participate effectively in this regard.

Friis-Hansen (1999) emphasized the need for taking indigenous knowledge on board when planning, developing, implementing and managing water resources. He argued that although experiences and knowledge of local people lack scientific explanations, they are a strong weapon in solving local problems. Research in local knowledge could ensure community participation, and indigenous/local knowledge could be used to facilitate development of water projects that are environmentally sustainable and meet national and community development objectives (Adams et al., 1994). It has been shown that the implementation of well considered, community accepted drinking water quality management procedures can effectively change an unacceptable water quality to one that satisfies drinking water specifications (Mackintosh and Colvin, 2003).

South Africa is a signatory to a number of international and regional agreements relating to management of water resources such as the SADC Protocol on Shared Waters and therefore the communities in the country need to be empowered to deal with pollution and water quality issues so that the country does not find itself in breach of those agreements. Developing a model which will enable communities to participate effectively in water quality monitoring and management will most likely go a long way in avoiding a situation where the country is found to be in breach of the agreements.

"Indigenous People's Kyoto Water Declaration" at the Third World Water Forum in Kyoto (Japan) in March 2003 was an attempt to bring to the forefront the plight of indigenous communities from around the globe. This study therefore was a tangible contribution towards allowing an indigenous community to have a voice over the management of their natural resources including water. Indigenous knowledge (IK) concerning the environment, is in many societies around the world, in danger of being lost, since western science has lately been controlling the development of environmental management practices to such a large extent (Wigrup, 2005).

It is therefore of the greatest importance to record and assess such IK before it becomes extinct. Luvuvhu Catchment covers part of Makhado Local Municipality and most parts of Thulamela Locality Municipality in Vhembe District. The two local municipalities are dominated by two indigenous groups i.e. the Venda and the Tsonga. Therefore by pulling out some aspects of their knowledge, attitudes and practices and integrating them into the national water quality framework, the study will contribute towards the preservation of this knowledge.

2. The Study Area

The Luvuvhu Catchment forms part of the Luvuvhu/Letaba Water Management Area which lies between $29^0$49'E and $31^0$55' E and $24^0$1'S and $29^0$49S. The Luvuvhu River and some of its tributaries (including the Mutshindudi and Mutale Rivers) rise in the Soutpansberg Mountains. It flows for about 200km through a diverse range of landscapes before it joins the Limpopo River near Pafuri in the Kruger National Park. Except for the Thohoyandou and Malamulele urban centres, the catchment mainly consists of rural settlements, commercial agriculture and the Kruger National Park. The major uses of water in the catchment are domestic water supplies followed by commercial agriculture and then environmental use.

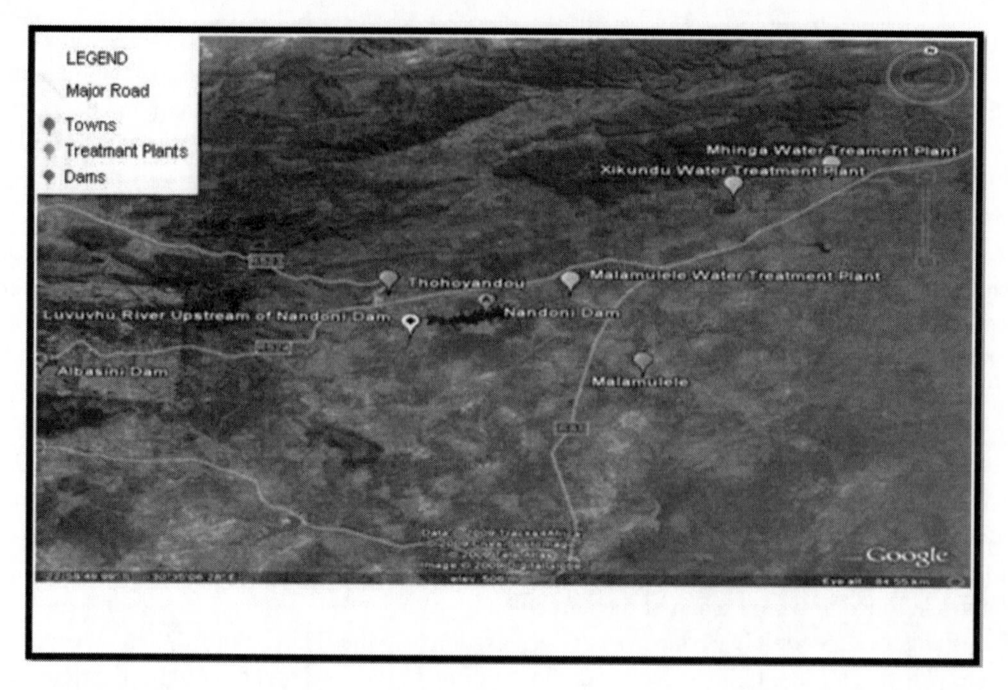

Figure 1. Map of the portion of the Luvuvhu River Catchment that comprised the study area showing treatment plants.

There are four water treatment plants in the catchment. These are Albasin, Malamulele, Xikundu and Mhinga. Albasin is located in a commercial area and it supplies water to Louis Trichardt which is in the Sand River Catchment and therefore the area around Albasin was not included in the study. The study area effectively covered the areas that are supplied with water from Malamulele, Xikundu and Mhinga treatment plants (Figure 1).

3. METHODOLOGY

The first objective of the study was aimed at proving that indeed there were pollution problems in Luvuvhu Catchment warranting investigation. The study had to prove community vulnerability to pollution problems in the catchment and that they faced potential risks to their health due to the pollution problems. The first action was to review any literature and official documents that could prove that there was potential vulnerability and health risk faced by communities due to pollution problems in the catchment. The second action was to analyse water quality from Luvuvhu River to prove that indeed the water in the river was polluted as suggested by the literature reviewed. Water quality results from four monitoring stations along the Luvuvhu River the period 2005 – 2007 were obtained from the Department of Water Affairs (DWA) and analysed to further confirm the vulnerability and risks faced by communities in the catchment. The water quality monitoring program in Luvuvhu River commenced in 2005 and at the time the data were obtained, complete data sets were only available for four stations between 2005 and 2007.

The third action was to monitor water quality from one street tap from each of the 15 villages participating in the study for 8 months to further confirm and fortify the case for community vulnerability and risks in the catchment. Microbiological analysis of water was carried out using Total Coliforms and *E. coli* as indicators.

The fourth action was to analyse water treatment efficiencies from the treatment plants that supplied water to the community in the catchment to find out if the quantities were adequate and the quality good for the target population. Treatment efficiency for a plant in relation to quantities of water processed was calculated from dividing the volume of treated water per day by the corresponding volume of raw water that went in and then multiplying the product by 100. The treatment efficiency in relation to quality was calculated from dividing the magnitude of a particular water quality parameter in treated water by the magnitude of the same parameter in raw water. Only Mhinga Water Treatment Plant had enough data to cover a period of two years. Therefore treatment efficiencies could only be calculated for the Mhinga Water Treatment Plant.

The fifth action was to collect and analyse data from health facilities in the catchment relating to waterborne diseases such as diarrhoea to really prove that the communities in the catchment were indeed at risk due to water pollution problems. The final or sixth action under the first objective was to find out the feelings and perceptions of the communities themselves relating to the quality and safety of their water supplies.

The second objective was to analyse and identify any gaps in the current water quality monitoring and management frameworks. This involved an extensive review of the policy, legal, institutional and organizational frameworks relating to water quality monitoring and management in South Africa. It also involved conducting interviews with practitioners and communities to solicit their views on the current frameworks.

The third objective sought to evaluate indigenous knowledge, attitudes, practices and perceptions related to water quality monitoring and management in Luvuvhu Catchment. This was achieved in two parts. The first part was achieved through interviews with around 8 000 members of the community from the catchment. Participatory tools and a structured questionnaire were used to collect data from the community. The second part evaluated the community's perception of water safety using turbidity as the first part had indicated that communities relied on the physical appearance of water to decide whether it was still suitable for use or not. The second part of the study therefore sought to identify the point at which members of the community would reject the use of water based on its physical appearance. A template was used to collect data from the respondents.

The fourth objective sought to evaluate community participation in water quality monitoring and management in Luvuvhu Catchment. This involved carrying out interviews with practitioners and members of the community and review of any related literature and official reports.

The fifth objective sought to evaluate the reasonableness and acceptability of the proposed model for effective community participation in water quality monitoring and management for the communities in Luvuvhu Catchment and other stakeholders. The proposed model was circulated to various academics and government officials to solicit for their inputs.

A survey was carried out among the community leadership in the 15 participating villages using a questionnaire. This included the Chief of the area, Village Heads, Civic Chairman and the Ward Committee Member from each village. The study sought to evaluate the perception of the leadership in relation to the proposed model, finding out if it was acceptable to them and capturing their contributions.

3.1. Water Quality Monitoring

Secondary water quality data collected by Department of Water Affairs (DWA) from their monitoring stations along the Luvuvhu River were used in the study. These were monthly total water quality data covering a period of two years starting from July 2005 to August 2007 to indicate the quality status. The quality of water from street taps in the villages participating in the study was monitored for a period of eight months. The results for Total Coliforms and *E. coli* were almost always negative and therefore the exercise was stopped after eight months.

3.2. Community Surveys

Two major community surveys were carried out to evaluate indigenous knowledge, attitudes, practices, perceptions relating to water quality monitoring and management. These studies involved people obtaining their water for different uses from the Luvuvhu River system. The communities mostly have access to reticulated water supplied by three treatment plants namely Malamulele, Xikundu and Mhinga. For the first study the sample involved a large number of people (over 8 000 out of 53 333 people) (15%) and the use of structured interviews to collect data from individuals proved to be very expensive. Therefore participatory tools were used to gather qualitative data which was then converted into quantitative data. Using the formula;

$$n = [(z^2 * p * q) + ME^2] / [ME^2 + z^2 * p * q / N].$$

This normally used in social science studies to determine the sample size, the ideal sample for the given study population would have been around 5000 respondents but the sample had to be increased to 8000 to improve on the accuracy of results.

The exercise was divided into a number of sessions with each session having 45 people. In a session the people were divided into three groups of 15 people each. Each group was assigned some themes relating to water quality monitoring and management and asked to work through them using relevant participatory tools and then write the findings on flip charts following instructions. The groups would then present at a plenary session and the Research Assistants (RA) would take down the responses from each group. The data were then converted into quantitative data as the responses were presented in terms of how many of the groups gave a particular response. This automatically converted the data into numbers which could be analysed statistically instead of qualitative statements. Structured questionnaires were used to interview government officials and other stakeholders at catchment level.

The second part of the study was based on the findings of the first one and aimed at finding the point at which the community would reject the use of water for various purposes based on the physical appearance of water. The perceptions of 1 000 people (2% of the total population) relating to safety of water based on the degree of turbidity were evaluated. Samples of water with known turbidity, chemical and microbiological values were shown to each respondent.

The respondent was then asked whether he/she would be able to use the water from each sample for various uses such as drinking, cooking, bathing and laundry. The responses were recorded on a template. A profile of turbidity values which the communities considered to be unsafe for each use was compiled. A table outlining the relationship between the level of turbidity and possible health impacts as given in the South African Water Quality Guidelines (1996) was used to interpret and make sense of the turbidity values at which respondents rejected the water for different uses.

3.3. Sampling

3.3.1. Water Sampling Points

Total water quality data from the following DWA sampling points was collected and analyzed;

- Downstream of Thohoyandou sewage works
- Downstream of Waterfall sewage works
- Downstream of Elim sewage works
- Vuwani Oxidation Ponds

3.3.2. Community Sampling

There are four water treatment plants in the part of the catchment where the study was done namely Albasini, Malamulele, Xikundu and Mhinga. Albasini supplies water to Louis Tritchadt which lies outside the catchment and therefore was not included in the study. Fifteen percent (15%) of villages being supplied by each of the remaining treatment plants were included in the study. The selection was done randomly. Each village was given a number and the number was written on a small piece of paper which was put in a container and the contents thoroughly mixed up before a paper was selected and the number recorded. The paper was then thrown back into the container. The process was repeated until 15% of the villages from each treatment plant were picked.

Multi - stage sampling was carried out to come up with the people who took part in the study. Once the villages were selected, 15% of the households in each village were selected to participate in the study. Random sampling was carried out to select the participating households. Each household in a village was assigned a number, which was then written on a small piece of paper. All the pieces of paper were then put in a container and shuffled thoroughly before the chief of the village was asked to pick a paper. The number on the paper was written down and the piece of paper thrown back into the container. The process was repeated until 15% of the households were selected. Three members from each selected household were then invited to take part in the study. If one of the required members was not available in a particular household, a member from the next household was invited to participate in the study.

3.4. Analysis of Data from Community Survey

Two sets of data were gathered from the community surveys. The first set concerned the data relating to the knowledge, attitudes, practices and perceptions (KAPP) relating to water quality management. The second set of data consisted of responses of community members to the water samples containing the different degrees of physical appearance of water (turbidity). In the first part of the KAPP study, communities indicated that they depended on the physical appearance of water to determine whether a body of water was still fit for human consumption. Turbidity was therefore used as an indicator of physical appearance of water in this study.

The analysis of data from KAPP study started with the conversion of qualitative data gathered through participatory approaches to quantitative data that is capable of being analysed by statistical methods. The responses from the different groups were then entered into EXCEL, a computer programme or spreadsheet before being transported into SPSS which analyses statistical data. SPSS stands for Statistical Package for Social Sciences and it organises quantitative research data to determine the relevance of variables associated with the research topic. PASW (Predictive Analysis Software) brand was used for this analysis. The information obtained from analysing the data from the KAPP study was subjected to further statistical analysis to obtain the mean and standard deviation.

The data from the turbidity/colour study was analysed manually. The information obtained was used to develop a profile of different degrees of turbidity/colour and the likely responses from members of the community to the water.

4. Results

4.1. Water Quality Monitoring and Contemporary Management Practice

Although South Africa has a comprehensive and sophisticated water quality monitoring and management programme backed by very elaborate and extensive policy, legal, institutional and technical frameworks, implementation at community level leaves a lot to be desired. While the national constitution guarantees all citizens the right to access sufficient water and a safe environment, the practical modalities on the ground make it impossible to achieve this. The way in which the water quality monitoring programme is implemented currently makes it hard for the country to guarantee the above constitutional provisions. While the policy, legal and technical frameworks are comprehensive enough to guarantee the above constitutional provisions, the structures of the institutional framework and lack of resources make it impossible to do that.

The water quality monitoring programme is currently being transformed from a centralised to a decentralised one. The transition phase also contributes to the weaknesses in the system. While the main actors in the old system were the Departments of Water Affairs and Health, the Water Services Act transfers the primary responsibility for managing the quality of water for domestic use to the Water Services Authorities. This has implied the reorganisation of the whole programme. DWA has had to shed the key responsibility of managing water quality for domestic uses to the WSAs. This meant transferring staff and resources to the WSAs and the transition is not yet completed and at the moment that is affecting water quality monitoring on the ground. As indicated in Figure 6 DWA at regional office level has three sections charged with water quality monitoring and management (Water Services, Water Quality Management and Geohydrology). The Water Quality Management section is headed by an Assistant Director. The Assistant Director in the case of Polokwane Regional Office is incharge of two Water Management Areas i.e. Limpopo WMA and Luvuvhu/Letaba WMA. The section is responsible for monitoring quality in surface waters and there are only ten officers to cover the whole of Luvuvhu/Letaba WMA. Luvuvhu River only, has 31 points to be monitored at least once a month.

The inadequate manpower has a serious impact on the quality of the service being rendered. The officers are not able to collect data every month due to other work commitments and therefore a lot of gaps exist in the data collected over the years. For example, for the period between July 2005 and June 2007 only four stations out of thirty one had complete sets of data for a year. Although the data was computerised and could be easily accessed by making a request through internet, it was not analysed. This brings into question the usefulness of the data collected as a tool for community interventions and decision making.

The geohydrology section in DWA Polokwane Regional Office is responsible for monitoring and management of groundwater quality. According to Mr. WH du Toit the Information Manager in the Water Resource Directorate at the Polokwane Regional Office, the National Groundwater Quality Monitoring Program has 55 monitoring stations in Limpopo Province which includes Luvuvhu Catchment and the data is available on Water Management System (WMS). The data obtained from DWA after request through internet also suffered from lack of completeness. For example, there is hardly a point with a complete set of data for a year on any given water quality monitoring parameter between 1995 and 2007. Some of the points were only monitored for only one month in a year. Consequently the data available over the internet is not analysed meaning that only experts can make use of it.

As already indicated the primary responsibility of monitoring quality of water for domestic use lies with WSAs. The role of DWA through the Water Services section is to support and regulate the WSAs. Therefore the district structure for DWA has remained with skeletal support staff just to support the work of the WSAs. In Luvuvhu Catchment the WSAs are Vhembe District Municipality which treats water and Thulamela and Makhado Local Municipalities which distribute the water to consumers. The water quality function is supposed to be carried out by the district municipality but Vhembe District Municipality has not completed setting up the water services structures and therefore DWA is still responsible for that function.

The water quality programme in Vhembe is carried out by staff based at four water treatment plants in the catchment by taking samples and forwarding them to a laboratory located in Sibasa. The samples are taken from the raw water supplying a treatment plant, from the treatment plant itself and from a designated point just outside the treatment plant. Other sampling points are situated along the distribution line for every 10 000 people using the system. Water can get recontaminated along the distribution system (McGhee, 1991). Therefore the above arrangement cannot guarantee that the consumers are receiving safe water. The laboratory caters for other local municipalities under Vhembe district but under different catchments such as Nzhelele and Sand. It has employed 8 technicians and according to Mr. Nemaunzeni (Control Engineering Technician) the laboratory is coping with samples from the whole district. This has made the sampling points to be very limited. If more samples were to be randomly added from community level, the laboratory may fail to cope with the workload.

The Department of Health through its Environmental Health section is tasked with monitoring water quality at the point of consumption and giving health and hygiene education at community level from the Drinking Water Quality Management Guide for Water Services Authorities (2005). The Environmental Health Services are currently in transition. They are being moved from the Department of Health to Vhembe District Municipality. At the moment an operational structure has not been worked out for the services but staff has just been seconded to work in areas under the local municipalities.

There are seven Environmental Health Officers (EHO) deployed under Thulamela Locality Municipality which covers most parts of Luvuvhu Catchment to serve an estimated population of 800 000 people (Vhembe IDP, 2009). This translates to a ratio of 1 EHO to 114 286 people to be served. The World Health Organisation recommends a ratio of 1 EHO to 10 000 people but South Africa relaxed this to 1 EHO to 15 000 people (Agenbag and Gouws, 2004). Figure 2 gives a comparison of functional Environmental Health Officers per population in South Africa between 2006 and 2007. Limpopo Province had 1 EHO for 32 000 people, which is more than the nationally recommended staffing levels. But the current staffing situation for the areas under Thulamela Local Municipality is far much worse as shown above. This has serious implications on the ability of the EHOs to carry out their mandate of monitoring water quality at the point of consumption and giving health and hygiene education to the community. The EHOs also have other duties in addition to water quality monitoring including; malaria control, food control and sanitation.

The EHOs themselves on the ground do not see water quality monitoring as a priority. One of the EHOs under Vhembe District Municipality Mr. Matshakatini said they only prioritised water quality monitoring during outbreaks of waterborne diseases such as cholera. He said when they were still under DoH they only monitored water quality when ordered to do so by their superiors but currently there is confusion since they do not have a structure yet to report to.

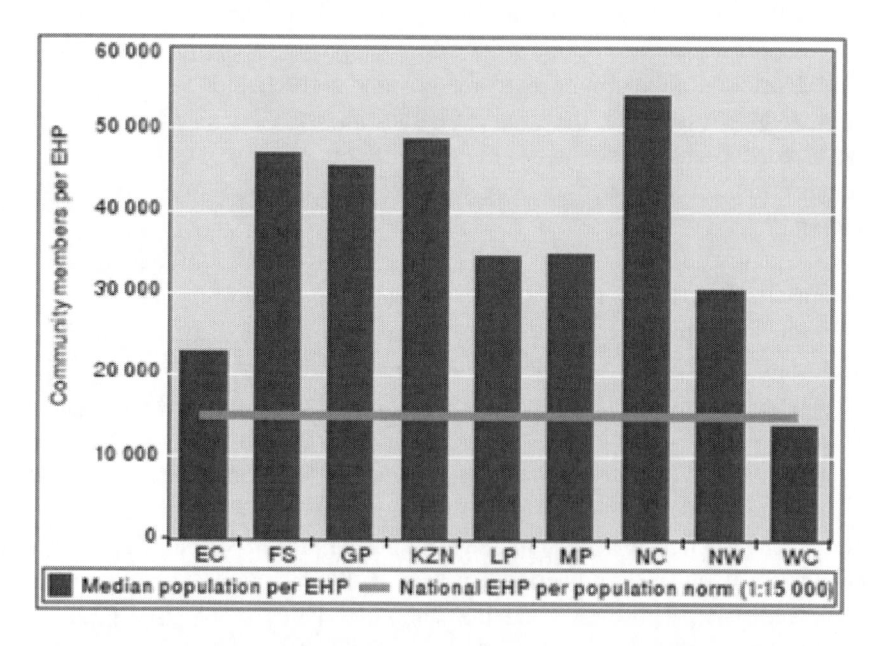

Figure 2. Comparison of functional Environmental Health Practitioners (EHP) per population in South Africa (2006 – 2007) (Agenbag, 2008).

There is very limited contact between the EHOs and the rural population they serve. Five of the seven EHOs under Thulamela Municipality operate from Thulamela Municipality offices in Thohoyandou and the other two are based in Malamulele. There are no officers based at clinics which are at community level. With such limited interface with communities, their ability to influence behaviour change towards water quality management through delivery of health and hygiene education is in doubt.

Although DWA has set up a water quality information system as required by the National Water Act and the information is readily available on the internet or through request to the department, the quality of the information and access by all the members of the public remains a concern. As already indicated, a lot of the data sets are not complete and the information is kept in the form of raw data which can only be useful to professionals who are able to analyse and interpret it.

DWA has put into place an audit system to monitor the water quality monitoring function of WSAs. Mr. Nemaunzeni who is in charge of the laboratory at Sibasa confirmed that once a month a private laboratory takes samples from designated water sampling points and analyses them for quality control purposes.

From the foregoing discussion, it is clear that although water quality monitoring and management is well organised on paper in South Africa, on the ground the programme is not likely to satisfy the provisions of the national constitution and ensure that every citizen in Luvuvhu Catchment is protected from using unsafe water. This is why waterborne diseases are still common. Therefore there is need for authorities to be innovative and bring other stakeholders including local communities on board.

4.2. Indigenous Knowledge and Community Perceptions Relating to Water Quality Monitoring

According to the Health Belief Model, for an individual to undertake recommended preventive health action, he/she needs to perceive a threat to his/her health, be simultaneously cued to action and his/her perceived benefits outweigh his/her perceived losses. Therefore for one to successfully mobilize communities in Luvuvhu Catchment to participate in water quality monitoring and management, one needs to understand their perceptions of water quality related threats/risks to their health. The study sought to evaluate indigenous knowledge and community perceptions in the catchment and utilized them as a basis for mobilizing the communities to participate in water quality monitoring and management projects.

There was a wealth of knowledge relating to water quality management among the communities in the catchment as provided in Table 4. The communities knew how water got polluted, with a high percentage of the groups (ranging from 91 – 99% mean = 93; Sample standard deviation (S) = 2.4) saying water got polluted through the introduction of foreign materials. The percentage of the groups who said they would recognize that water had been polluted through its physical appearance ranged from 61 – 68% (mean = 60; S = 1.5). The percentage of the groups that said they would employ a combination of conventional and traditional methods to treat water ranged from 79 – 89% (mean = 85; S = 2.9). An example of a conventional method mentioned was the use of chlorine products to disinfect water at household level, while the traditional methods included putting tree branches with fresh leaves in the water to be treated, allowing the water to settle for a while before using it, filtering water through sand and boiling water before using it. The communities were aware that poor home hygiene could contaminate water, with the percentage of the groups participating in the study who confirmed this ranging from 83 – 89% (mean = 86; S = 2.3). For the groups that said they would stop using a source and report to the local leadership if they suspected that the source was polluted the percentages ranged from 69 – 77% (mean = 73; S = 2.4). The communities were also aware of the fact that consumption of polluted water affected the gastro – intestinal system with the percentage of the participating groups saying this ranging from 81- 88% (mean = 86; S = 2.4).

The communities generally said they would reject using water for various purposes based on the aesthetic quality of the water more than any other reason. For drinking water, 64 – 71% (mean = 66; S = 2.3) said they would reject it for aesthetic reasons and 71 – 79% (mean = 75; S = 1.9) said they would reject water for bathing for the same reason. For cooking, 61 – 67% (mean = 63; S = 2.3) said they would reject water for aesthetic reasons while 82 – 89% (mean = 85; S = 1.8) said they would do the same for laundry purposes.

The community's perception of water safety was evaluated using turbidity as an indicator of poor quality in water. Although the same communities had said they would recognize that water had been polluted through its physical appearance, their perception of unsafe water using physical appearance (turbidity) as an indicator differed significantly from acceptable scientific value. For example, about 35% of the participants said they would use water for drinking with turbidity of 92 NTU. This is way above the maximum limit of 5 NTU recommended in SANS 241. According to South African Water Quality Guidelines (Volume 1: Domestic Water Use) any water with a turbidity value above 10 NTU carries an associated risk of disease due to infectious disease agents and chemicals adsorbed onto particulate matter. This study demonstrated the point made in the Health Belief Model in that, now that it is clear the communities in Luvuvhu are at risk because they tolerate water with very high turbidity, the health and hygiene education by EHOs can be effective in changing this behaviour.

Since the communities felt that the reticulated water supply was not safe in terms of smell, taste, and physical appearance, they were advised to treat water at the point of use with simple technologies such as using chlorine compounds or boiling water before use in the short term. In the long term, the communities were advised to engage authorities at the treatment plants so that the treatment processes can be improved upon.

4.3. Community Participation in Water Quality Monitoring and Management

South Africa has very comprehensive policy, legal, technical and institutional frameworks that are either meant or can be used to promote community participation in water quality monitoring and management.

The various policies and pieces of legislation discussed in chapter two encourage and facilitate participation of communities in development activities including water quality monitoring and management. The National Constitution guarantees the citizens the right to be consulted in any activities that affect them and since water quality affects public health it becomes imperative that they be consulted on it.

South African Water Quality Guidelines (Volume 1: Domestic Water Use) give a tabulated relationship between the concentration of a water quality constituent and the possible risk to human health as illustrated for pH and turbidity in Tables 1 and 2 respectively. This tabulation makes it possible and easy to involve members of the public in water quality monitoring and management. For example if members of the community were trained in the use of simple equipment such as pH and turbidity meters and given Tables 1 and 2 they would be able to take samples, analyse them and interpret the results for themselves.

Table 1. Effects of pH on Aesthetics and Human Health

pH Range (pH units)	Effects
<4.0	Severe danger of health effects due to dissolved toxic metal ions. Water tastes sour
4.0-6.0	Toxic effects associated with dissolved metals, including lead, are likely to occur at a pH of less than 6. Water tastes slightly sour
Target Water Quality Range 6.0-9.0	No significant effects on health due to toxicity of dissolved metal ions and protonated species, or on taste are expected. Meta ions (except manganese) are unlikely to dissolve readily unless complexing ions or agents are present. Slight metal solubility may occur at the extremes of this range. Aluminium solubility begins to increase at pH 6, and amphoteric oxides may begin to dissolve at a pH of greater than 8.5. Very slight effects on taste may be noticed on occasion
9.0-11.0	Probability of toxic effects associated with deprotonated species (for example, ammonium deprotonating to form ammonia) increases sharply. Water tastes bitter at a pH of greater than 9
>11.0	Severe danger of health effects due to deprotonated species. Water tastes soapy at a pH of greater than 11

Table 2. Effects of Turbidity on Aesthetics and Human Health

Turbbidity Range (NTU)	Effects
Target Water Quality Range 0-1	No turbidity visible No adverse aesthetic effects regarding appearance, taste or odour and no significant risks of associated transmission of infectious micro-organisms. No adverse health effects due to suspended matter expected
1-5	No turbidity visible A slight chance of adverse aesthetic effects and infectious disease transmission exists
5-10	Turbidity is visible and may be objectionable to users ar levels above 5 NTU. Some chance of transmission of disease by micro-organisms associated with particulate matter, particularly for agents with a low infective dose such as viruses and protozoan parasites
>10	Severe aesthetic effects (appearance, taste and odour). Water carries an associated risk of disease due to infectious disease agents and chemicals adsorbed onto particulate matter. A chance of disease transmission at epidemic level exists at high turbidity

Table 3. Colour coded classification system (Quality of Domestic Water Supplies Series 1998, Volume 1: Assessment Guide)

Blue [B]	Class 0	*Ideal water quality* – suitable for lifetime use.
Green [G]	Class I	*Good water quality* – suitable for use, rare instances of *negative effects.*
Yellow [Y]	Class II	*Marginal water quality* – conditionally acceptable. Negative effects may occur in some sensitive groups
Red [R]	Class III	*Poor water quality* – unsuitable for use without treatment. Chronic effects may occur
Purple [P]	Class IV	*Dangerous water quality* – totally unsuitable for use. Acute effects may occur.

The Quality of Domestic Water Supplies (Volume 1: Assessment Guide) gives a colour coded classification system as shown in Table 3, which is created after analysing the quality of water at a particular locality for a period of time, for example a year or two. The use of colour is innovative in that even illiterate members of the public can follow and understand the quality of their water resources and the implication to their health. For example they are taught that if the colour displayed is blue then it means they can use the water without any precautions but if it is red then they need to treat the water first and if it is purple then they should never use the water. This framework would empower communities to seek explanations from experts and participate in any remedial action.

DWA also publishes Water Services Authority Awareness Pamphlet and a Consumer Awareness Booklet on water quality issues. These if properly distributed go a long way in keeping members of the public informed and allow them to contribute to water quality monitoring and management. The problem is that all these opportunities offered by the different frameworks are not being utilised at the moment. The data generated from water quality monitoring activities is reserved for use by researchers and other professionals instead of being shared with or used for the benefit of the affected communities.

All water quality monitoring programmes run by DWA and the WSAs at the moment do not involve communities at any stage. DWA does not involve communities in its monitoring of surface waters in the catchment. According to Mr. William Moasefoa, the former Assistant Director (water quality) at the Polokwane Regional Office, communities do not even know where the 31 monitoring points along the Luvuvhu River are located. They do not explain to the affected communities the reasons for taking water samples. In fact according to him "*the communities think we are members of the Zion Church collecting water for their prayers".* They do not give feedback to communities concerning the results of the monitoring done along the river. The same situation prevails in the ground water quality monitoring programme according to Mr. WH du Toit, the Information Manager in the geohydrology section at the Polokwane Regional Office.

The monitoring of quality for domestic water supplies done by the DWA district office in Vhembe district does not involve communities also. According to Mr. Nemaunzeni (Control Eng. Technician) they do not disclose the designated points where they collect samples to the communities because "*people will be bothering us about why we concentrate on sampling while the service itself is not consistent".* The results are not shared with communities but compiled for the benefit of higher offices. The only time the results are used locally is when there is a failure in the water quality system in which case the laboratory advises those in charge of the water supply system to shut down the supply and supply the villages from tankers. Even when the communities do not understand the explanation they are advised to be patient while the system is being sorted out.

Communities confirmed in a survey that, they were not involved in water quality monitoring and management activities in their villages. Although 91 – 98% (mean = 94; S = 2.4) of the groups interviewed said they knew that their water supplies were monitored/tested, only 3 – 8% (mean = 5; S = 1.5) mentioned communities as participating in the process. The percentage of the groups that said they got any feedback from the authorities on the results of the water samples that are taken from their villages ranged from 3 – 9% (mean = 5; S = 2.2). The communities were aware that there were water quality problems in their areas with 89 – 97% (mean = 93; S = 2.3) of the groups supporting this view. The percentage of the groups that felt that the water quality problems in their areas needed to be monitored regularly ranged from 92 – 99% (mean = 95; S = 2.1) and 91 – 98 (mean = 94; S = 2.2) said they are to be involved in the monitoring. The percentage of the groups that said they were willing to contribute financially towards monitoring of their water ranged from 87 – 97% (mean = 93; S =3) and they were willing to pay amounts ranging from at least R10 – R150/month towards this. More people were willing to pay R20/month, with the percentage ranging from 51 – 75% (mean = 55; S =1.8). Communities held meetings to discuss development including water quality issues as confirmed by 94 – 98% (mean = 96; S = 2.3) participating groups. The percentage of the groups that said the water quality issues discussed at these meetings were related to water quality monitoring and management ranged from 70 – 78% (mean = 74; S = 2.6). The communities acknowledged that they got external assistance whenever they had problems with water quality, with 88 – 96% (mean = 92; S = 3.4) of the participating groups confirming this. The percentage of the groups that said the external assistance came from government/municipality system ranged from 62 – 72% (mean = 67; S =2.9) while those that said the assistance came from NGOs ranged from 29 – 38% (mean = 33; S = 1.9).

In terms of the Health Belief Model, the communities in Luvuvhu Catchment appreciate risks of using polluted water, appreciate the benefits of monitoring water quality and are ready to take action to eliminate the risk. This could be the appropriate community where the government could start experimenting with meaningful involvement and participation of communities in development activities including management and protection of water resources. The government can start experimenting with bottom – up approaches that can lead to true empowerment of the communities in managing and protecting water resources in Luvuvhu Catchment. This is expected to bring about meaningful participation of communities in water resources management.

Luvuvhu Catchment falls under the Luvuvhu/Letaba WMA and the CMA for this WMA is in the process of being formed. The process of setting up the water management institutions as is underway. A proposal for its establishment was produced in early 2010 and it includes structures that will in theory promote the participation by communities from grass root level to WMA level (Figure 3). The CMA will take over most of the functions currently being performed by the DWA Regional Office in Polokwne except for the regulatory ones.

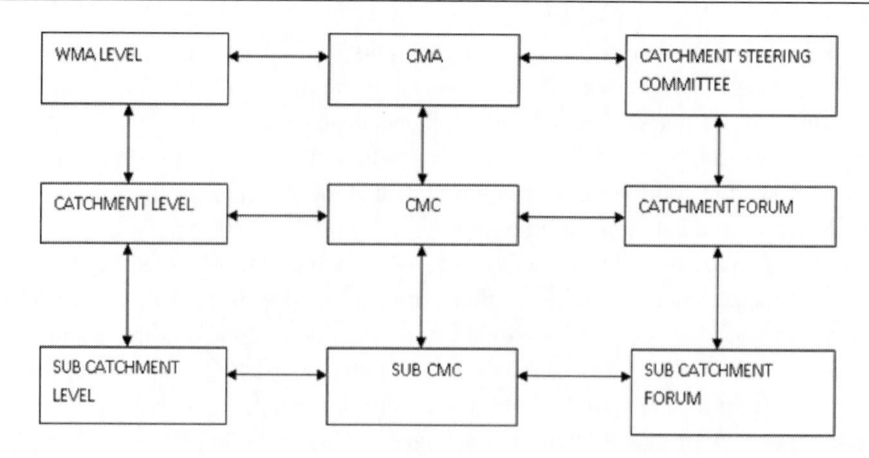

Figure 3. Proposed catchment management structures for Luvuvhu/Letaba Water Management Area.

The immediate problem with the proposed structure (Figure 3) is that, it does not connect with local government administrative structures such as municipalities, wards, villages and sub villages. This means that it cannot articulate the feelings and perceptions of the common public but concentrate on those who are considered to be major water users such as farmers, industry, etc. The second weakness lies on how the structures were established. DWA contracted a private company to conduct "*public participation*" meetings and then establish the structures. The contract was for six months with predetermined outputs. The schedules were very tight and the private company had little time to deliver. Therefore the whole objective on the part of the contracted company became meeting the targets instead of allowing communities to understand the whole concept and then making suggestions on how the whole issue should be conducted. At the end most of the meetings were poorly attended and major decisions were taken by very few people on behalf of the whole community.

The next problem relates to sustainability. When the private company completed its work and left the connection between the community and DWA was broken since there was no one to follow up on issues that were raised in the meetings.

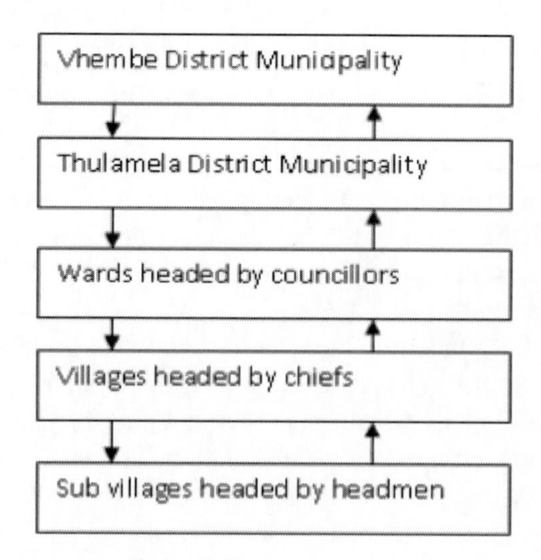

Figure 4. Local government structures in Luvuvhu Catchment.

The introduction of structures that will change fundamentally the way government engages with communities would have required protracted, committed and sustainable communication with communities to allow them time to adapt and adjust to the changes. Instead of contracting out the task to private companies the government should have allowed the decentralisation processes of water services to Vhembe District Municipality to be completed and then build capacity within the proposed Institutional Social Development (ISD) unit at that level.

The municipality would have the capacity to carry out the task since its structures are already linked to the communities as illustrated in Figure4. Locally based NGOs could have been involved in mobilizing the communities and establishing the structures since these would remain for a long time in the community. They would offer the communities support as they adapt and adjust to the changes.

4.4. Water Scarcity and Failure by Service Providers to Supply Adequate Amounts of Water

Although Luvuvhu Catchment is viewed as the only catchment in South Africa that potentially has surplus water, the water is not adequate to satisfy all users. The Thulamela Local Municipality IDP (2008), showed there was a backlog of 93 121 households without access to reticulated water supplies in the year 2008. When these households are finally connected to the water supply system the surplus water is likely to be reduced. Luvuvhu Catchment is already transferring water to Louis Tritchardt in Sand Catchment and some areas in Letaba Catchment.

Studies carried out by students from the University of Venda in the catchment including those by Masidiri (2008); Nemadodzi (2008); Malume (2010); Mvundlela, (2010) have shown that at household level communities experience severe water shortages. The communities only receive water for a limited number of days in a week and sometimes go on for weeks without getting any water. Mvundlela (2010) determined that the per capita use of water in Tshidembe Village which is part of the catchment was 17 l/c/d and was lower than the figure given in the RDP document of 25 l/c/d. Although the three treatment plants supplying water to the population within the catchment had large design treatment capacities, their treatment efficiencies were relatively low and not adequate for the target population. For example, the Mhinga treatment plant has a design capacity of 3 mega litres (ML) but its treatment efficiency ranges from 30 to 90% (900 - 2700 m^3) with a mean of 60% (Figure 5).Therefore, at its lowest it produces 900m^3/day.

According to the DWAF Limpopo Water Services Regional Bulk Infrastructure Grant document for the years 2008/2009 and after, the design domestic consumption levels for different communities is as shown in Table 12. Using the guidelines for basic service (16 l/c/d, 25 l/c/d and 35 l/c/d), the minimum water demand for the villages under Mhinga treatment plant would be 800, 1250 and 1750 m^3/day respectively. At 30% efficiency, the plant *"produces"* 900m^3/d, barely enough to cover a supply required for survival (16 l/c/d) of 50000 people for a day (800 m^3/d) especially considering that there are always unaccounted for water losses along the distribution line. If the supply is operated at standard level (25 l/c/d) which is supported by the Reconstruction and Development Programme (RDP) document the supply deficit becomes even bigger.

Table 4. Design domestic consumption levels: DWAF (2008)

Service level	Domestic consumption l/c/d		
Scenario	Basic	Higher	Urban
Survival	16	50	70
Standard (DWAF)	25	90	150
Higher	35	120	200

l/c/d = litres/capita/d.

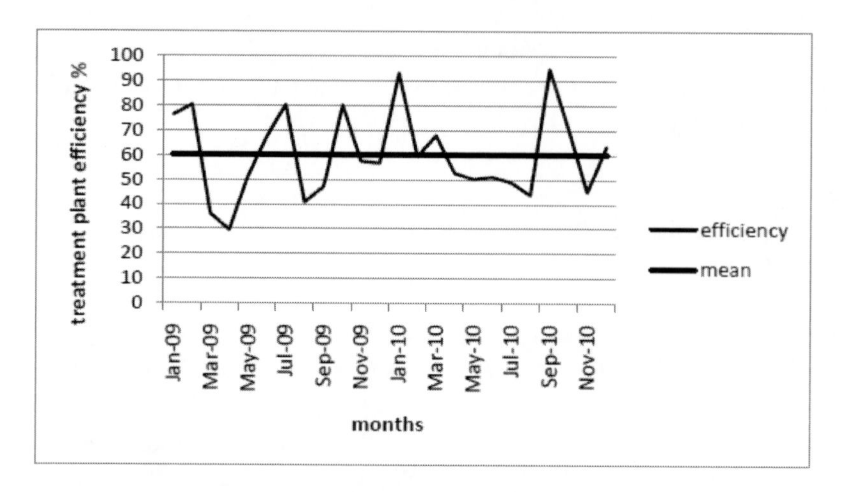

Figure 5. Water treatment efficiency at the Mhinga treatment plant.

As the daily water demand exceeds the supply, the authorities resort to water rationing tactics to even out the demand. Each sub village is supplied with water for less than 5 days in a week. This forces communities to resort to the use of other sources including the unprotected ones.

A community survey carried out in the area also confirmed the fact that not all people in the area had access to a reticulated water supply all the time. The percentage of the groups that confirmed that they sometimes used water from a combination of protected and unprotected sources for drinking ranged from 20 – 29% (mean = 24; S = 2.6) while 67 – 75% (mean = 68; S = 17) and 61 – 69% (mean = 64; S = 16) said that they used a combination of protected and unprotected sources for bathing and cooking respectively.

Failure to access adequate water of sufficient quality increases community vulnerability to water quality related problems. When people are not supplied with adequate amount of water, they turn to other alternative sources of water which may cause diseases (Dungumaro, 2007). Each year about 4 million people die of diarrhea and more than 800 million people in the world are malnourished due to insufficient water (Cosgrove and Rijsberman, 2000).

4.5. Exposure to Polluted Water

The second source of vulnerability faced by the communities in Luvuvhu Catchment relates to the potential use and exposure to polluted water. Although literature on pollution problems in the catchment is limited, review of available information relating to land use patterns in the catchment showed a high potential for water pollution occurring in the catchment. The upper part of the catchment is dominated by irrigated agriculture which uses fertilizers and pesticides with high potential for polluting water. The middle part of the catchment is dominated by human settlements. Most of the settlements are rural and use pit latrines for their sanitation.

The high density of pit latrines poses a threat to the quality of groundwater in the catchment. Contamination of groundwater with human waste can raise nitrate levels in water. Figure 6 shows the results of groundwater quality monitoring from four points located in various parts of the catchment over two years. The nitrate levels in some stations at some points exceeded even the maximum allowable limit for nitrates given in SANS 241 of 20 mg/l. SANS 241 require that nitrate – nitrite concentration in water should remain below 10 mg/l under Class I limits and range between 10 and 20 mg/l under Class II limits provided the consumers are not exposed for more than 7 years.

Nitrate levels from four stations along the Luvuvhu River monitored for a period of two years also exceeded both the minimum and maximum limits in Thohoyandou and Waterfall Sewage Outflow in some months (Figure 7).

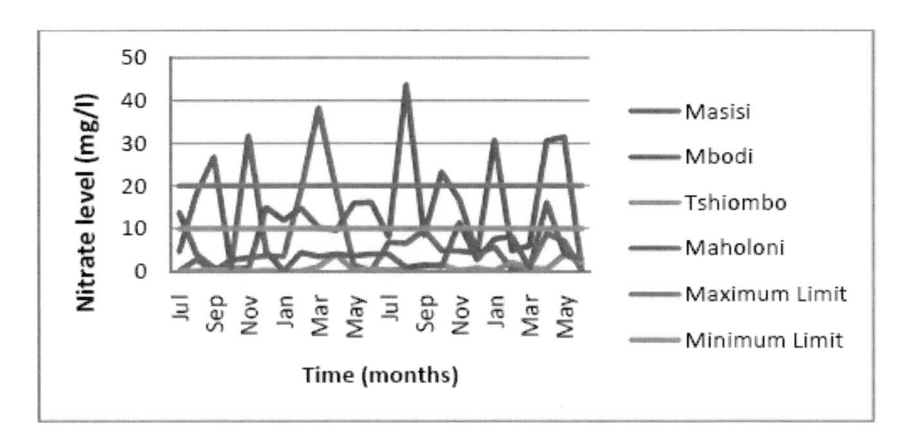

Figure 6. Nitrates levels in ground water from four monitoring points in Luvuvhu Catchment (2005 – 2007).

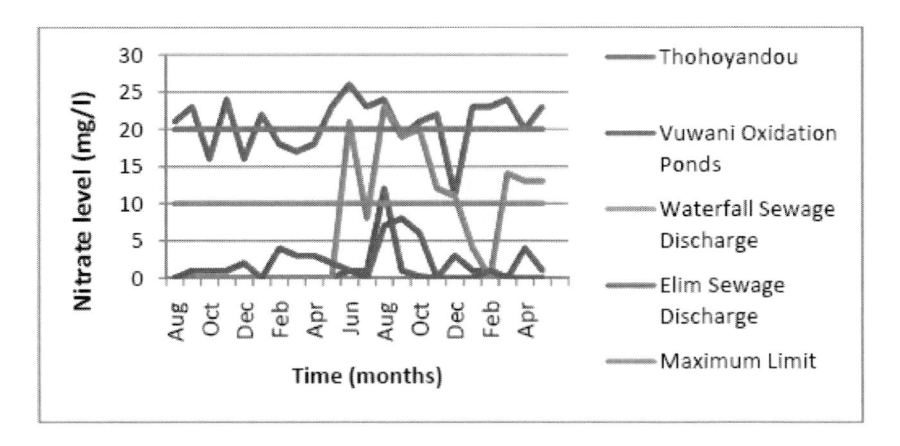

Figure 7. Nitrate levels from four stations along Luvuvhu River (2005 – 2007).

In Thohoyandou, the nitrate level was always above the minimum acceptable limit. This could be due to the influence of irrigated agriculture in the upper part of the catchment, contribution from the sewage treatment facilities upstream of Thohoyandou or contamination from groundwater. Groundwater can carry nitrogen (in the form of nitrate) into surface water bodies through recharge and spring discharges (West, 2001).

Turbidity values in all the four stations analysed, were mostly found to be way above even the maximum limits given in SANS 241 of 5 NTU (Figure 8). This could be attributed to anthropogenic activities in the catchment such as agriculture and human settlements (urban and rural) which have altered land cover and led to generation and introduction into water of particulate matter that causes high turbidity in water. The middle part of the catchment has two peri - urban centres of Thohoyandou and Malamulele and these have the potential of causing pollution problems associated with raised turbidity values in water courses. Urban settlements contribute towards water pollution through alteration of land cover, urban drainage, phosphorous and sewage generation. The paved surfaces in urban areas tend to increase the production of dissolved salts, suspended solids, and nutrients (Ahearn et al., 2005). Sewage facilities can also discharge sewage that is not adequately treated and raise turbidity levels in water.

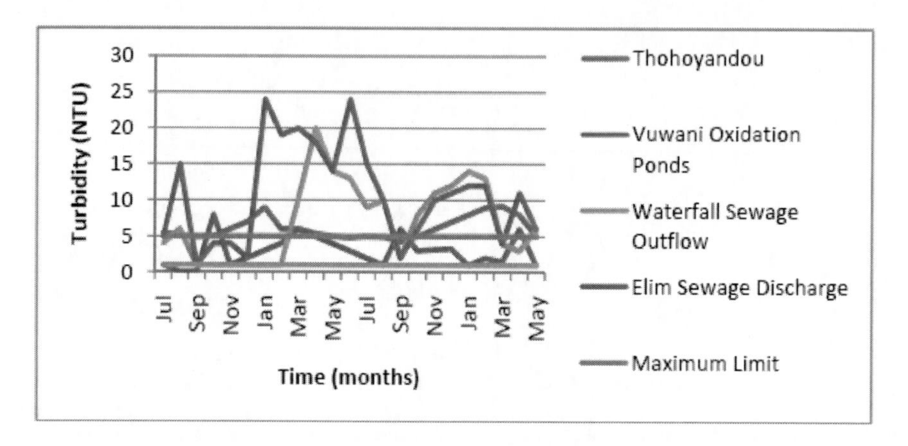

Figure 8. Turbidity values from four stations along the Luvuvhu River (2005 -2007).

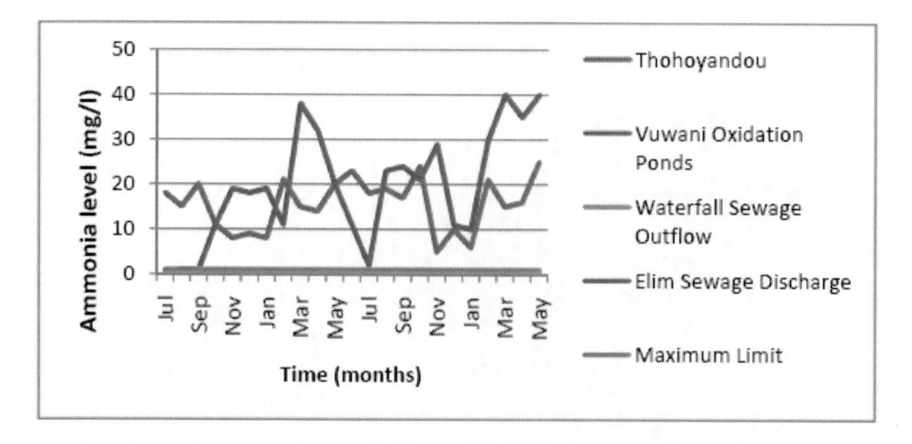

Figure 9. Ammonia (NH_3) levels from four stations along Luvuvhu River (2005 – 2007).

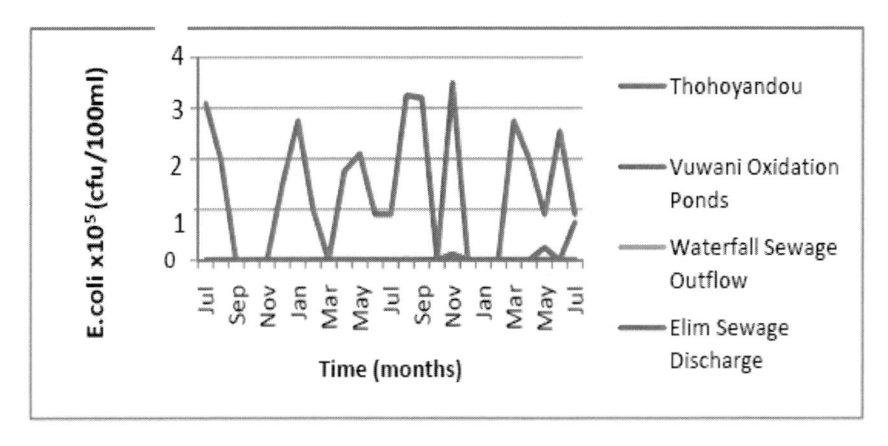

Figure 10. Results of *E. coli* monitoring along Luvuvhu River (2005 – 2007).

Results of Ammonia monitoring from two stations along Luvuvhu River (Thohoyandou and Vuwani Oxidation Ponds were way above the maximum limit for ammonia given in SANS 241 (Figure 9). The recommended limits for ammonia in SANS 241 are 1 mg/l Class I operational limits and 1 – 2 mg/l for Class II limits despite the duration of exposure. Concentrations exceeding 10 mg/l are found in raw untreated sewage, since ammonia concentrations tend to be elevated in waters where organic decomposition under anaerobic conditions takes place. The sewage facilities at Vuwani could be malfunctioning and releasing raw sewage into the river. The areas around Vuwani and rural areas upstream of Thohoyandou have livestock which could also contribute to the high levels of Ammonia in the water.

The discharge of raw sewage into the water could also account for the high numbers of *E. coli* indicator bacteria recorded at the Thohoyandou monitoring station (Figure 10). According to SANS 241, 95% of all water samples (count/100ml) analysed should yield no count of *E. coli* for water to be declared safe.

Most of the samples from Thohoyandou monitoring point show viable counts of *E. coli* up to over 350000/100ml. This shows heavy pollution with material containing faecal matter, strengthening the suspicion that the sewage ponds upstream at Vuwani are discharging raw sewage into the river. The water needs to be treated properly because there is a possibility of pathogens passing through treatment and disinfection processes and posing a danger to public health.

The COD values at all the four stations monitored along the Luvuvhu River mostly exceeded by far the maximum limit given in SANS 241 (Figure 11). COD is a component of Dissolved Organic Carbon (DOC).

According to SANS 241 DOC concentration in water should not exceed 10 mg/l for Class I and range from 10 to 20 mg/l under Class II limits provided the consumers are not exposed for more than 3 months. The results from the four stations indicate a possibility of serious pollution from a rich source of organic carbon.

This further strengthens the possibility that sewage facilities along the river are discharging raw sewage into the Luvuvhu River. The treatment system needs to be thorough and reduce the COD to manageable levels. Excessive COD not only causes problems with aesthetics in the treated water but could promote the formation of trihalomethanes (THMS) which are carcinogenic substances.

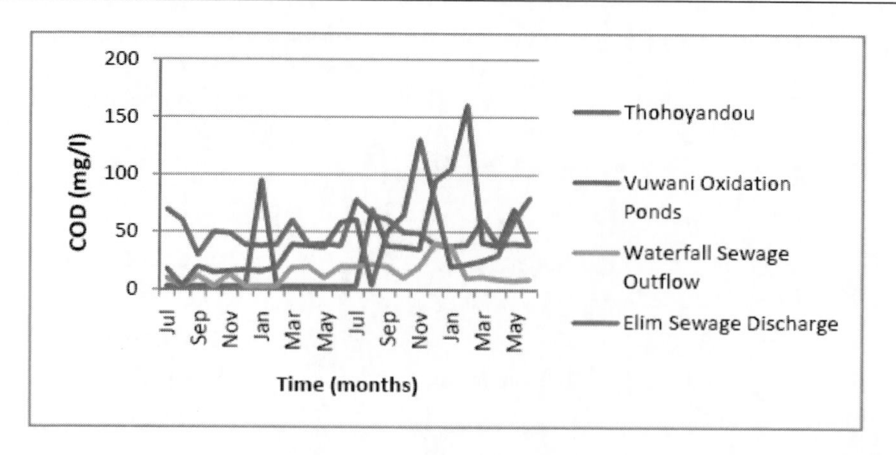

Figure 11. Results of COD monitoring along Luvuvhu River at four monitoring sites (2005 – 2007).

Review of literature and official documents revealed that there were other potential sources of water pollution besides the ones discussed above. Roadsides and rooftops from the two urban centres were implicated as sources of metals and salts which could pollute water when it rains.

The lawns and other garden trimmings from the urban centres contribute nutrients such as phosphorous into the water courses. The geology of the catchment has the potential to contribute to pollution of water. Literature reviewed by Odiyo et al. (2009) showed that, the geology of the Soutpanseberg Mountains contains minerals such as copper, iron, refractory flint, salt, sillimate and coal. Since the Luvuvhu River starts from this mountain, it means there is a potential for the minerals to pollute the water in the river. While Tshikondeni coal mine may not be causing any pollution problems at the moment according to DWA (2004b), literature has shown that mines can cause problems when they cease to operate (van Zyl, 2002).

Review of literature also revealed that the common practice by local communities to dig and open up the ground to obtain materials such as quarry stones and pit sand can cause water pollution. Oxidation of sulphide minerals, especially pyrite, during weathering can cause Acid Mine Drainage (AMD) and release iron (Fe), sulphate (SO_4), acid (H^+) and trace elements, such as nickel (Ni) and arsenic (As), into the aquatic environment (Pope et al., 2006b). Masidiri (2008) found that water from boreholes in the Tshimbupfe area in the catcthment had very high *E. coli* count and suggested that, this could have been caused by poor water point hygiene. The boreholes were not fenced off from animals and they could have licked the spouts and contaminated them. But critically, this shows that water at the point of collection is vulnerable to contamination putting the health of the community at risk.

4.6. Weaknesses in the Contemporary Water Quality Monitoring and Management Practices

The third source of vulnerability emanates from the weaknesses in the contemporary water quality monitoring and management practices. While the quality of water from the treatment plants was generally acceptable, treatment failures did occur during the rainy season. For example, analysis of treatment efficiency data from the Mhinga plant between

January 2010 and February 2011 shows a failure in the removal of turbidity in treated water (Figure 12). Turbidity in treated water in December 2010 reached 22 NTU and exceeded by far the allowable limit of 5 NTU given in SANS 241.

Turbidity has significant implications on human health. Excessive turbidity or cloudiness in drinking water is aesthetically unappealing and may also represent a health concern (Fox and Tversky, 1995). It can provide food and shelter for pathogens and can promote their regrowth in the distribution system, leading to waterborne disease outbreaks (Fox and Tversky, 1995). Although turbidity is not a direct indicator of health risk, numerous studies show a strong relationship between removal of turbidity and removal of protozoa. There is a relationship between turbidity removal and pathogen removal. Low filtered water turbidity can be correlated with low bacterial counts and low incidences of viral diseases (LeChevallier et al., 1991). Positive correlations between removal (the difference between raw and plant effluent water samples) of pathogens and turbidity have also been observed in several studies. Data gathered by LeChevallier and Norton in 1992 from three drinking water treatment plants using different watersheds indicated that for every log removal of turbidity, 0.89 log removal was achieved for the parasites *Cryptosporidium* and *Giardia.*

The particles of turbidity can also provide "shelter" for microbes by reducing their exposure to attack by disinfectants. Microbial attachment to particulate material or inert substances in water systems has been documented by several investigators and has been considered to aid in microbe survival (Marshall, 1976).

Even the communities expressed dissatisfaction with various issues surrounding the quality of their reticulated water supplies. While the percentage of the groups that said the taste of the water from taps was very good ranged from 51 – 59% (mean = 54; S = 2.4), 41 – 49% (mean = 45; S = 2.2) said the taste of the water was just good, meaning there was a significant proportion of the communities that were not completely happy with the taste of the water.

The same applied to the question of smell of the water where 43 – 49% (mean = 46; S = 1.8) of the groups were not entirely happy with it saying it was just good. The percentage of the groups that said the water had floating particles soon after collection ranged from 72 – 79% (mean = 75; S = 2.2) and 68 – 78% (mean = 74; S = 2.4) said the containers where the water was kept were stained brown after some time.

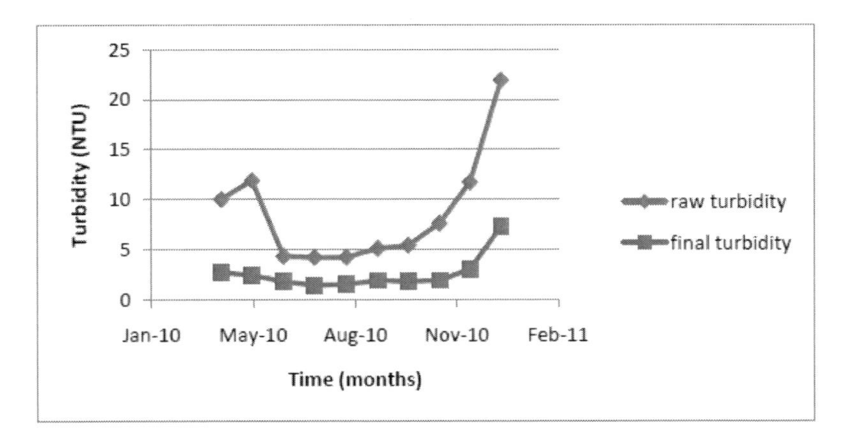

Figure 12. Turbidity removal efficiency at the Mhinga treatment plant.

Although most of the results from a monitoring exercise for microbiological quality from street taps in selected villages included in the study were satisfactory, there were occasional failures for both Total coliform and *E. coli* counts in some of the villages.

According to SANS 241, 95% of all samples (count/100ml) during monitoring should have no sign of *E. coli.* In some villages *E. coli* counts were recorded in more than 5% of the samples.

Since the results from most of the villages were satisfactory, it would suggest that the problem is not with the treatment plants but water could be getting contaminated at the point of collection or along the distribution pipes. The street pipes are not fenced off and allowing animals to lick the taps. This could compromise the hygiene of the tap and contaminate water. This serves to emphasize the need for water quality to be monitored at the point of collection to reduce the risk of communities getting sick from consuming water that has been contaminated after treatment.

Statistics obtained from four clinics in the study area, suggest a correlation between the rainy season and the increase in the incidences of diarrhoeal diseases (Figs. 13; 14; 15; 16). Although there was no official threshold to declare an outbreak when diarrhoea cases exceeded a certain value, the shapes of the graphs drawn from the statics obtained from the clinics showed that there was an enormous rise in cases during the rainy season. From 2008 to 2010 the peaks of the graphs were always between September and January or between January and April. This further confirms that the health of the public is under threat from waterborne diseases.

It suggests that people are using water from unprotected sources such as streams and pools during the rainy season when surface water is easily available. The statistics confirm the findings of the community survey that said between 20 and 29% of the people used a combination of protected and unprotected sources of water for drinking purposes sometimes. The diarrhoea cases could not be attributed to food sources because according to the Health Statistics Manual from the Department of Health, food poisoning can only be suspected when four members of a family fall ill at the same time. The cases are recorded and reported separately according to Ms Khangale, the Health Information Officer at the Vhembe district Department of Health offices.

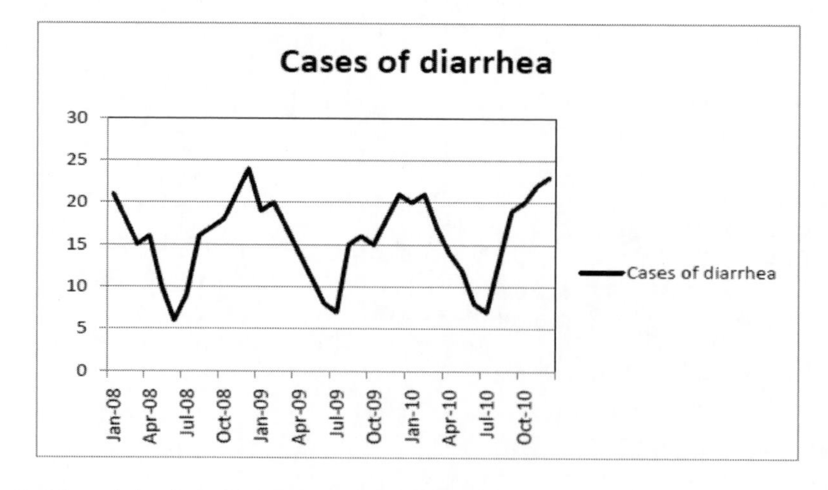

Figure 13. Health statistics from Mphambo Clinic (2008 -2010).

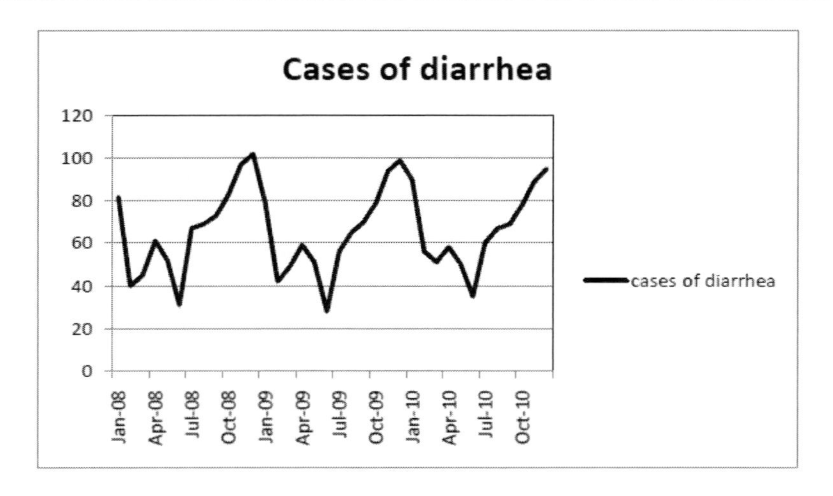

Figure 14. Health statistics from Mavambe Clinic (2008 – 2010).

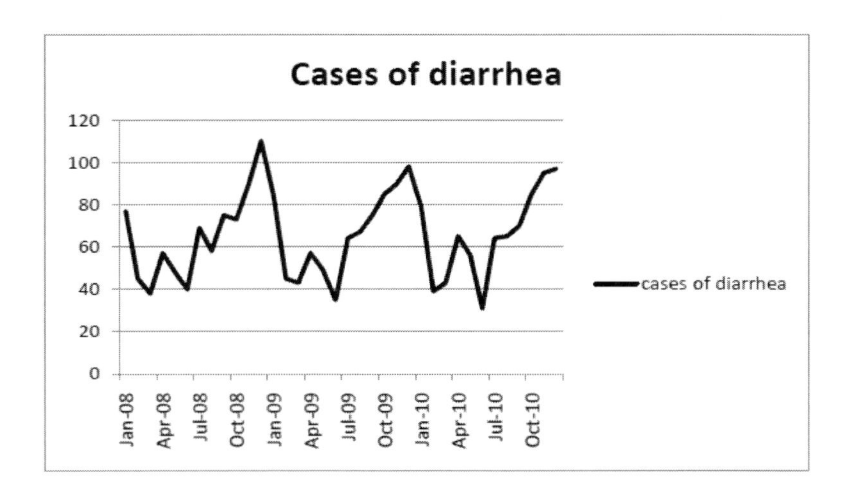

Figure 15. Health statistics from Tshikonelo Clinic (2008 – 2010).

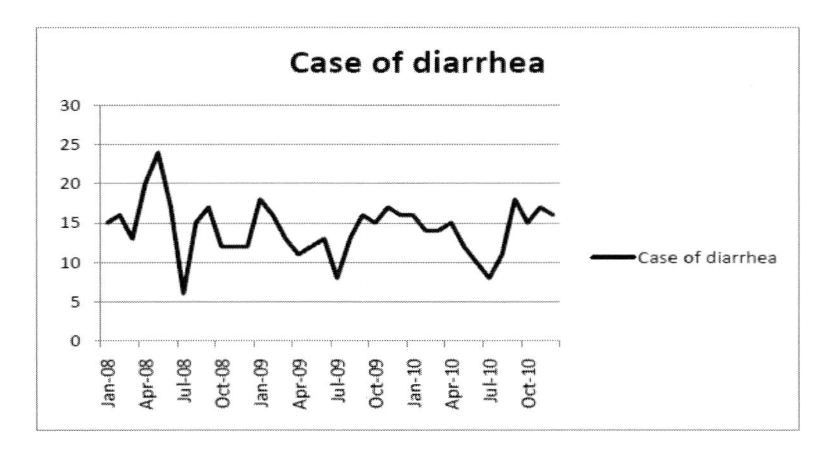

Figure 16. Health statistics from Xigalo Clinic (2008 – 2010).

4.7. Fragmented Structure of the Decentralised Health Services

The other source of vulnerability is related to the fragmented decentralised structure of the health services in Vhembe district. The Environmental Health Services who are responsible for investigating outbreaks of diseases including waterborne diseases have been decentralised to Vhembe District and its local municipalities. The clinics which are the ones that collect statistics of diseases have remained with the Department of Health. According to Mr. Mukwevho a Senior Environmental Health Officer at Vhembe District Municipality, there is a problem with the flow of statistics between the DoH and the municipality. Sometimes the DoH sends the statistics to their Head Office and Environmental Health Services can only access it through the mother Department of Local Government. This creates serious time lapses between an outbreak occurring and the reaction time by Environmental Health Services and increases community vulnerability to water quality problems.

CONCLUSION

The first objective which relates to providing proof that water quality problems exist in Luvuvhu Catchment and are worth receiving attention from the authorities was achieved. The potential water pollution sources in the catchment include the geology of the catchment, irrigated agriculture in the upper part of the catchment, rural human settlements in the middle part of the catchment, livestock and wildlife in the middle and lower parts of the catchment, mining activities, motor vehicle emissions and peri – urban areas of Thohoyandou and Malamulele with all the water pollution associated activities such as generation of both solid and liquid waste. The results obtained from DWA stations along the Luvuvhu River confirmed that water pollution was a threat to the health of the people who depended on the Luvuvhu River system for water supply. Parameters such as turbidity, nitrates, COD and *E. coli* exceeded by far the safety limits given in SANS 241. Results of ground water quality monitoring also confirmed this position with nitrate levels exceeding the limits in SANS 241. Results of a monitoring exercise conducted on selected street taps in the 15 study villages also confirmed the vulnerability and risks faced by the communities when some of the taps failed the *E. coli* tests. People resort to using unprotected surface water sources which become abundant during the rainy season making them vulnerable to waterborne diseases such as diarrhoea. The study proved that communities in the catchment are vulnerable to water quality problems and that water quality problems in the catchment are worth receiving the attention of authorities and researchers.

The second objective was to analyse the water quality monitoring frameworks in South Africa and identify any gaps. Although South Africa has a well developed and extensive policy, legal and technical framework relating to water quality monitoring and management, Luvuvhu Catchment has not developed the institutional capacity to adequately monitor and manage the quality of water resources in the catchment. The water quality monitoring services are in transition with key functions and staff being moved from DWA and DoH to Vhembe District Municipality as the Water Services Authority. This transition is currently affecting the level and quality of services offered because Vhembe District Municipality has not developed its structures to carry out this function. Although DWA still assists with monitoring the quality of water for domestic purposes, they only monitor the quality of raw water just before it enters the treatment plant and the quality of treated water at designated points after it leaves the treatment plant. This is not adequate to guarantee the safety of the consumers since water can get contaminated during distribution and needs to be ultimately monitored at the point of use.

Environmental Health Officers who are supposed to monitor water quality at point of use are currently in transition and are not able to carry out this function. Although DWA monitors water in the Luvuvhu River and ground water in the catchment, the monitoring is mostly irregular with many months going unmonitored in a year. The data generated from the monitoring activities although readily available on internet, is not analysed and not easily accessible to members of the public. The water quality programme in Luvuvhu Catchment currently fails to fulfil the requirements of the national constitution that every citizen has a right to a safe environment.

The partial decentralisation of health services from government to the municipalities is increasing community vulnerability to water quality problems. The fragmented health structures where the clinics are run by the Department of Health while the Environmental Health Officers who are responsible for investigating incidences of disease outbreaks are in the municipalities make it impossible for the system to react timeously when poor water quality causes problems within communities.

The third objective was to evaluate indigenous knowledge, attitudes, practices and perceptions and community participation in water quality monitoring and management. The communities felt that water pollution was a problem in their areas and that water quality needed to be monitored regularly. They had knowledge relating to various aspects of water quality such as how pollution of water happens, how to identify water pollution and improve water quality. Although they said they would recognise that water had been polluted through its physical appearance, their perception of water safety based on different degrees of turbidity did not match the expected scientific requirements. The communities accepted to use water with turbidity values as high as 92 NTU and yet the recommended maximum limit of turbidity in water for domestic use in SANS 241 is 5 NTU. These communities lack knowledge on water safety and are exposed to the risk of contracting diseases. They need education to raise their awareness to a level where they can make at least scientifically acceptable judgements.

The study established that, there is no meaningful participation by communities in the catchment in water quality monitoring and management despite the fact that there is comprehensive policy, legal and technical framework that could promote community participation. The institutional framework and the way water quality monitoring and management is carried out at the moment do not promote participation by communities. A top – down approach is employed in water quality monitoring and management with the provision of the service being viewed as purely technical domain to be carried out by the experts only.

The current practice where DWA engages private companies as consultants to engage communities in development issues e.g. establishment of CMAs should be discouraged because it is not sustainable and does not deliver the desired change at community level. Private companies are profit oriented and do not allow for the time required for communities to move through all the phases in the resistance to change continuum. Therefore as they pull out the initiative collapses since communities will have nobody to continue helping them building confidence in the initiative.

The current approach to water quality monitoring and management in the catchment does not promote an integrated approach. The programme is implemented vertically by DWA mostly and at times Department of Health and the municipalities are involved. Other critical players like communities and government department which deal with land use related functions such as Departments of Agriculture and Environment are not involved. This has serious implication of sustainable use of water resources since it is the land uses that impact on the quality of water.

The extension staff involved in water quality monitoring and management view communities as passive recipients of the service who have nothing to contribute in terms of ideas and materials. Although terms like stakeholders are used in the water sector in South Africa to refer to communities, they are just words while the practical reality on the ground is completely different.

Communication with communities is monologic (one sided) with "*experts*" depositing their new and educated ideas into the *"empty minds"* of communities without expecting any feedback from them. Although community participation has been adopted in policy and legal documents and accepted as the official approach to development, the current implementation system does not provide for dialogue where the communities can also make their own contribution or seek clarification at least. For example, if there is water quality failure from one of the treatment plants, the staff just responds by shutting down the supply and rectify the situation without discussing with the communities.

The fifth objective was to assess the reasonableness and acceptability of the proposed model to the communities. The communities currently feel left out in water quality monitoring and management activities and they view the proposed model as a way to get them involved. They view the model from a wider perspective within the democratic agenda in the country.

The communities are even willing to contribute towards the financing of the proposed community based water quality monitoring programme. The community leadership believe that the proposed model will empower communities to participate in water resources management at that level.

6. Recommended Conceptual Model for Community Empowerment in Water Quality Monitoring and Management

6.1. Conceptualised Participatory Community Based Water Quality Monitoring and Management Model

In response to the weaknesses in the current water quality monitoring and management model pointed out in chapter 4, an alternative conceptual community based water quality monitoring and management model has been proposed and shown to be reasonable and acceptable. It is anchored on three frameworks which allow it to achieve what the current model has failed to achieve.

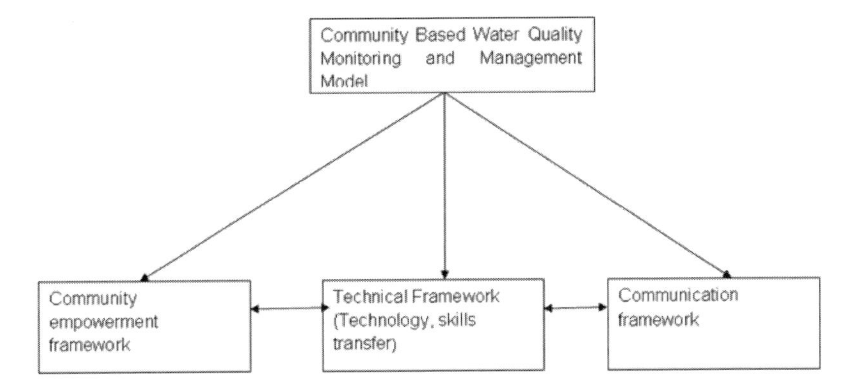

Figure 17. Community based water quality monitoring and management conceptual model.

The three frameworks are technical, community empowerment and communication frameworks (Figure 17). The current model is anchored from the top by DWA and the municipalities. Although the current policy and legal frameworks support community participation, the institutional framework is not designed to do so. There is no opportunity in the current model to empower communities through education and skills provision which are cornerstones of the proposed model.

6.2. Technical Framework

The proposed main features of the technical framework include simple technologies and a GIS spot map showing flashy spots which require more attention. For example, Figure 18 shows villages in the catchment with stand pipes that recorded some failures during monitoring for bacteriological quality over a period of eight months. These villages would therefore receive more attention in monitoring than the other villages that performed well.

Figure 18. GIS spot map showing villages that had stand pipes that failed microbiological tests during monitoring.

The framework proposes simple technologies to monitor water quality at community level. The first line of water quality monitoring at community level has been proposed to rely on human senses such as taste, smell, sight and touch. Hrudey et al. (1999) found evidence that an aesthetic problem (an unpleasant odour, taste and colour) is usually translated into a potential health risk. Gingras et al. (1999) showed that the taste of water and its source (lakes, rivers, groundwater aquifers) influence perception on water quality. If the relationship between these sensory parameters, their possible causes and the health implications as outlined in various pieces of scientific literature can be formalised as shown in Table 5, then communities can be trained to use their natural senses to monitor water quality.

Indigenous knowledge and practice can inform the technologies developed under this framework. Koocheki (2007) found that indigenous communities have valuable knowledge systems that can be incorporated into strategies for environmental management and that the knowledge systems are an essential cultural and technological element of human societies. The communities in Luvuvhu Catchment depend on physical appearance of water to decide whether the water is polluted or not. This fact can be built upon to ensure that simple technologies to measure turbidity levels in water are developed. The practice by the communities that, whenever they suspect that, there is a problem with their water supply, they report the problem to the municipality/government, can be built upon to form a community based surveillance system. The national water quality monitoring and management can be improved significantly if the communities acted as part of the "alert" system which will draw the attention of the authorities to trouble spots in time to prevent any disaster.

The major challenge that African countries continue to face according to Kamara (2004) is how to reconcile indigenous knowledge and modern science without substituting each other, respecting the two sets of values, and building on their respective strengths. Therefore in the case of Luvuvhu Catchment, this can be achieved by integrating the dependency of communities on physical appearance of water to detect pollution into the national water quality monitoring and management framework.

According to (Hens, 2006) indigenous knowledge is context specific and what works successfully in one location for one community may not for another. It is unique to a particular culture and acts as the basis for local decision making in agriculture, health, natural resource management and other activities and it is embedded in community practices, institutions, relationships and rituals. Therefore, there is need to develop technologies that will help perpetuate the uniqueness of the knowledge and practices related to water quality monitoring and management in Luvuvhu Catchment. This should motivate communities to participate effectively in water quality monitoring and management.

The second line of monitoring is proposed to use simple equipment and technologies that support the reasons for the sensory judgements. Instead of measuring particular water quality constituents, the technologies measure the effects of those constituents in water and confirm the sensory judgements. Measuring of physical and chemical water quality parameters such as pH, turbidity and residual chlorine can explain such things as the taste, smell, physical appearance and sense of touch of water, link it to possible causes and health implications. The health implications of concentrations/levels of each parameter are given in tabular form under the South African Water Quality Guidelines (Volume 1: Domestic Water Use) (DWAF, 2002). The technologies used to measure such parameters as turbidity, pH and microbiological quality (H_2S strip test) are relatively cheap and easy to use. Rural communities can be trained to use them effectively with relative ease. For example the cost of pH meter is estimated at R6 579 – 00 as shown in a quotation sent to the University of Venda by Labotec (Pty) Ltd (Appendix C1).

Therefore if the kits are distributed to Malamulele, Tshikonelo, Xikundu and Mhinga as the main villages in the study area and the sub villages under each village share a kit then for pH only about R27 000 – 00 is needed for capital equipment. This is relatively cheap when compared with treatment of people in hospitals when they fall sick from waterborne diseases such as cholera, especially considering that sometimes they lose their lives. According to DWAF (2005), there are various potential sources of funds that can be invested in community based water quality monitoring and management as shown in Table 18. In addition to that, the communities indicated their willingness to contribute towards the cost of monitoring and managing water quality at their own level. This should add to the potential sources of funds for the proposed program.

Experiences from elsewhere have shown that rural communities, if given necessary training can handle and use simple technologies to monitor water quality. Community based water quality monitoring has been experimented on in other parts of the world and the results have been good and used to influence environmental policy.

Table 5. Summary of common water quality manifestations, possible causes and health implications (DWAF, 2005)

Manifestation	Possible Causes	Health Implications
Sour taste	Low pH which could be due to pollution from anthropogenic activities or addition of excess chlorine during disinfection	pH below 6 will encourage corrosion of cadmium which can be toxic to human beings
Bitter taste	High pH which could be due to pollution or addition of excess lime during treatment of water	pH around 12 encourages corrosion of lead which has adverse effects on human health including growth retardation in infants
Soapy touch	High pH	
Salty taste	Chlorides above 400 mg/l	Excessive chlorides encourage corrosion of metals which can be toxic to humans. Will cause nausea between 2000m/l and 10000 mg/l and cause vomiting at 10000 mg/l and above
Colour	Oxidation of ferrous and manganous salts by iron bacteria Aluminium hydroxide floc due to aluminium concentrations in excess of 0.1 – 0.2 mg/l remaining in water during the water treatment process Blooms of cyanobacteria and other algae in reservoirs	Makes the water unsightly but also could be due to other toxic chemicals which would be injurious to public health Some cyanobacterial products such as cyanotoxins are also of direct health significance.
Excessive turbidity, or cloudiness	Pollution from various sources including sewage systems Blooms of cyanobacteria and other algae in reservoirs	Turbidity can provide food and shelter for pathogens Promotes regrowth of pathogens in the distribution system, leading to waterborne disease outbreaks.
Odours	Presence of sulphur compounds leading to the formation of hydrogen sulphide Presence of decaying organic material	Odours could indicate pollution from chemicals that could be harmful to human health

In the 1990s a rural community in the Philipines worked side by side with researchers, non-governmental and governmental workers over a five year period to develop science based indicators for water quality monitoring relevant for developing environmental policy (Deutsch et al., 1997). The results prompted the Lantapan Municipal Council to incorporate community based water testing and some of the research findings and recommendations into their Natural Resource Management Plan. In Umgeni, South Africa a GREEN (Global River Environmental Education Network) water project involved school children and community groups in monitoring water quality using simple scientifically valid methods (O'Donoghue et al., 1993). The methods included the use of litmus paper, to test for pH and then taking action to remedy the situation e.g. lobbying local administrative structures to build better toilets, step up health education and rehabilitate wetlands.

The Department of Civil Engineering at the University of Cape Town in South Africa is currently involved in collaboration with a number of international organisations in a project called Aquatest Project seeking to develop a combination of low-cost water quality field kit and a cellphone based data collection tool (Rivett, 2011). The aim of the project is to support basic drinking water quality monitoring at supply level, as well as the collection of results in a digital format. Although the implementation of the Aquatest project at Hantam Municipality (Eastern Cape, South Africa) is in its early stages, it has already provided some valuable lessons. The most important of these is that there is a clear need for systems that improve the ability of staff in the field to conduct operational monitoring. However, in order to be acceptable, systems should be well integrated with existing municipal processes and structures. Therefore the proposed community based water quality monitoring model in Luvuvhu Catchment provides an opportunity for staff to improve on their monitoring coverage at community level since the communities will act as an *"extension"* of the staff and provide preliminary surveillance services.

The results of the community based surveillance activities can then be plotted on a GIS map as done in Figure 18 to help the under staffed and poorly resourced water quality monitoring system in Vhembe District to focus its surveillance on the trouble spots. The health and hygiene education can be directed and focussed onto those areas that need it more than the others.

The community based water quality monitoring and management programme can serve as part of an *"alert system"* described below;

- Alert Level I (no significant risk to health): Routine problems including minor disruptions to the water system and single sample non-compliances (Internal WSA response only).
- Alert Level II (potential minor risk to health): Minor emergencies, requiring additional sampling, process optimisation and reporting/communication of the problem (Internal WSA response only).
- Alert Level III (potential major risk to health): Major emergencies requiring significant interventions to minimise public health risk (Engagement of a designated Emergency Management Team).

The community based water quality monitoring and management programme can form part of Alert Level I (no significant risk to health) which deals with routine problems including minor disruptions to the water system and single sample non-compliances requiring internal WSA response only. When monitoring programmes at community level raise an alert signal, the WSA responds and investigates further. If there is a potential minor risk they activate Alert Level ll which deals with minor emergencies requiring additional sampling, process optimization and reporting/communication of the problem. This also requires internal WSA response only. If the potential risk to health is a major one, they activate alert level lll which deals with major emergencies requiring significant interventions to minimise public health risk. This requires engagement of a designated Emergency Management Team.

6.3. Community Empowerment Framework

The community based water quality monitoring and management model can serve as a model for empowering communities in development issues. The fact that the communities can conduct preliminary water quality monitoring using their senses and confirm that with scientific equipment, boosts their self esteem and gives them confidence when discussing with the "*experts*". When the results from the community based monitoring system are interpreted against Tables 1, 2 and 5 the communities will not only be able to relate the results to possible impacts on their health but will be motivated to take up remedial actions to mitigate against the situation as envisaged in the Health Belief Model.

The possible remedial actions might include treating water at the point of use through such methods as boiling, disinfection with chlorine products, allowing sedimentation to take place before using the water or sand filtration. The most important issue is that this model will be able to provide evidence at a local level that the health of a community is at risk and as stated by Redding et al., (2000) individual perceptions of risk play a critical role in influencing the probability of adopting protective behaviour to prevent illness. According to Dogaru et al. (2009) perceptual ability heavily depends upon the amount of perceptual practice and experience that the subject has already enjoyed, implying that perception is a skill that can be improved tremendously through judicious practice and experience. Therefore by continuously monitoring water quality at their level the communities will improve their perceptions of risk related to use of water of poor quality and cue themselves to take action as stated in the Health Belief Model.

The corrective action taken by the community can also include engaging the authorities on the need to take action beyond the capacity of the community. For example if turbidity levels in reticulated water are higher than expected, then it could mean that the filtration process at the treatment plant could be defective. The monitoring results will empower communities to take up the issue with those responsible for the treatment plant. As stated by Mackintosh and Colvin (2003) the implementation of well considered, community accepted drinking water quality management procedures can effectively change an unacceptable water quality to one that satisfies drinking water specifications.

The community based water quality monitoring and management model can empower the communities to deal with not only issues relating to water resources management but environmental management issues in general in line with integrated water resources management principles. As stated by Ferreyra et al. (2007) integrated water resources management is one of the major bottom-up alternatives that emerged during the 1980s in North America as part of the trend towards more holistic and participatory styles of environmental governance. It aims to protect surface and groundwater resources by focusing on the integrated and collaborative management of land and water resources at a watershed scale.

For example if the pH of a water body is lower than expected, by referring to Tables 5 and 2 the communities can come up with possible causes and sources of pollution which they can refer to the authorities for further investigation with confidence since they will be backed by the monitoring results. An environmental management expert wanting to introduce a programme on soil conservation to a community can start with a project monitoring total suspended solids (TSS) and work with communities throughout the project cycle. This will allow the communities to perceive and understand the extent of the problem in their area and make it easy for the expert to mobilise the communities to take action.

While the use of simple technology empowers communities to monitor the quality of their water resources, a supportive environment needs to be created for effective empowerment to take place. An environment that allows the voice of the communities to be heard needs to be created. Bottom – up development approaches need to be adopted when dealing with issues relating to water quality monitoring and management. This will lead to sustainable use of the limited water resources.

As stated by Haddad (2007), sustainability is the capacity to maintain services and benefits both at the community and institutional levels without detrimental effects even after external assistance has been phased out. Key aspects of sustainability include empowerment of local people, self-reliance and social justice (Haddad, 2007). These reflect principles of equity, accountability and transparency. One way to incorporate these principles into real-life management is to move away from conventional forms of water governance, which have usually been dominated by top-down approaches, and professional experts in the government and private sector. The movement should be towards the bottom-up approach, which combines the experience, knowledge and understanding of various local groups and people.

Given the right environment and orientation communities can participate effectively in different aspects of water resources management including water quality monitoring and management. The case study from the Philippines referred to in subsection 6.2., did not only help communities to acquire new skills but empowered them to participate in the management of the quality of their water resources. For example a community leader who wanted to tap water from some mountain spring and convey the water to the community, requested the services of the community water monitor to determine the bacterial level of the water prior to making the final decision of installing pipes. The tests revealed that some of the springs had unsafe levels of coliform bacteria and alternative sources were found, saving the government funds and minimising the risk of waterborne diseases.

A study from northern Thailand with villages participating in the management of their watersheds and other villages not participating showed that the villages participating had higher income after completion of the project (Empandhu et. al., 1996). Although they had higher opportunity costs due to the time spent for meetings and seminars, the villagers taking care of their own resources benefited from improved water, soil and forest quality and could make use of it through higher yields and thus through a higher income (Empandhu et al., 1996).

While the above cases show that communities can participate effectively in development, the Food and Agricultural Organization (FAO) has carried out controlled experiments to prove that output from farming activities can be improved through community involvement and participation in the planning of the project. The case studies outlined as case studies 1 and 2 below illustrate the potential for community participation and farmer training in improving harvests.

Case study 1: Start – up of participatory community planning in Mexico
From: FAO. 1997. Communication for Rural Development in Mexico: In Good Times and Bad. By Fraser, C. and Restrepo-Estrada Rome

What might be coined FAO's first concerted venture into participatory planning by intended beneficiaries of a project, occurred in Mexico under the PRODERITH (Programme of Integrated Rural Development in the Tropical Wetlands Project), funded by the World Bank and technically backstopped by the Development Support Communication Branch. The first phase ran from 1978 to 1984 and was concerned with improving agricultural development in Mexico's wetlands that make up 23 percent of the country's total land area. Prior to PRODERITH, a large-scale integrated rural development project had been launched in the wetlands which drained 83 000 hectares, built roads, new villages, schools and medical centres, yet was never successful.

The peasants never identified with it nor did they use or maintain the infrastructure properly. This was attributed to "a lack of effective mechanisms for the participation of the beneficiaries". The objectives of the new project, budgeted at US$149 million, were to increase agricultural productivity in the tropical wetlands, improve the living standards of peasant families, and conserve natural resources. People in the targeted communities were involved in the planning process from the start. The mechanism to do this was imbedded in a Central Rural Communication Unit created for the project. It worked principally with video and support print materials to cover three types of communication needs: a) situation analysis and participatory planning with peasants, b) education and training for peasants, and c) information for project coordination and management.

Outreach field units were set up to work with communities. Video was used to record local people's attitudes and perceived needs and then played back to individual communities as a basis for promoting internal dialogue about its past, present and future, and options for improvement.

People began to articulate the realities of their situation, their priorities, and what they felt capable of doing. This was followed up with a synthesis of collective perceptions and elaboration of a "local development plan" for project concentration. During its implementation, video was also extensively used for orientation and farmer training in a wide range of agricultural and rural development topics.

At the end of its first phase in 1984 incomes of some 3 500 families in a 500 000 hectare zone had increased by 50 percent over 1977 levels. And perhaps most significantly, it had put in place a methodology for replication in a second phase involving 650 000 people in an area covering 1.2 million hectares.

The World Bank considered PRODERITH to be among the most successful projects they had supported up to that time, attributing much of its success to the participatory approach adopted by the communication units in synthesizing community development priorities, with follow-up skills training for farmers in its implementation. As for the bottom line, the communication component for the first phase absorbed only 1.2 percent of the total costs, while the internal rate of return, a measure of the economic success of the project, was 7.2 percent higher than originally foreseen.

Case study 2: Comparison of Inputs and Outputs of ten IPM versus ten Non-IPM Rice Farmers in West Sumatra, Indonesia
From: FAO .1993. IPM Farmer Training: The Indonesian Case, Jogyakarta: FAO – IPM Secretariat

A controlled study was conducted in West Sumatra, Indonesia, during the wet season of 1992-1993 (December to May). The study compared costs of rice farming inputs and outputs among ten farmers who had participated in IPM farmer field schools during the previous wet season with practices and outputs of l0 farmers who had never participated in FFSs. The two groups of farmers were matched by location, farm size and land tenure. The only treatment variable was the IPM-FFS training.

Observations and discussions with both sets of farmers were held on a weekly basis. IPM training had stressed "Growing a Healthy Crop" (improved seed varieties, balanced fertilization, proper plant spacing in straight rows), Conservation of Beneficial Insects" (low pesticide use), and Weekly Field Observations to Determine Management Actions. The foregoing training focus was determined to be the major difference between IPM and non-IPM farmers. The comparative results on a number of key variables based on actual harvests are tabulated below.

Overall, the IPM farmers achieved 21 percent more rice harvest yield on a per hectare basis (6.9 tons versus 5.7 tons), for 97 percent of production costs, when compared to their non-IPM farmer counterparts. The significantly lower "input" costs for IPM farmers were largely attributed to minimal usage of commercial pesticides. Labour costs were also slightly lower for IPM farmers, possibly because of better management actions.

The above two case studies prove that community involvement and participation can improve results in a project and should be adopted in water quality monitoring and management in Luvuvhu Catchment. The first step will be to make adjustments to the national water quality monitoring information system to allow it to incorporate data generated by the community based water quality monitoring system. This should not pose serious challenges since the community based system will be using scientifically sound and accepted field kits.

For example there will be no variation in the way experts measure turbidity and the way the communities will do it. The only difference is that the new model will not only allow for more water quality monitoring points to be included in the national data base but also for "*real – time*" monitoring to take place at community level. Allowing the data from community level to be incorporated into the national system, amounts to transmitting the community's voices throughout the system to the top. The authorities at different levels will make decisions based on data that reflects the real conditions at community level.

The second action to encourage true or effective community participation will be for authorities to finalise the establishment of the catchment management system in Luvuvhu/Letaba WMA since the proposed structures are likely to enhance community participation in water quality monitoring and management as shown in Figure 3. The structure would need to be linked to local government structures at community level so that it can articulate the opinions of the majority of the people. Links have been created between Figure

3 and Figure 4, resulting in Figure 19 so that information from the grass root level can flow through to the national level.

Variable	Average Budget for 10 IPM Non-IPM Farmers (In Rupiahs)	Average Budget for 10 IPM Trained Farmers (In Rupiahs)
Pre-Harvest Labour/Ha	414 660	384 656
Harvest Labour/Ha	657 730	659 851
Total Inputs/Ha	163 268	139 819
Total Production Costs/Ha	820 998	799 670
Total Output in Kg/Ha	5 741	6 953

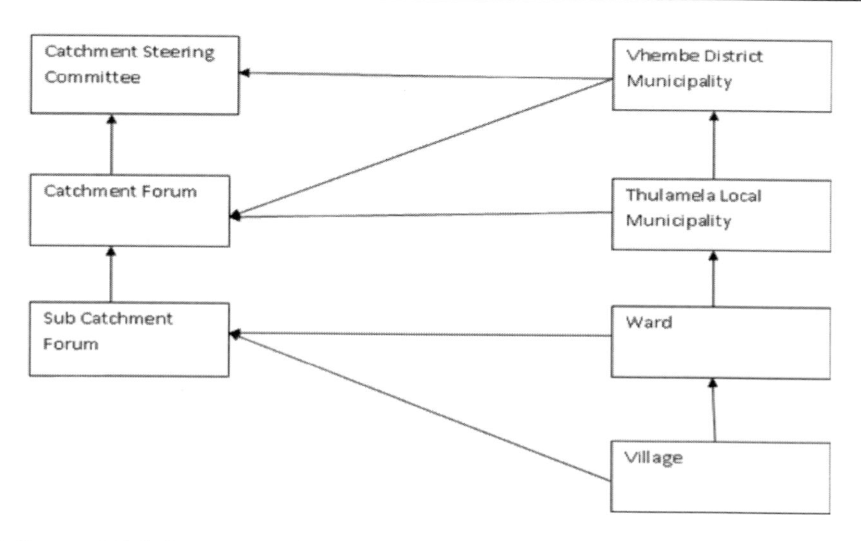

Figure 19. Proposed links between catchment management and local government structures.

Each sub village will operate as the smallest unit of water quality monitoring. A set of field kits for monitoring, for example pH, turbidity, residual chlorine and microbiological quality of water can be issued to a chief and all the sub villages under that chief can share the same set. Each village will select at least two volunteers who will be trained by experts from DWA/Vhembe District Municipality in the use of the kits. The volunteers will submit the monitoring results to the staff at the nearest treatment plant for onward transmission to the district level and incorporation into the national water quality information system. Rapport between the volunteers and the staff in - charge of treatment plants should be encouraged so that if the results of monitoring at community level are not satisfactory, the volunteers can easily bring this to the attention of the staff at the treatments plants. The staff will either sort out the problem if it is within their capacity or report to Vhembe District Municipality as the WSA responsible for treating domestic water. This will be done as part of Alert Level framework as discussed above.

The results from each of the sub villages should be discussed at the weekly meetings held at the Tribal Office and chaired by the chief to allow the rest of the community to be aware of the state of their water supply. The elected Ward Committee members from each village will then take the results to the monthly Ward Committee meetings chaired by the Councillor.

Decisions are then made in this committee. From this point two routes should be taken. First, the Councillor as a member of the local municipality should take the results to municipal meetings where the results and any suggestions from each ward should be discussed and actions recommended. The results will then follow the normal local government reporting structures up to national level. The second route is that, the councillor who should be a member of the sub Catchment Forum should take the results for discussion at the sub Catchment Forum. The chiefs should be part of the sub CF so that they can represent their villages. The members from the sub CFs who attend the Catchment Forum present the results from each sub catchment for discussion. The Catchment Forum is made of all stakeholders in the water sector at catchment level. The results will then be taken up and discussed at WMA level by the CMA and the Catchment Steering Committee before being passed onto the national level. This should empower the communities to have *"a voice"* in how their supplies are managed and deepen the democratic agenda within the catchment.

The third action will require investing time and resources to bring about a mindset change at all levels. Currently the accepted norm of service delivery is that the government is responsible for everything. The communities are passive recipients of services. The approaches to development are top – down oriented and all the decisions are made from above and transmitted down to the communities. Communities on the other hand believe that their duty is to demand for services from above and wait for them to be provided. Therefore for the proposed model to work there will be a need for change of mindset from all levels starting from policy and decision makers to implementers and communities. Everyone has to change and accept that communities are capable partners who can take responsibility for their actions. Everyone involved has to be taken through the five phases of adapting and overcoming resistance to change. All the people involved in the community based water quality monitoring and management programme need to accept and believe in the SARAR philosophy. Therefore, there will be need to ensure that mindset activities are carefully planned and implemented at all levels.

Advocacy meetings and workshops should be conducted for policy makers (politicians, directors and community leadership, etc.) while intensive training on development and leadership theories should be carried out among implementers (managers and extension workers). This process should ensure that, it assures all involved of the community's capacity and capability to implement and manage the proposed community based water quality monitoring and management programme. All role players, especially extension workers need to be assured that the proposed programme will not lead to any loses (jobs, allowances, status, etc.) on their part to avoid generating resistance among them.

6.4. Communication Framework

For effective community participation in water quality monitoring and management to take place, an appropriate communication framework has to be developed. The communication framework should place the targeted communities at the centre of its three key components as shown in Figure 20. The three fundamental components of the communication framework are education and communication, research and extension.

The Health Belief Model dictates that before one attempts to change the behaviour of individuals, one needs to understand how they perceive the situation under discussion. Hens

(2006) states that indigenous knowledge is context specific and what works successfully in one location or for one community may not work for another. Koocheki, (2007) further states that indigenous knowledge is unique to a particular culture and acts as the basis for local decision making in agriculture, health, natural resource management and other activities and it is embedded in community practices, institutions, relationships and rituals.

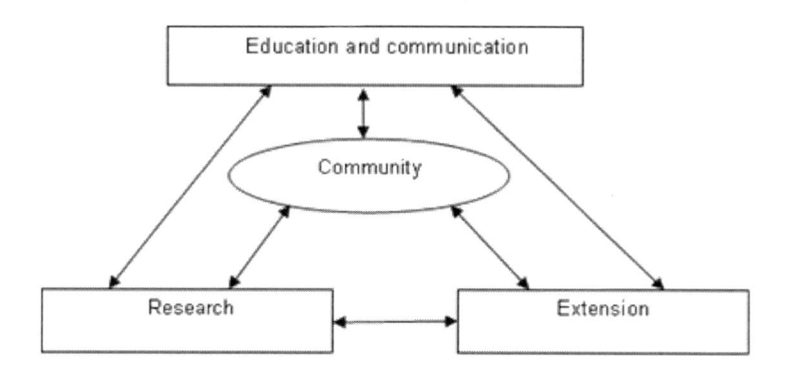

Figure 20. The Knowledge Triangle: Adopted from FAO (2003).

As already discussed under the empowerment framework, individual perceptions influence the chances that an individual will take action against a perceived risk and that perceptions can improve with experience. Therefore research will play a critical role in shaping the whole water quality monitoring and management agenda at community level. Besides the government agencies that are involved in water quality monitoring and management, Luvuvhu Catchment has academic institutions such as the University of Venda, Madzivhandila Agricultural College, Techniven etc. These can play an important role in carrying out research which will focus on the education and communication component of the framework. The trademark of this framework will be the dialogue between the communities and the extension staff. Before the extension staff go out to give health and hygiene education, the researchers should collect relevant data and analyse it and use it to come up with the content of the health and hygiene education topic. In other words any education programme will act as a *"reply"* to a community request for information.

While the academic institutions can play an important role of research, it remains critical to realise that the extension workers themselves through their engagement with communities collect a lot of data that can guide their education and mode of communication. The extension workers should always use participatory tools to engage communities as these allow the communities to discuss the issues amongst themselves and reach consensus before giving feed back to the extension workers. People should always be divided into groups and given a theme to work on and give feedback to the meeting.

The education and communication component is responsible for supplying information to the communities. This involves simplifying all the data accumulated by researchers or extension staff to a level that the communities where the data was collected can understand it and use it. For example, the water quality monitoring results collected by DWA staff and the academic institutions in the catchments can be classified and translated into colour codes as shown in Table 9. These will allow even those who are not educated to understand the results and the implication to public health.

There are many ways of giving out information as outlined in section. Most of the villages in Luvuvhu Catchment have electricity and it is possible to use electronic based methods of communication such as videos which show villagers how other villagers elsewhere have performed in similar projects or dissemination of information through the internet. The researchers and extension workers can simplify water quality monitoring results from each village and upload them onto internet and villagers through their representatives can download the results from internet cafes and discuss them at their meetings. India has used internet successfully in community education programmes (Bailur, 2007). Simplified journalistic approaches can be utilized to disseminate information. For example weekly values of turbidity, pH, and residual chlorine can be published on the wall at the tribal office. Interpersonal communication methods such as extension workers explaining water quality monitoring results at community meetings or the community members explaining results to each other remain viable options.

Community radios have played a significant role in improving community perception of different development scenarios. University of Venda has a radio station which is not a public broadcaster and can play the role of a community radio in Luvuvhu Catchment. For example researchers at the University can simplify results of water quality monitoring done in the catchment over the years and disseminate these through Univen FM as a way of provoking debate around water resources management in the catchment. Univen FM can also broadcast regular updates on the quality of water in different villages.

Efficient management of existing limited water resources in terms of quality and quantity requires real time flow of information both from managers to users and users to managers. These processes can be made effective and efficient by using latest available open source technology such as cell phones (Khan, 2010). Mobile phone networks extend over most of South Africa and have an untapped potential for data capture and water quality management. In 2008, South Africa had 45.68 million subscribers which came only second in Africa to that of Nigeria with 45.89 million subscribers (Moorgas et al., 2010). Luvuvhu Catchment is therefore in a position to speed up the dissemination of water quality monitoring by employing simple technologies to monitor water quality at community level and then uploading the results onto a cell phone for dissemination. One such technology is the Electronic Mobile Water Application (eMWAP) an innovative concept harnessing the use of appropriate field kits and mobile phones to support monitoring and management of water quality in remote areas.

The eMWAP application works via the process of using simple field test kits, such as the H_2S strip test or World Water Monitoring Day Test Kits. The results of the test are then immediately loaded onto a mobile phone which gives instant feedback to the user and to the manager/authority, who can then ensure that problems are identified and resolved without delay. eMWAP currently captures data for pH, turbidity, electrical conductivity, H_2S strip test and free chlorine residual (any determinants can be added on request).

eMWAP helps to:

- Capture water quality analysis related information in remote places by means of a mobile phone
- Communities receive instant feedback on their water quality even though would not have immediate access to a laboratory

- Understand what issues exist by receiving direct results
- Capture water quality results on a database without the need for internet access
- View data on a dedicated website that gives you the number of samples collected, compliance and failures
- Inform your sampler, manager, authorities of any immediate water quality issues
- Know and understand the health status of rivers in your area
- Inform municipalities of water services related issues, e.g. burst water pipes

The extension component of the communication framework deals with issues of training, imparting and transfer of skills. The communities will need to be trained and given skills to monitor quality at their level. They need to be trained on how to use their natural senses to monitor water quality and interpretation of these sensory judgements in a scientifically acceptable manner. They need to be trained to use simple water quality monitoring field kits and how to interpret the results. The ratio of EHOs to the people to be served is very low and therefore the available manpower cannot offer quality extension service. Vhembe District Municipality has also not appointed staff for water quality monitoring purposes.

To overcome manpower shortage, there is a need to select and train community volunteers to give health and hygiene education and to monitor water quality at their villages. In the Nzhelele area under Makhado Local Municipality, the Environmental Health staff has created waste management fora at community level to mobilise communities to clean up the area. These fora report any waste management problems to the staff based at Siloam Hospital. The same concept can be employed in water quality monitoring activities in Luvuvhu Catchment.

REFERENCES

Adams, W.M., Potkanski, T.P. and Sutton, J.E.G. (1994) Indigenous farmer managed irrigation in Sonjo, Tanzania. *GeogrlJ.* 160: 17-32.

Agenbag, M., Gouws, M. (2004) Redirecting the role of environmental health in South Africa. In: *Paper presented at 8th World Congress on Environmental Health,* Durban, 2004 Feb 23-27.

Agenbag, M. (2008) The Management and Control of Milk Hygiene in the Informal Sector by Environmental Health Services in South Africa. *Masters Thesis in Environmental Health at the Central University of Technology,* Free State, Bloemfontein, South Africa, 2008. Unpublished.

Ahearn, D.S., Sheibley, R.S., Dahlgren, R.A., Anderson, M., Jonson, J. and Tate, K.W., (2005) Land use and land cover influence on water quality in the last free-flowing river draining the western Sierra Nevada, California, *Journal of Hydrology* 313 (3): 10 – 13.

Anderson, B. A., Romani, J.H., Phillips, H.E., Wentzel, M., Tlabela, K., 2008. Exploring Perceptions, Behaviors and Awareness: Water and Water Pollution in South Africa www.hsrc.ac.za/Publication-Keyword-924. phtml.

Bailur S., (2007) The complexities of community participation in ICT for development projects: the case of "our voices". *Proceedings of the 9th International Conference on Social Implications of Computers in Developing Countries,* São Paulo, Brazil, May 2007.

Barwell, L., (2006) Global Change and Ecosystems Research in South Africa: Opportunities for Partnership with Europe2006 www.csir.co.za/nre/docs/ NREARR2009_FINAL.

Deutsch, M. W. (1997). *Antibiotic use in necrotizing pancreatitis*; 122:356-61. (AN : 97259146).

DWAF, (2002) Water Supply and Sanitation Policy: *White Paper,* Cape Town www.dwaf.gov.za/documents.

Department of Health, (2005a) *Diarrhoea and Typhoid outbreak in Delmas under control,* Sept 2005. www.hsr.iop.kcl.ac.uk/prism/tag/download/ DHFast.

Department of Health, (2005b) *Protecting what's underneath the tap* www.hsr.iop.kcl.ac.uk/prism/tag/download/DHFast-forwardingPCMH.pdf.

Dogaru, D., Zobrist J., Balteanu D., Popescu C., Sima M., Amini M. and Yang, H., (2009) Community Perception of Water Quality in a Mining-Affected Area: A Case Study for the Certej Catchment in the Apuseni Mountains in Romania. *Environmental Management* 43:1131–1145.

Dungumaro, W.E, and Madulu, N.F., (2002) Public Participation in Integrated Water Resources Management: the Case of Tanzania: 3rd WaterNet/Warfsa Symposium *'Water Demand Management for Sustainable Development'*, Dar es Salaam, 30-31 October 2002.

Dungumaro, W. E., 2007. Socioeconomic differentials and availability of domestic water in South Africa: *Physics and Chemistry of the Earth*, Parts A/B/C, Issues 15 -18; 1141 – 1147.

DWAF, (2006). National Water Resource Policy (NWRP) Department of Water Affairs Forestry: Pretoria www.dwaf.gov.za/documents.

DWAF, 2005. Luvuvhu/Letaba WMA: Internal Strategic Perspective: APPENDICES. Report No. P WMA 02/000/00/0304. Department of Water Affairs and Forestry Directorate: *National Water Resource Planning* (North).

Empandhu, D, Lakhaviwattanakul, T. and Kalyawongsa, 1996. A Case of Successful Participatory Watershed Management in Protected Areas of Northern Thailand. *In case studies of people's participation in Watershed Management in Asia.* Part II : Sri Lanka, Thailand Vietnam and Philipines, ed. Sharma, P.N Field Documents No. 5 Kathamandu, Nepal.

FAO, (1993). IPM Farmer Training: *The Indonesian Case,* Jogyakarta: FAO – IPM Secretariat. Published by Republic of Indonesia, IPM Secretariat, Jakarta; pp 512.

FAO, (1997). Communication for Rural Development in Mexico: In *Good Times and Bad.* By Fraser, C. and Restrepo-Estrada Rome www.fao.org/ DOCREP/003/X8002E/x8002e00.htm - 9k.

FAO, (2003) The state of world fisheries and aquaculture: *Highlights of special FAO Studies Editorial Group FAO Information Division* www.fao.org/DOCREP/003/X8002E/x8002e00.htm - 9k.

Ferreyra, C., de Loë, R. C. and Kreutzwise, R. D., (2007) Imagined communities, contested watersheds: Challenges to integrated water resources management in agricultural areas: *Journal of Rural Studies;* 304(24) 423 – 478.

Fox, C. R. and Tversky, A., (1995). Ambiguity Aversion and Comparative Ignorance. *The Quarterly Journal of Economics*, Vol. 110, No.3. 585–603.

Friis-Hansen, R., (1999) The Socio-economic dynamics of farmers` management of local plant genetic resources – A framework for analysis with examples from Tanzanian case study. *CRD Working Paper* 99.3.

Genthe, B. and Steyn, M., (2006) Good Intersectoral Water Governance – A Southern African Decision-Makers Guide: *Chapter on Health and Water* www.researchspace.csir.co.za/dspace/bitstream/.../1/Genthe_2006_d.pdf.

Gingras AC., Gygi S. P., Raught B., Polakiewicz R. D., Abraham RT., Hoekstra MF., Aebersold R., Sonenberg N, (1999). Regulation of 4E-BP1 phosphorylation: a novel two-step mechanism. *Genes Dev.* Jun 1;13(11): 1422-37.

GWP, (2000). Towards water security: *A framework for action* www.gwptoolbox.org/index2.php?option=com...id...

Haddad, B ., (2007) The professional and intellectual challenges of sustainable water management, in *proceedings of the 3rd Dubrovnik conference on sustainable development of energy*. Water and Environment system. eds. Singapore: World scientific publication.

Hemson, D., and O'Donovan, M. (2005). *Putting numbers to the scorecard: Presidential targets and the state of delivery*. In S. Buhlungu, J. Daniel, R. Southall and J. Lutchman (Eds.), State of the nation: South Africa, 2005-2006 (pp. 11-45). Cape Town: HSRC Press.

Hens, L., (2006) *Indigenous Knowledge and Biodiversity Conservation and Management in Ghana*: Human Ecology Department, Vrije Universiteit Brussel, Laarbeeklaan 103B-1090 Jette, Belgium: www.vub.ac.be/MEKO.

Hirji, R., Mackay, H., Maro, P., (2002) *Defining and Mainstreaming Environmental Sustainability in Water Resources Management in Southern Africa.* SADC, IUCN, SARDC, World Bank, Maseru, Harare, Washington, DC.

Hodgson, K. and Manus, L., (2006) *A drinking water quality framework for South Africa)* ISSN 0378-4738 = Water SA Vol. 32 No. 5 (Special edn. WISA 2006) ISSN 1816-7950).

Kamara, J., (2004) *Indigenous knowledge in natural disaster reduction in Africa,* periodic publication by UNEP/GRID-Arendal.

Kauzeni A. S. and Madulu N. S., (2000) Review of development programmes, Projects in Serengeti, *Consultant report to SIDA*, Daressalam.

Koocheki, A.A., (2007) Indigenous knowledge in Agriculture with Particular Reference to Saffron Production in Iran: *I International Symposium on Saffron Biology and Biotechnology* May 20-26, 2005, Tehran (Iran).

LeChevallier, M.W., Norton, W.D. and Lee, R.G. (1991). "Giardia and Cryptosporidium in Filtered Drinking Water Supplies." *Applied and Environmental Microbiology.* 2617-2621.

McGhee, T.J. (1991). *Water Supply and Sewerage*, 6th Ed., McGraw-Hill, New York, 602.

Mackintosh, G. and Colvin C., (2003) Failure of rural schemes in South Africa to provide potable water. Environmental Geology, CSIR Research Space > *General science, engineering and technology.*

Malume, F. (2010) Evaluation of Water Demand in Makonde Village in Thulamela Local Municipality of South Africa and the Implications on Water Resources Management, *Unpublished honours dissertation.* School of Environmental Sciences University of Venda.

Marshall, M. M., 1976. Waterborne protozoan pathogens. *Clinical Microbiology Reviews* Vol 10, No. 1, 67-85.

Masidiri, F., (2008) Evaluation of the Quality of Water Supplies and the Implications on Domestic Uses in Malamulele area of Thulamela Local Municipality in South Africa,

Unpublished honours dissertation, School of Environmental Sciences, University of Venda.

Moorgas, S., Naidoo,V., Wensley, A., Mackintosh, G. and Charles, K. (2010) The use of mobile based DWQ monitoring and management technology for remote areas. Feedback on Kwa Zulu–Natal pilot initiative, 2nd DWQ Conference 7a4Paper www.ewisa.co.za/.../ DWQUALITY CONFERENCES.

Mvundlela, M.S., (2010) Evaluation of Actual Domestic Water Consumption in Tshidembe Village under Thulamela Local Municipality in South Africa (*Unpublished in honours dissertation*, School of Environmental Sciences, University of Venda.

Oberholster, P.J., Botha, A.M. and Cloete, E., (2007) Using a battery of bioassays, benthic phytoplankton and the AUSRIVAS method to monitor long-term coal tar contaminated sediment in the Cache la Poudre River. Colorado. *Water Res.* 39, 4913 -4924.

Nemadodzi, N., (2008) Evalution of Water Poverty in Mhinga Village in Thulamela Local Municipality of South Africa, *Unpublished honours dissertation*, School of Environmental Sciences, University of Venda.

Odiyo, J. O., Makungo R. and Muhlarhi, T. G. (2009) Investigating the impacts of geochemistry and agricultural activities on groundwater quality in the Soutpansberg fractured rock aquifers, *Paper presented at the groundwater conference on pushing the limits held at Somerset West,* Cape Town, South Africa, 16 – 18 November 2009.

O'Donoghue., Habel., Normal., Michael., Maddox., Marion., (1993) Myth, ritual and the sacred. *Introducing the phenomena of religion*, (Underdale: University of South Australia, 1993).

Ong'or, D.O., (2005) Community Participation in Integrated Water Resource Management: *The Case of the Lake Victoria Basin:* Department of Agriculture, Moi Institute of Technology.

Pope J., Singh B. and Thomas D., (2006b) Mining related environmental database for West Coast and Southland: *Data structure and preliminary geochemical results,* AUSIMM 2005 Annual Conference: Auckland, 7p.

Redding, C.A., Rossi, J. S., Rossi, S. R., Velicer, W. F. and Prochaska, J.O. (2000) Health Behavior Models. *The International Electronic Journal of Health Education*, *3* (Special Issue):180-193.

Tsiho S., (2007) *Water Pollution in Southern Africa* www.whois.ws/whois_ index/g/domain.

Moyo, N. and Mtetwa, S., (2002) Water Quality Management and Pollution Control. In Hirji, J., Johnson, P., Maro, P. and Matiza Chiuta, T. (eds) 2002. *Defining and Mainstreaming Environmental Sustainability in Southern Africa*. SADC, IUCN, SARDC, World Bank: Maseru/ Harare/ Washington DC.

Silberbauer, M. J., Rossouw, J. N., Kamish, W., Chetty, K., Hadebe S. and Wildemans D., (2001) Building Capacity in Water Quality Modelling in South Africa: Report on a U. S. - South Africa Binational Commission training visit by staff of Ninham Shand, Department of Water Affairs and Forestry and Umgeni Water to the United States of America, September 2001 *Report number*: N/0000/00/USA/0601.

Van Zyl, (1999) Impact on Diffuse Pollution? *Proceeding of the international conference on diffuse pollution in Australia centres*. exeter.ac.uk/cws/ publications.

Van Zyl, B., (2002) Challenges for industry in reaching mine closure. *Paper presented at the WISA Mine Water Division,* Mine Closure Conference, Randfontein, 23–24 October.

West, D., (2001) Nitrates in Ground Water: *A Continuing Issue for Idaho Citizens:* Ground Water Quality Information Series No. 1: Idaho Department of Environmental Quality: State Ground Water Program 1410 N: www.deq.idaho.gov/media/471641-nitrates_issue_citizens.pdf.

Wigrup, I., (2005) The Role of Indigenous Knowledge in Forest Management : *A Case Study from Masol and Sook Division*, West Pokot, Kenya: University essay from SLU/Dept. of Silviculture ex-epsilon.slu.se:8080/.

In: Water Quality
Editor: You-Gan Wang
ISBN: 978-1-62417-111-6

Chapter 8

THE FATE AND PERSISTENCE OF THE ANTIMICROBIAL COMPOUND TRICLOSAN AND ITS INFLUENCE ON WATER QUALITY

Teresa Qiu[1,3], Christopher P. Saint[2,*] and Mary D. Barton[1]
[1]School of Pharmaceutical and Medical Sciences, University of South Australia, Australia
[2]SA Water Centre for Water Management and Reuse, University of South Australia, Australia
[3]Water Quality and Environment, South Australian Water Corporation, Australia

ABSTRACT

The Earth's water resources are coming under increased stress from the combined pressures of quality and quantity and wastewater has an increasingly important impact on these issues. The quality of treated wastewater entering both the marine and freshwater environments can greatly influence the health of natural ecosystems. In addition, a shortage of freshwater supply in drought affected countries has led to an escalation in the use of treated wastewater ("reclaimed water") for the purpose of water supply augmentation for agricultural, industrial and even potable uses. The fate and persistence of chemical contaminants in wastewater is of considerable concern when considering the potential environmental and health impacts of discharges, be they to the open environment or captured for reuse. Of the many types of anthropogenic compounds found in wastewater, pharmaceutical and personal care products (PPCPs) make up a significant portion of encountered pollutants. Of these, triclosan is one of the prevalent compounds known to be recalcitrant to removal or inactivation in wastewater treatment processes. This chapter deals with the significance of triclosan in the wastewater cycle, its persistence and fate, and examines the microbiological processes involved in the removal of this compound. The significance of microbial resistance to triclosan is also considered and how this relates to other types of microbial resistance to antibiotics, and we discuss the potential role of wastewater as a source for the creation of new antibiotic resistant strains of bacteria.

[*] Corresponding author.

1. INTRODUCTION

A recent study estimated that 80% of the world's population is exposed to high levels of threat to water security (Vörösmarty et al., 2010). This is compounded by the fact that many of the world's most densely populated areas are a significant distance from readily available sources of freshwater.

Wastewater is not only a significant source of pollution in the environment but is now increasingly being viewed as a resource with regard to providing an alternative water source or "reclaimed water". The greatest potential sources of industrial and domestic wastewater lie in our cities, so in communities with well developed capture and treatment systems it is a conveniently located source of water, primarily for irrigation and industrial purposes. Most Western countries have well established sewer networks and advanced treatment plants that can produce water of a good quality, however it is known that inorganic and some organic contaminants can enter the environment post-treatment. This is an issue in terms of potential environmental degradation but also increasingly of concern when one considers re-use of treated wastewater with regard to its intended use, something often referred to as "fit for purpose". For example, reducing the amount of treated effluent released to sensitive marine environments is of course a good thing but we also need to be mindful that we are not merely reducing an environmental effect in one location and creating a new issue elsewhere through inappropriate re-use.

Pharmaceutical and Personal Care Products (PPCPs) refers, in general, to any product used by individuals for personal health or cosmetic reasons or used by agribusiness to enhance growth or health of livestock. PPCPs comprise a diverse collection of thousands of chemical substances, including prescription and over-the-counter therapeutic drugs, veterinary drugs, fragrances, and cosmetics. The use of PPCPs is on the rise, for instance in the USA between 1999 and 2009 the annual prescription rate for these compounds virtually doubled from 2 billion to 3.9 billion (Tong et al., 2011). PPCPs largely enter the environment through wastewater disposal whose original source is human or animal excretion or trade waste from industry deposited directly into the environment or via a sewerage and treatment network. PPCPs include the types of compounds referred to as endocrine disrupting compounds (EDCs) and although effects on a variety of vertebrates have been demonstrated a link is still to be made regarding the adverse human health effects of these compounds.

There is also concern about the fate and persistence of medical and veterinary antibiotics in the environment. It is well documented that low level consistent exposure of microorganisms to antibiotics can bring about increased levels of resistance. While the causes of this effect are well documented in medical and veterinary applications, environmental exposure (for example in the wastewater treatment process) and its significance with this regard are largely unknown. Likewise the fate and persistence of antibiotics in the environment and their influence on the overall burden of resistance traits and genes and their dissemination is poorly understood.

A crucial factor in implementing the widespread use of reclaimed water is the economic considerations when designing and operating treatment facilities to provide such water. Whereas, in years to come, the value of any source of water may become so great as to essentially neutralise the economic considerations; at present water reuse schemes are generally expensive to construct, operate and maintain. The water industry and related health authorities are conservative in their approach towards the level of treatment required prior to use and rightly so – as the onus has to be on providing the science to demonstrate that reduced levels of treatment do not result in increased levels of risk. Therefore, there is much interest in researching the various treatment options that will provide a satisfactory level of removal of contaminants such as PPCPs. There is interest in monitoring removal by existing wastewater treatment processes and by alternative, and possibly more cost effective processes, such as biofiltration.

In this chapter we consider the fate and persistence of a major PPCP, namely triclosan. Triclosan has become widely used as an antiseptic and is of significant environmental concern. We consider the history of its use and its mode of action and why there are concerns regarding its continued use. For the reasons mentioned in this introduction we are also interested in the presence of triclosan in wastewater, its removal by treatment processes and persistence in the environment when these processes fail to remove it completely. With regards to biodegradation we discuss what is known about the biochemical and genetic processes that may be involved in triclosan dissimilation. Finally, we consider the significance of triclosan resistance and biodegradation with regard to the role that wastewater and other aquatic systems may play in the spread of antibiotic resistance.

2. What Is Triclosan and Why Is It Used?

2.1. General Properties of Triclosan

Triclosan (*TCS*, $C_{12}H_7Cl_3O_2$, *Irgasan*) is a synthetic broad spectral antimicrobial compound (*General Chemistry Online, 2004*) with a molecular weight of 289.5. Figure 1 below shows the molecular structure of this compound. It belongs to a group of 2-hydroxydiphenyl ethers that is a class of compounds showing antimicrobial activity.

Triclosan was originally introduced to health care settings as a surgical scrub in 1972, and was first included in toothpaste as an additive in Europe in 1985. Since then, triclosan has been used as an active antimicrobial component in many PPCPs, which include detergents, soaps, deodorants, cosmetics, lotions, toothpastes, mouthwashes and plastics for hospital and household use. In recent years, there has been a remarkable increase in the popularity of the addition of antimicrobial chemicals to consumer products, and triclosan is commonly used in products ranging from chopping boards, pizza-cutters, mop handles and mattresses to bowling ball finger inserts and baby toys.

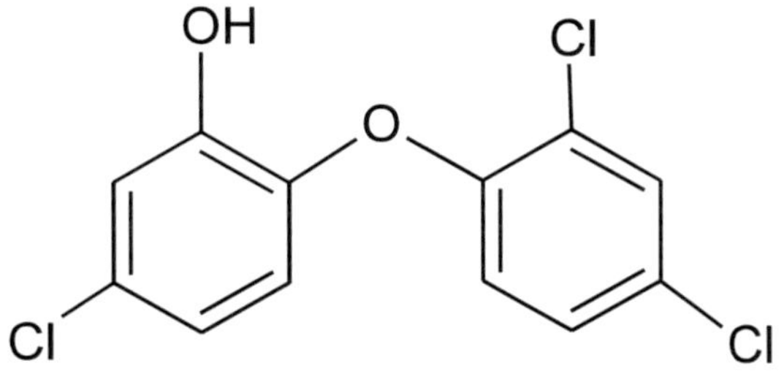

OH Cl O Cl Cl

Figure 1. Molecular structure of triclosan.

Triclosan exhibits a broad spectrum and clinically significant antimicrobial activity (Regos et al., 1979, Vischer and Regos, 1974). It is acutely or chronically toxic to aquatic organisms (Tatarazako et al., 2004), but has limited antiviral and antifungal properties (Samsøe-Petersen et al., 2003).

The solubility of triclosan in water is poor under acidic and neutral conditions, and this poor solubility significantly complicates the study of triclosan. However, triclosan is fat-soluble, i.e. soluble in most organic solvents, and it is also water soluble in alkaline conditions (pH>7). Triclosan is chemically stable, and has a decomposition temperature of 280 °C (Triclosan MSDS).

2.2. Antimicrobial Properties and Mechanisms of Action of Triclosan on Bacteria

Triclosan is active against many, but not all, types of Gram-positive and Gram-negative bacteria. It is bacteriostatic at low concentrations, but at higher concentration it is bactericidal. Many studies have been done to evaluate the effectiveness of triclosan as a disinfectant for dentifrices and hand-wash products as well as its use in clinical antibacterial settings (Barry et al., 1984, Faoagali et al., 1999, Fine et al., 1998, Stephen et al., 1990, Walker et al., 1994, Webster, 1992). Triclosan was found to be an efficient active component in toothpaste to prevent the development of plaque on teeth and in hand washes and scrubs for general hygiene (Aiello et al., 2004, McBain et al., 2003). One of the most significant findings about the biocidal efficiency of triclosan has been its effectiveness in the control of outbreaks of methicillin-resistant *Staphylococcus aureus* (MRSA) (Webster et al., 1994, Zafar et al., 1995).

The efficiency of the antimicrobial activity of triclosan is dependent on several factors. A bacterial population kinetic study (Gomez Escalada et al., 2005b) showed that generally bacterial growth was delayed under triclosan exposure and the extent of delay corresponded to the concentrations to which bacteria were exposed. Interestingly, instead of showing a longer delay effect at higher concentrations, the study showed a longer delay at lower concentrations. This is probably due to the low solubility of triclosan. It was observed in this study that triclosan tends to crystallize from solution at high concentration (i.e. higher than saturation concentration, triclosan solubility in water 0.01g/L at 20°C). Once the crystallization process is initiated, the formed crystal will become a nucleus to promote further crystallization even in an unsaturated solution. This will result in soluble triclosan in the solution being present at a concentration lower than the saturated level. This explains the reason for longer growth delay at low concentrations, where less crystal nuclei are formed and more soluble triclosan is present in the solution. Only soluble triclosan is bio-available and this accounts for the more efficient biocidal effect.

Bacterial growth phase is another key factor that influences the efficacy of triclosan action. Bacteria in the log phase of growth are more susceptible to triclosan compared to bacteria in stationary phase, and washed cells were found to be least affected (Tabak et al., 2007).

Concentration (soluble component) is another very important factor in determining the lethality of triclosan. At high concentrations, triclosan was found to be lethal regardless of the growth phase of the bacterial population and this lethal activity could be effective within a time period as short as 15 seconds (Gomez Escalada et al., 2005b). The effect of triclosan at high concentration is positively related to contact time and triclosan concentration. However, this effective triclosan concentration (represented by Minimal Inhibitory Concentrations) varies among different isolates (Braoudaki and Hilton, 2004a, Gomez Escalada et al., 2005a, McBain et al., 2004, Slayden et al., 2000, Suller and Russell, 2000).

These facts indicate that the mechanisms of triclosan bactericidal activity are complicated and concentration related. Until a specific target was identified in 1998, triclosan had long been thought to act as a non-specific biocide by attacking the bacterial cell envelope and rendering it too porous to retain nutrients, which leads to the death of cells. The publication of McMurry and co-worker's study on the genetic target of triclosan (McMurry et al., 1998), namely *fabI*, brought the study of mechanisms of triclosan resistance into a new era. This study showed that triclosan acts on a specific target in the bacterial fatty acid biosynthetic pathway, namely the NADH-dependent enoyl ACP (acyl carrier protein) reductase (FabI). Subsequent studies investigated various areas of composition, kinetics and genetics of this pathway (Heath et al., 2000, Heath et al., 1999, Kuo, 2003, Levy et al., 1999, McMurry et al., 1999, Parikh et al., 2000, Slayden et al., 2000, Stewart et al., 1999).

Although the capacity of triclosan to target the fatty acid biosynthesis pathway has been well established in recent years, triclosan biocidal action by targeting FabI in the bacterial fatty acid synthesis pathway can only form part of the story of triclosan activity. This mechanism applies only when the concentration of triclosan is not high enough to kill the bacteria immediately, as the inhibition in bacterial fatty acid synthesis results in the inhibition of bacterial proliferation. At this low triclosan concentration even when a complete inhibition of fatty acid synthesis is achieved, triclosan will only inhibit the growth of organisms, but will not kill them (Gomez Escalada et al., 2005a, Gomez Escalada et al., 2005b). Recent studies have shown that the inhibitory factor is actually a ternary complex of triclosan, NAD+ and InhA; therefore, triclosan can only act as a slow binding inhibitor of InhA due to the nature of its requirement to form a complex with NAD+ (Kapoor et al., 2004, Kuo, 2003). However, we can speculate that with the bacteriostatic effect, if lipid synthesis was prevented from taking place over a long time, and no cell membranes could be renewed, the bacteria would eventually die. Thus the bacteriostatic activity of triclosan at lower concentration can still be used as a strategy in hygiene or bacterial growth control, yet it is not enough to achieve an immediate lethal effect at this concentration range. Obviously there are some other more efficient lethal actions of interest, which might be independent or related to fatty acid biosynthesis, that remain to be elucidated.

3. Why Are There Concerns Regarding the Use of Triclosan?

3.1. Use of Triclosan and Resulting Concerns

Triclosan is widely used as a constituent in PPCPs, and so is commonly discharged into the water and wastewater environment via sewage and discharge of waste. Triclosan, at various concentrations, has been detected worldwide in the water and wastewater environment (Boyd et al., 2003, Jackson and Sutton, 2008, Kuster et al., 2008). As discussed later in Section 4.1 triclosan concentrations vary from 0.02 μg/L to 0.21 μg/L in surface waters, and much higher levels of 0.1-21.9 μg/L were detected in sewage plant influents worldwide (Chau et al., 2008, Lishman et al., 2006, McAvoy et al., 2002, Miege et al., 2008, Nagulapally et al., 2009, Okumura and Nishikawa, 1996). Although these concentrations may not be significant enough to select triclosan resistance, the concentration of triclosan in the digested sludge from wastewater treatment plants ranges from 0.5 to 15.6 μg/g (dry wt) (McAvoy et al., 2002), which may be close to the MICs of some bacteria. In addition, a much higher concentration of triclosan is expected to be present in the sewer immediately downstream of households as well as major triclosan users, such as medical clinics. Concerns have thus arisen in the water and wastewater management area that the biocidal property of triclosan against many bacteria and the acute toxicity to algae and other aquatic organisms will significantly influence the balance of the natural ecology in water environments. Another key concern arising because of the presence of triclosan at sub-lethal levels is the potential for selection of bacterial resistance to this bis-phenol (Thompson et al., 2005, Xia et al., 2005).

3.2. Bacterial Resistance to Triclosan

3.2.1. Triclosan Resistance and Adapted Resistance in Bacteria

Many bacteria, such as *Neisseria subflava, Prevotellanigrescens, Porphyromonas gingivalis* and *E. coli, Salmonella, Staphylococcus* etc, are intrinsically susceptible to triclosan at low concentrations (0.1–3.9 mg/L) (Copitch et al., 2010, Randall et al., 2004), whereas *Lactobacillus* and *Streptococcus mutans* were less susceptible (MIC range 15.6–20.8 mg/L) (Randall et al., 2004). Most *Pseudomonas* strains are found to be resistant to triclosan even at high concentrations (Chuanchuen et al., 2002, Ellison et al., 2007).

Acquired resistance to triclosan has also been widely observed across bacterial genera (Ledder et al., 2006), and the trait varies in different species. A highly significant reduction (400-fold) in triclosan susceptibility was achieved in a laboratory *E. coli* strain when exposed to triclosan, whereas only minor (two-fold) decreases in triclosan susceptibility occurred for *Prevotella nigrescens* and *Streptococcus* strains (Randall et al., 2004).

In a recent study, forty environmental and human isolates were tested for effects of chronic triclosan exposure on antimicrobial susceptibility; however only 5 isolates gained acquired resistance (Ledder et al., 2006).

The inconsistency seen in studies on acquired resistance in bacteria under prolonged exposure of triclosan suggests that this adaption is not through a common pathway shared by all bacteria, but it is confined to certain bacterial species that share similar specific pathways.

3.2.2. Mechanisms of Resistance to Triclosan

A recent study investigated the prevalence of decreased susceptibility to triclosan and a panel of antibiotics in over 400 human and animal isolates of *Salmonella enterica*. Seventy four percent of the isolates were shown to be susceptible to triclosan at a concentration of 0.12 mg/L. Of the 111 isolates showing growth on media containing 0.12 mg/L, only 16 strains showed consistent triclosan resistance, but mostly at a low level (MIC 0.25-0.5 mg/L). Only 1 isolate had MIC of 4 mg/L. Of these strains showing reduced triclosan susceptibility, 56% were also multidrug-resistant, and expressed higher efflux activity compared to the strains showing reduced susceptibility only to triclosan. No specific mutations within *fab*I were detected for these multi-drug resistant (MDR) strains (Copitch et al., 2010). This indicates that triclosan resistance maybe a result of generic resistance mechanisms that can be promoted by a multidrug efflux system. However, this investigation was restricted to *Salmonella* strains and mechanisms in other bacteria could be significantly different.

Pseudomonas is intrinsically resistant to triclosan, and the mechanisms involved in triclosan resistance as well as cross resistance to other antibiotics have been studied extensively. So far, the most commonly accepted mechanisms of reduced susceptibility to triclosan and cross-resistance to antibiotics in *Pseudomonas* are mutation of *fabI* and over-expression of a Resistance Nodulation Division (RND) efflux system with more emphasis on the latter.

3.2.3. Resistance via Biodegradation?

Theoretically, if bacteria have the ability to decompose triclosan at a specific concentration, then they will automatically be resistant to this compound at this concentration, as for biodegradation to occur, bacteria have to physically interact with triclosan without being killed. Resistance to antibiotics can occur in several ways: by mutations that render the site of antibiotic action inaccessible, or alter the target site of action so the antimicrobial agent cannot bind, or by the action of inactivation enzymes carried by the target bacterium.

Therefore, if one considers biodegradation a type of resistance this can only be proffered via an enzymatic pathway. Depending on the pathways endowed the breakdown (inactivation) may also provide a source of energy for the host microorganism thus imparting a dual benefit. The advantage of such a mechanism over mere resistance is that it provides the microorganism with a selective advantage allowing it to boost its population; this in turn would actually result in greatly enhanced triclosan removal from the immediate environment. However, it is worth noting that the intermediates formed during the biodegradation process could also be toxic and so may demonstrate inhibition to the growth of the degraders (Aranami and Readman, 2007, Canosa et al., 2005). In this situation, triclosan biodegradation would actually function as an inhibitory mechanism for these bacteria.

4. Triclosan in the Environment

4.1. Detection of Triclosan and its Effects on the Water and Wastewater Environment

Presence of triclosan in water environments has been reported worldwide (Boyd et al., 2003, Chau et al., 2008, Federle et al., 2002, Jackson and Sutton, 2008, Lishman et al., 2006, McAvoy et al., 2002, Sabaliunas et al., 2003, Samsøe-Petersen et al., 2003, Singer et al., 2002, Xia et al., 2005). Table 1 below summarizes the levels of triclosan reported recently in water and wastewater environments.

It can be seen that triclosan is largely present in the wastewater environment and is removed during wastewater treatment processes, which is indicated by the significant drop of triclosan concentrations between the influent and effluent of sewage plants. Furthermore, when discharged into a natural water body, the triclosan-containing effluent from sewage plants is also diluted in the surface water by natural dilution and the self-purification capacity of the natural water body. The resulting concentrations of triclosan in surface water after the wastewater treatment process and the natural process thus fall far below inhibitory concentration levels and thus will not have a lethal effect on environmental microorganisms and other ecosystems.

However, the predicted no effect concentration (PNEC) of triclosan for algae is as low as 50 ng/L (Singer et al., 2002). Therefore, even after all the natural and artificial removal processes, the triclosan levels listed in these studies are still significantly above this PNEC level.

Therefore, the presence of triclosan in the natural water environment as well as in sewage plants may potentially have an impact on the ecological environment (Tatarazako et al., 2004). A recent study (Thompson et al., 2005) highlighted the necessity for further investigations into the effect of residuals, at concentrations far below inhibitory concentration, on bacterial populations and their role, if any, in the continued problem of antibiotic resistance.

Concentrations of triclosan in local sewage collection systems leading to the sewage plant could be much higher than those listed as influent concentrations of triclosan in Table 1 due to the well documented adsorptive property of this substrate (Federle et al., 2002, Heidler and Halden, 2007, Singer et al., 2002). The concentration of triclosan measured in sewage plant influent may therefore be lower than that in the local sewer system, because of the adsorption of triclosan onto the surface of organic material, such as polyvinyl chloride (PVC) pipes. Therefore, the sub-lethal concentration of triclosan immediately or shortly after discharge from homes could act as an inducer of the expression of bacterial multidrug resistance determinants, such as multidrug efflux. The potential for triclosan to select for antibiotic resistant bacteria has been addressed by scientists all over the world. Various studies have reported an increase of triclosan resistance and a possible reduction in susceptibilities to antibiotics after exposure to triclosan (Braoudaki and Hilton, 2004b, Chuanchuen et al., 2001, Schweizer, 2001).

In the natural water environment, apart from the potential selection of antibiotic resistant bacteria, the presence of triclosan has also raised various other concerns about the impact that triclosan could have on the ecological environment. For example, triclosan is acutely or chronically toxic to some aquatic organisms, as in the case of acute toxicity to algae at trace levels, i.e. PNEC of 50 ng/L (Tatarazako et al., 2004).

Table 1. Occurrence of triclosan in the surface water and wastewater environment as reported in the literature (Chau et al., 2008, Lishman et al., 2006, McAvoy et al., 2002, Miege et al., 2008, Nagulapally et al., 2009, Okumura and Nishikawa, 1996)

	Surface Water (μg/L)	Sewage Plant	
		Influent (μg/L)	Effluent (μg/L)
Europe	0.04-0.21	0.6-21.9	0.11-1.1
USA	0.021	3.8-16.6	0.2-2.7
Japan	0.05-0.15	NA	NA
China	0.004-0.117	NA	NA
Canada		<4.01	<0.324
France			0.09-0.43

Triclosan is also found to have a high potential for bio-accumulation (Coogan et al., 2007). In a Japanese study, a range of 0.89-2.5 μg/kg of triclosan was detected in fish tissues (Okumura and Nishikawa, 1996), and a range of 0 to 2100 μg/kg in lipid was found in human breast milk samples (Dayan, 2007). Despite the lack of knowledge on how this accumulation of triclosan will affect human or environmental health, it is significant enough to cause an alert on public health grounds.

Recently, the formation of dioxin has been reported in a natural water body as a result of chlorination of triclosan in the presence of ultraviolet light (UV) (Aranami and Readman, 2007, Chung et al., 2007, Lores et al., 2005, Son et al., 2007, Yu et al., 2006). These much more toxic triclosan derivatives will have greater impacts on the aquatic fauna.

The presence of triclosan in the wastewater environment may have similar impacts as that in the natural water body as described above. In addition, it may also have significant impact on the performance of the wastewater treatment plant.

Engineered wastewater treatment processes represent artificial enhancement of biological processes that occur spontaneously in the natural water environment. It employs a biological process as the principal treatment process, where the behaviour of the microbiological community dictates the performance of the processes. In particular, for conventional wastewater treatment processes, the nutrient removal efficiency relies largely on the activity of microorganisms in the activated sludge bio-reactors. Therefore, a selective pressure from the presence of the biocidal effect of triclosan will potentially provide an inhibitory impact on the biological environment and affect the bacterial community profile as well as the viability of the microorganisms. This will affect the efficiency of the treatment process, particularly when triclosan enters the process as a matter of shock load, when a high level of triclosan enters within a short period of time.

There are limited studies investigating the effect of triclosan on the performance of the wastewater treatment process. A bench-top study showed that the load of triclosan at levels expected in household and manufacturing wastewaters is unlikely to upset sewage treatment processes (Federle et al., 2002).

However, it is noteworthy that in this study, the sewage treatment process tested was a triclosan pre-acclimated system (dosed with additional triclosan), which was evaluated after the adaptation to triclosan had already occurred. Therefore, the result does not reflect the real situation.

A recent study (Heidler and Halden, 2007) revealed that a large proportion of particle-active triclosan (composed of 80±22% on average of the total triclosan inflow) was bound to particulates and eventually sequestered and accumulated in excess sludge in the wastewater treatment process.

This study therefore raised concerns about the practice of applying large quantities of triclosan-containing sludge to soils used for animal husbandry and crop production, and consequent concerns over the ecological, environmental and health impacts. However, in this study, no information on sludge digestion was given; therefore it is hard to determine if the concern is valid. In another study (McAvoy et al., 2002) investigating triclosan in digested sludge, it was revealed that triclosan was adequately removed from digested sludge prior to disposal for land application.

To conclude, the presence of triclosan in the water and wastewater environment represents a potential risk to public health and the natural environment. The efficient removal through sewage treatment plants is thus critical to mitigate the risk.

4.2. Wastewater Treatment Process and Triclosan Removal Efficiency

A conventional wastewater treatment process train includes preliminary treatment, primary treatment, secondary treatment and sludge treatment (Wiesmann et al., 2007). The preliminary treatment normally includes screens and a silt settling basin, which removes large objects and a proportion of the suspended solids. Most of the suspended solids are removed through the primary sedimentation basin, where a proportion of organics that attached to the suspended solids are also removed.

Secondary treatment is considered as the main process for removal of organics and other nutrients. Conventional secondary treatment includes aeration and sedimentation. During the start-up of a new system, the indigenous microorganisms in the source water are concentrated; the microorganisms that are capable of utilizing organics and other nutrients in the source water proliferate and outgrow other microorganisms. By working synergistically, the dominant microorganisms gradually evolve the ability to utilize the organic compounds in the source water as carbon sources and thus treat the water.

In the aeration basin, aerobic microorganisms utilize oxygen, organics and other nutrients in the water to proliferate. As a result, biomass increases in amount and the water is treated. Downstream to the aeration basin, biomass is settled in the secondary sedimentation basin. A proportion of the settled biomass is removed from the water treatment system before further treatment in the sludge treatment process prior to disposal, while another portion of the settled sludge from the secondary sedimentation basin is recycled into the aeration basin as inoculum to facilitate the biological process in the aeration basin. Excess sludge from the secondary sedimentation tank is digested using aerobic and/or anaerobic processes to remove organics and nutrients through nitrification.

Triclosan may be removed by wastewater treatment processes together with other organic compounds in the water. Therefore its removal efficiency is related to the overall organics removal efficiency of the treatment process. Table 2 lists a brief summary of reports on triclosan removal from wastewater treatment plants in different geographic locations. It is assumed that these wastewater treatment plants accepted influent streams composed of mostly domestic wastewater and they all operated well. In all processes, Biological Oxygen Demand (BOD) and Total Suspended Solids (TSS) removal were steady and efficient.

From the data listed it can be seen that a good removal rate (over 90%) of triclosan may be achieved by most secondary biological treatment processes. Of the commonly used wastewater treatment processes, conventional activated sludge treatment demonstrated reliable performance for triclosan removal with an effluent concentration of around 1μg/L even at a high triclosan load in the influent.

A high removal percentage of 95% can be achieved in most activated sludge processes. This is probably due to the high Hydraulic Retention Time (HRT) and Sludge Retention Time (SRT), commonly seen in most conventional activated sludge processes. In some activated sludge processes designed for enhanced removal of phosphorus and nitrogen, the HRT can be maintained up to 12-18 hours and the SRT up to 15-30 days. Trickling filters and biological contactors normally have less HRT, and therefore are slightly less efficient in triclosan removal. Other parameters of operational conditions like temperature at the Waste Water Treatment Plant (WWTP) will also influence the efficiency of the treatment processes. Nevertheless, efficient removal of triclosan from wastewater can be achieved in most secondary treatment processes (as shown inTable 2).

Table 2. Triclosan Removal Efficiency from Sewage by Wastewater Treatment Processes (Bester, 2003, McAvoy et al., 2002, Sabaliunas et al., 2003)

Type of Works	Percentage removal	Influent conc. (μg/L)	Effluent conc. (μg/L)	Source
Activated Sludge	95%	21.9	1.1	UK
Activated Sludge	83%	5.4	0.9	UK
Activated Sludge	98%	4.7	0.07	USA
Activated Sludge	96%	1.2	0.05	Germany
Activated Sludge	95%	5.2	0.2	USA
Activated Sludge	96%	10.7	0.41	USA
Trickling filter	58%	3.8	1.6	USA
Trickling filter	86%	16.6	2.1	USA
Trickling filter	83%	15.4	2.7	USA
Trickling filter	79%	2.8	0.6	UK
Trickling filter	95%	7.5	0.34	UK
Biological contactor	58-96%		0.15-1.1	

However, we can also see from this table that downstream of a successful triclosan removal treatment process, there are still hundreds or thousands of nano-grams of triclosan per litre of effluent, and this concentration will still cast an influence on the natural water environment as discussed previously. Therefore, investigations into triclosan removal mechanisms in these wastewater treatment processes are necessary in order to develop strategies for an enhanced triclosan removal during these processes or to identify some advanced supplementary treatment process to ensure that the residual concentration of triclosan in the wastewater environment has minimum adverse impact on the natural environment.

4.3. Mechanisms Involved in Triclosan Removal from Activated Sludge Wastewater Processes

4.3.1. Flow of Triclosan in Activated Sludge Wastewater Treatment Process

Triclosan enters wastewater systems in the form of soluble triclosan in the aquatic phase and also particulate-active mode. A molecular level of adsorption and desorption process occurs at all times, and the ratio of soluble and particulate triclosan remains kinetically balanced according to the property of the carrying water and the performance of the treatment during the whole process.

Triclosan in wastewater will undergo a biological treatment process as described previously, where physical adsorption and biological degradation work synergistically for the removal of triclosan. At the biological treatment process, primary wastewater is mixed with activated sludge at the beginning of the process, where soluble triclosan is adsorbed on activated sludge. This adsorption is expected to be effective due to the polarity of triclosan. At the same time, the large percentage of triclosan in particle form will be mixed and trapped into activated sludge. The active component of activated sludge is the consortium of microorganisms that are able to utilize the contaminants in water as nutrients. Under optimal conditions provided by the bioreactor, these microorganisms will utilize triclosan and other organics to metabolize and proliferate. During this process, triclosan is biodegraded, and the remaining triclosan will be separated and sediment to the bottom of the reactor and is disposed as excessive sludge at the end of the water treatment process. The excessive sludge will then undergo an aerobic and anaerobic digestion, where contaminants in the sludge are further degraded prior to disposal.

4.3.2. Variable Contribution of Different Triclosan Removal Mechanisms

Despite the fact that the triclosan removal rate in most activated sludge treatment processes remains consistently high (95-98%) (Bester, 2003, McAvoy et al., 2002, Sabaliunas et al., 2003), the mechanisms of this removal have been subject to debate. In a recent mass balance study on triclosan removal from conventional wastewater treatment processes (Heidler and Halden, 2007), an overall of 50±19% of the triclosan entering the plant was found to remain in the excessive sludge after the whole treatment process, and only 48±16% of the triclosan intake was biodegraded or transformed through other mechanisms. However, in a bench-top based study (Federle et al., 2002) on continuous activated-sludge systems, biodegradation and removal of triclosan radio-labelled with C^{14} was monitored, and the results showed that, over a series of triclosan concentrations tested, 94-99.3% of triclosan in this system underwent primary biodegradation, and only 1.5-4.5% was adsorbed to the waste solids. The conflict in this literature suggests that although the mechanisms involved in triclosan removal are similar, the dominant mechanism may vary in different systems, and they may to some extent influence the removal efficiency.

For an existing wastewater treatment plant, the efficiency of treatment depends on lots of factors such as property of wastewater influent, property of activated sludge and operational conditions including aeration rate, temperature etc. The adaptation of activated sludge to a specific wastewater influent is very important for a successful treatment of this wastewater. As described previously, hydraulic retention time (HRT) will substantially influence the efficiency of secondary wastewater treatment.

4.3.3. Biodegradation as a Primary Mechanism to Be Promoted for Triclosan Removal

All studies on the efficiency of removal of triclosan from wastewater treatment systems are based on the concentration of triclosan residual in the water phase of the effluent after solids/liquid separation occurs in the secondary sedimentation process. Because of its polarity, triclosan is readily absorbed into activated sludge, which guarantees a good removal rate of triclosan (over 90%) (Bester, 2003, McAvoy et al., 2002, Sabaliunas et al., 2003) seen in most conventional activated sludge processes.

However, better or complete removal of triclosan requires faster biodegradation rate and longer HRT. Fast biodegradation also leads to a rapid and sustained decrease of triclosan concentration in the excess sludge, where biodegradation also occurs. As discussed before, a kinetic balance of triclosan exists between the water phase and the sludge phase, so more triclosan from water will be absorbed into the sludge phase, if the degradation of triclosan in the sludge phase takes place at a higher rate. This will consequently reduce the triclosan concentration in the treated water.

This theory is consistent with published experimental results (Federle et al., 2002), which described that adaptation of activated sludge to triclosan favoured its removal and it is related to the overall treatment efficiency of the system. Based on the above theory, variable efficiency in the removal of triclosan in treatment processes described in the previous section could be largely attributed to the properties of activated sludge and variable HRTs in different systems. These two factors determine the extent of biodegradation that occurs in the sludge. The better acclimatised sludge has more degradation capacity, and decomposes triclosan more efficiently, thus achieving a better removal in a system running at a specific HRT. For one system with pre-established activated sludge composition, the longer the HRT, the more biodegradation in sludge will occur, and a higher removal rate will be achieved.

This theory also explains the conflicting results of studies on the efficiency of triclosan removal from excess sludge in the sludge treatment process. A recent mass balance assessment of triclosan removal from the wastewater treatment process (Heidler and Halden, 2007) revealed that despite the good removal rate of triclosan from the liquid phase, a large proportion of particle-active triclosan (composed of 80±22% on average of the total triclosan inflow) was bound to particulates and eventually sequestered and accumulated in excess sludge in the process. In contrast, in another study (McAvoy et al., 2002) investigating triclosan in digested sludge, it was revealed that triclosan was adequately removed from digested sludge particularly when using an aerobic digestion process. Removal of triclosan from digested sludge under anaerobic conditions was much less efficient.

The above comparison indicates that an aerobic biodegradation of triclosan took place in the McAvoy study, where a good triclosan removal was achieved in the sludge treatment; whereas no or minimum biodegradation of triclosan occurred in the Heidler investigation and triclosan accumulated in the excess sludge. It becomes obvious that biodegradation is a key mechanism for assured triclosan removal efficiency in the wastewater and sludge treatment process. Additionally, it is also a more environmentally friendly and sustainable mechanism that should be promoted in the triclosan treatment process.

As described above, triclosan biodegradation occurs naturally as well as in wastewater treatment plants. However, the performance varies between plants. Understanding of the mechanisms involved in triclosan biodegradation will assist in providing information for process optimization, improve triclosan removal efficiency, and consequently reduce the total quantity of triclosan present in the environment.

5. The Biochemistry and Genetics of Triclosan Degradation–What is Known?

5.1. Microorganisms Involved in Triclosan Biodegradation and Possible Pathways

Transformation of triclosan in the natural water environment involves photolytic degradation (Aranami and Readman, 2007), chlorination (Canosa et al., 2005), and methylation (Coogan et al., 2007). Apart from these degradation pathways that occur naturally in the water environment, a small scale laboratory based investigation was carried out on sonochemical degradation of triclosan in various environmental samples including seawater, urban runoff and influent domestic wastewater, as well as pure and saline water (Sanchez-Prado et al., 2008). A complete removal of triclosan was achieved in 120 minutes of reaction from most samples spiked with 5 μg/L of triclosan, except the influent domestic wastewater for which only 60% of removal was achieved within 180 minutes. No chlorinated and other toxic by-products were detected under the conditions studied (Sanchez-Prado et al., 2008).

Recently, triclosan biodegradation in soil has been investigated under aerobic and anaerobic conditions (Ying et al., 2007). In this study, a gradual decrease of triclosan concentration in non-sterile soil was observed from 1.01mg/kg at the beginning of the experiment to 0.08 mg/kg over 70 days of the investigation period. Interestingly, a consistently higher dehydrogenase activity was also observed in the triclosan incorporated soil compared to the trichlocarban incorporated soil, which showed much poorer removal of trichlocarbon. Aerobic triclosan biodegradation was also discussed in a study on triclosan removal during the wastewater treatment process (McAvoy et al., 2002). Much higher triclosan contents in the anaerobically digested sludge than found in aerobic sludge indicated a much more efficient triclosan biodegradation under aerobic conditions.

A wastewater consortium was found to be able to degrade triclosan (Hay et al., 2001). Two bacterial strains of *Pseudomonas putida* and *Alcaligenes xylosoxidans* subsp. *denitrificans* were isolated from soil and they were demonstrated to inactivate triclosan in liquid and on solid substrates and to use triclosan as a sole carbon source (Meade et al., 2001). Biotransformation of triclosan by fungi is also reported. The fungus *Pycnoporus cinnabarinus* was found to methylate the hydroxyl group of triclosan during cultivation, and the fungus *Trametes versicolor* metabolized triclosan and produced three metabolites, namely 2-O-(2,4,4'-tichlorodiphenyl ether)-β-D-xylopyranoside, 2-O-(2,4,4'-trichlorodiphenyl ether)-β-D-glucopyranoside, and 2,4-dichlorophenol (Hundt et al., 2000).

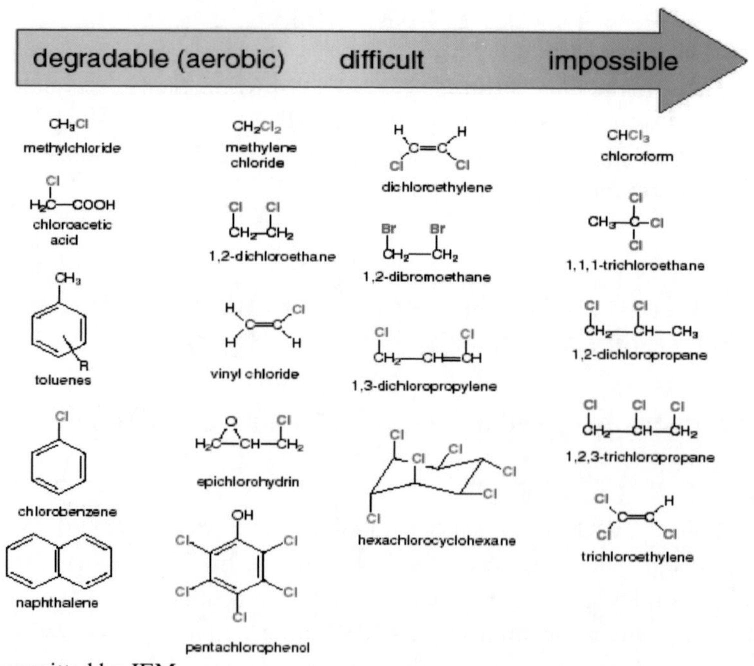

Reproduction permitted by JEM.

Figure 2. Degradability of chlorinated organic compounds by biological metabolism (Janssen et al., 2005).

The degradability or recalcitrance of chlorinated organic compounds was reviewed by Janssen (Janssen et al., 2005), where chlorinated compounds were classified by their degradability as illustrated in Figure 2.

Based on these theories and studies, triclosan could be considered as a moderately difficult compound that could be selectively degraded by bacteria after prolonged exposure.

Dehalogenation, irrespective of the mechanism (hydrolysis, oxidation or reduction) is an important step in the biodegradation pathway of a halogenated compound. Often the loss of halogen from a halogenated compound as a result of dehalogenase activity initiates a potential catabolic pathway for this compound, although the toxicity of intermediates of this pathway could also be key determinants in the maturity of this pathway, particularly for compounds carrying multiple halogen groups (van Hylckama Vlieg, 2000). Chlorinated aliphatic compounds can be utilized by both aerobic bacteria and strict anaerobic bacteria as growth substrates (Altenschmidt and Fuchs, 1992, Biegert and Fuchs, 1995, Caldwell et al., 1999, Leutwein and Heider, 1999, Li et al., 2010, Lovley and Lonergan, 1990, Pries et al., 1994). Some compounds, such as chloroacetate and 2-chloropropionate can be directly utilized by bacteria in pure cultures, while others, like di-chlorinated allyl alcohol and tri-chloroallyl alcohol, were found to be involved in a co-metabolisation pathway (van der Waarde et al., 1994). Chloroacetate was found to be only co-metabolized by one *Pseudomonas* strain, which can only degrade this compound in the presence of another readily metabolisable carbon source. By comparison, one proteobacteria from the r-subgroup was isolated and was successfully adapted to live on trichloroacetate as the only carbon and energy source (Laturnus et al., 2005, Yu and Welander, 1995). Interestingly, this bacterium cannot utilize monochloroacetate or dichloroacetate as the sole carbon source (Yu and Welander, 1995). Much less data is available on degradation of chlorinated compounds by anaerobic bacteria, partly due to the difficulty in culturing and studying anaerobes. An inducible enzyme was found to be responsible for transferring the methyl group of chloromethane onto tetrahydrofolate and therefore provide methyl-tetrahydrofolate as a substrate in the acetate synthesis pathway in a strict anaerobe (Meßmer et al., 1996). Some reductive dehalogenases were isolated and identified in anaerobic bacteria, such as *Dehalospirillum multivorans* and *Desulfomonile tiedjei* (Louie et al., 1997, Neumann et al., 1995).

In addition, it is well known that aromatic hydrocarbons and their substituted derivatives must be modified into o-diphenols through different peripheral reactions before ring cleavage can take place (Galli, 1996). Aromatic ring cleavage occurs through intradiol or extradiol ring cleavage pathways and a few homologous catechol 2,3-dioxygenases in different bacterial isolates have been cloned and characterised (Arensdorf and Focht, 1994, Bartilson and Shingler, 1989, Di Gioia et al., 1998, Kitayama et al., 1996, Moon et al., 1995). Following ring cleavage products are further metabolised via the TCA cycle.

5.2. Isolation and Growth of *P. citronellolis* F12 on Triclosan

We have performed enrichments from activated sewage sludge of triclosan degrading bacteria. Five hundred mL of activated sludge was added to a 1 litre flask along with 1g of triclosan (2g/L final concentration). The flask was incubated at 35°C with shaking at 160 rpm. The disappearance of triclosan was followed using HPLC-UV and when levels reduced to c.200mg/L the flask was augmented with an additional 1 g of triclosan.

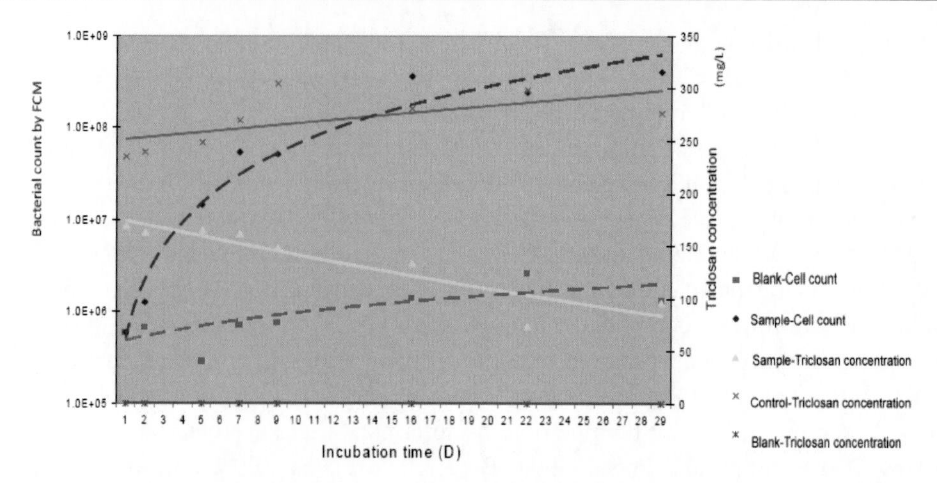

Figure 3. Growth of *P. citronellolis* F12 in Triclosan minimal medium containing 300mg/L of triclosan.

This process was maintained for 3 months following which colony isolation was performed on minimal agar containing 0.05% casamino acids and 1g/L triclosan. An isolate (F12) was obtained from this enrichment that was identified by 16S rRNA gene sequencing as *Pseudomonas citronellolis* and named *P. citronellolis* F12. Members of the genus *Pseudomonas* are fast-growing nutritionally versatile bacteria that are capable of utilizing a wide variety of carbon sources (Lloyd-Jones et al., 2005), and are capable of the degradation of environmentally important organic compounds (Wackett, 2003).

Figure 3 shows the growth curve for strain F12 in triclosan minimal medium broth. An increase in bacterial cell count and a corresponding decrease of triclosan concentration in the culture medium was observed over a 29 day period as monitored by flow cytometric cell counts. For the two control series, there was no bacterial growth evident in minimal medium without triclosan and killed bacterial cells inoculated into triclosan minimal medium showed a zero viable cell count and a constantly high concentration of triclosan in the culture medium. This indicated that the decrease of triclosan in the culture was due to the growth of isolate F12.

5.3. Tentative Identification of Biodegradation Intermediates

In this study we applied HPLC-UV, LC-MS and GC-MS to the identification and quantification of triclosan in liquid medium and detection of putative intermediates respectively. The putative triclosan biodegradation intermediates or by-products were tentatively identified as 2',4'-dichlorophenyl 4-chloro-6-oxo-2,4-hexadienoate, 4-methylene but-2-en-4-olide (or proto-anemonin), 2,4-dichloro-phenol (Figure 4) and an unknown compound having close mass spectrum to trichlorobenzene. The identified intermediates are all possible breakdown products of triclosan. *Ortho*-cleavage of the hydroxyl substituted ring of triclosan could yield 2',4'-dichlorophenyl 4-chloro-6-oxo-2,4-hexadienoate, subsequent hydrolysis 2,4-dicholorophenol and dehalogenation of the remaining 3-chloromuconic acid would yield protoanemonin. However, at this stage this breakdown pathway is speculative and needs to be supported by further high level analysis of intermediates, possibly by nuclear magnetic resonance analysis.

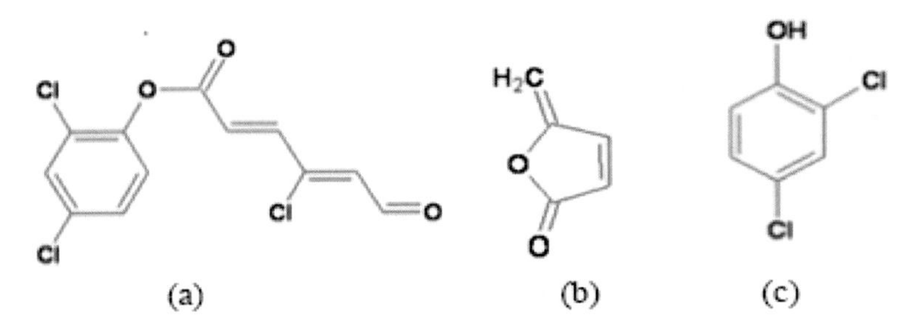

Figure 4. Putative triclosan biodegradation intermediates or by-products. (a) 2',4'-dichlorophenyl 4-chloro-6-oxo-2,4-hexadienoate. (b) 4-methylene but-2-en-4-olide (or proto-anemonin). (c) 2,4-dichloro-phenol.

5.4. Genes Involved in Triclosan Catabolism are Plasmid Encoded

It has been known for over 30 years now that genes involved in the catabolism of xenobiotic compounds can be associated with extra-chromosomal genetic elements known as plasmids (Downing and Broda, 1980). In fact, with regard to what is understood about the degradation of chlorinated organics, genes for their dissimilation are almost always plasmid encoded.

The interest in bacterial catabolic plasmids that are naturally present in soil or water bodies has grown over this time, due to the increasing amounts of toxic halogenated organic pollutants released to the environment in the last few decades (Bernhard et al., 2008, Feakin et al., 1994, Fewson, 1988, Haro and de Lorenzo, 2001, Kobayashi et al., 2009, Okpokwasili and Olisa, 1991, Sinton et al., 1986).

Plasmids, as well as other mobile genetic elements in environmental bacteria, have been demonstrated to play an important role in the acquisition of new catabolic functions. Through the evolution of different catabolic pathways such bacteria are important in the degradation of a range of environmental pollutants (Nojiri et al., 2004).

We undertook an examination of *P. citronellolis* F12 to ascertain whether its ability to degrade triclosan was directed by genes encoded on a catabolic plasmid. A plasmid was isolated from F12 using standard DNA extraction procedures (Qiu, 2012) and subjected to restriction enzyme analysis and gel electrophoresis, a plasmid of c.17kb in size was identified (Figure 5).The plasmid in the wild type F12 strain was designated pF12.

Plasmid associated genes are often lost at relatively high frequencies by a process known as plasmid curing. Essentially, this involves growing the host strain under non-selective conditions that promote plasmid loss. The maintenance and replication of plasmid DNA in the bacterial cell becomes a burden under non-selective conditions and cells that have spontaneously lost the plasmid gain a competitive advantage and can become amplified in the population as a whole due to their preferential growth rate. These "cured" cells can be isolated and identified in the laboratory. Growth at elevated temperature is known to be effective in removing plasmids from their hosts and we ascertained that strain F12 grew slowly at 46°C but not at all at 48°C. The strain was inoculated into LB broth and subcultured 3 times allowing for growth through at least 30 generations. Screening of subsequent colonies derived from nutrient agar revealed that approximately 10% of the colonies no longer grew on triclosan.

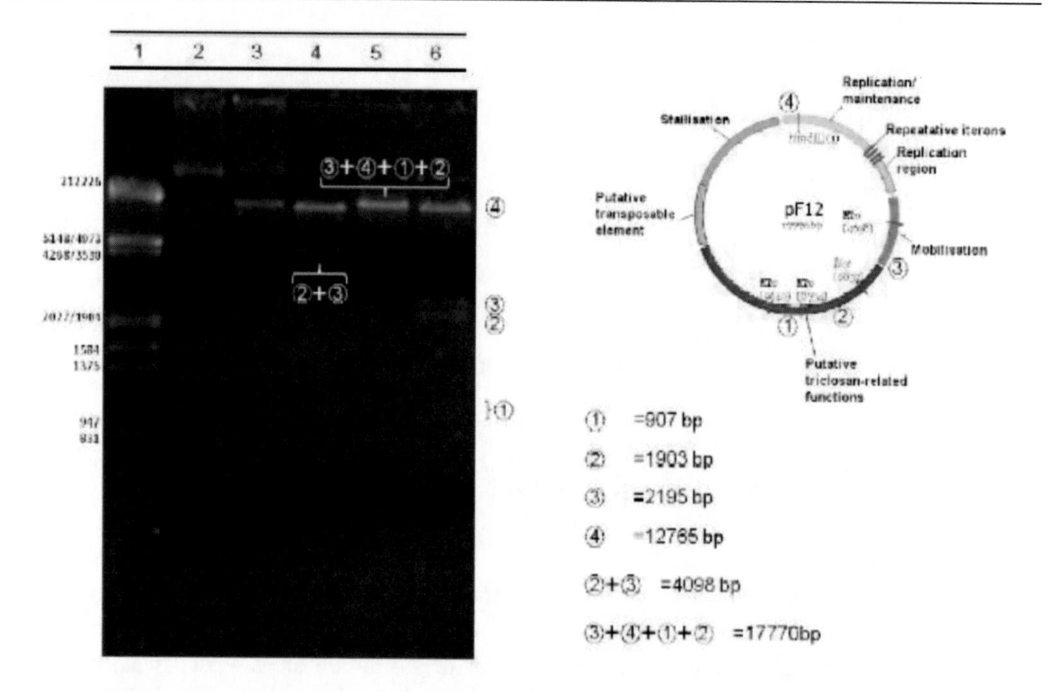

Figure 5. Characterisation of pF12. From left to right: Lane 1 DNA molecular weight marker III (0.12-21.2kbp) (λDNA•EcoRI and Hind III digested); Lane 2 Plasmid preparation from F12; Lane 3 Mock digestion of pF12 (no enzyme but nuclear free water was added); Lane 4 EcoRI digestion of pF12; Lane 5 NotI digestion of pF12; Lane 6 EcoRI and NotI double digestion of pF12.

Screening of a selection of these colonies by DNA extraction and restriction enzyme analysis revealed that all had lost the17Kb plasmid previously present in the F12 strain. One of these isolates was confirmed by 16S rRNA gene sequencing as being otherwise identical to the F12 strain. Subsequently, pF12 was completely DNA sequenced using standard procedures (Qiu, 2012) and found to be 17.77Kb in size.

The DNA sequence derived for pF12 was compared to known sequences within the DNA databases and a preliminary map outlining the functional regions of the plasmid constructed (Figure 5). The catabolic region involved in triclosan degradation is situated in a region just under 6 Kb in size. Analysis so far has revealed this region could encode several hypothetical proteins with known homologies to enzymes involved in catabolic functions. However, it is unlikely that this region could encode the entire suite of catabolic enzymes required for the complete mineralisation of triclosan. This constitutes the first report of the isolation and preliminary characterisation of a plasmid directly involved in triclosan dissimilation.

6. Triclosan and Selection of Antimicrobial Resistant Bacteria

6.1. Evidence for Triclosan Induced Multiple Resistance

Apart from concerns about intrinsic and acquired resistance to triclosan in various bacterial species, there is increasing concern that resistance to triclosan could select for resistance to clinically important antimicrobials. In 1991 a study reported that methicillin-resistant *Staphylococcus aureus* with low-level resistance to triclosan that had been isolated from patients receiving daily triclosan baths were also resistant to mupirocin (Cookson et al., 1991). Levy and his group (2002) investigated the link between mutations in efflux pump related genes and significant increases in triclosan resistance in *Esherichia coli*. Efflux pumps are associated with resistance to a diverse range of antimicrobials including fluoroquinolones, tetracyclines and glycylcyclines, chloramphenicol and florfenicol, macrolides, lincosamides and ketolides, β-lactams, aminoglycosides and oxazolinidones (Poole, 2005). However, studies demonstrating a clinical link between triclosan resistance and resistance to other antimicrobials are rare. In laboratory studies Chuanchuen *et al (2001)* found that *Pseudomonas aeruginosa* strains resistant to triclosan because of the presence of specific efflux pumps were also resistant to a number of antimicrobials. Similarly, resistance to quinolones was found to be induced in *Stenotrophomonas maltophilia* by use of triclosan and the up-regulation of efflux pumps (Hernández et al., 2011).

However, Cottell *et al* (2009) did not find any increased resistance to triclosan-tolerant strains of *E. coli*, *Staphylococcus aureus* or *Acinetobacter johnsonii*. In fact they noted that triclosan-tolerant strains of *E. coli* were more susceptible to gentamicin than wild-type strains. Similarly, Suller and Russell (2000) could not replicate the work of Cookson et al. (1991) and were unable to demonstrate any link between triclosan resistance and resistance to mupirocin or other antimicrobials. Other workers have reported similar findings with *S. aureus* and MRSA (Yazdankhah et al., 2006). It should be noted here that efflux pumps are not the only mechanism of resistance to triclosan, with mutations in Fab1, an enzyme in the fatty acid biosynthetic pathway, also important. In contrast to the *S. aureus* studies, there is a recent report of sublethal exposure of *Listeria monocytogenes* to triclosan resulting in decreased resistance to gentamicin and other aminoglycosides (Christensen et al., 2011). The detailed molecular mechanisms remain to be elucidated.

However, of concern is a very recent report of an apparent link between triclosan resistance and antibiotic resistance in a clinical setting. Strains of *Pseudomonas aeruginosa* isolated from an epidemic in immunocompromised patients in an oncohaematology ward were of identical genotypes and showed identical resistance patterns (resistant to fluoroquinolones and aminoglycosides) as well as high level resistance to triclosan. The source of infection was found to be a common soap dispenser which at the time of the outbreak contained triclosan soap (D'Arezzo S et al., 2012). So while the evidence is still scant it seems prudent to be mindful of the potential for selection of multiple antimicrobial resistance in pathogens from exposure to triclosan.

6.2. Could Wastewater Play a Role in the Amplification of Antibiotic Resistant Bacteria?

Antibiotic resistant bacteria and antibiotic residues have been found in wastewater (Baquero et al., 2008). Many of the studies concentrated on faecal organisms and/or human pathogens. Enterococci resistant to tetracyclines and fluoroquinolones and faecal coliforms resistant to amoxicillin, tetracyclines and ciprofloxacin were found by Ferreira da Siva et al., (2006, 2007). Lefkowitz and Duran (2009) reported resistant strains of *E. coli* in the different stages of wastewater treatment, with significant resistance detected to tetracyclines, streptomycin and beta-lactams. They noted that the percentage of isolates with multiple resistance increased through the treatment process. Faecal coliforms resistant to fluoroquinolones and trimethoprim-sulphamethoxazole and enterococci (principally *E. faecalis*) resistant to vancomycin were found by Nagulapally et al. (2009) and vancomycin resistant enterococci have also been detected in sewage sludge (Sahlström et al., 2009). Studies of hospital sewage treatment plants have yielded extended-spectrum β-lactamase producing organisms (Prado et al., 2008, Uyaguari et al., 2011) and gentamicin resistant strains of *E. coli* (Jakobsen et al., 2008).

Studies of environmental organisms in wastewater have also revealed microbial resistance. An increase in the percentage of isolates resistant to clinically important antibiotics such as amoxicillin/clavulanate, chloramphenicol and rifampicin (and multiple resistance) through the steps in treatment has been reported in *Acinetobacter* spp. (Zhang et al., 2009) and ticarcillin and quinolone resistance has been reported in *Aeromonas* spp. (Figueira et al., 2011) and fluoroquinolone resistance in *Pseudomonas* spp. (Schwartz et al., 2006). No studies have investigated transfer of antimicrobial resistance genes in wastewater. However, a number of recent reports have studied the presence of plasmids carrying antibiotic resistance genes, including class 1 and 2 integrons (Akiyama et al., 2010, Pellegrini et al., 2011, Pignato et al., 2009).

If mobile resistance plasmids carrying resistance genes are present in the wastewater microbial flora, then the presence of antimicrobial residues and/or triclosan (Saleh et al., 2011, Schweizer, 2001) is likely to drive the transmission of these plasmids (and their resistance genes) to other bacteria. This could contribute to the dissemination of antibiotic resistant bacteria into the environment.

CONCLUSION

Microorganisms are highly adaptive and by a process of oxidation or reduction most xenobiotics can be degraded, although rates are very much influenced by temperature and generally proceed at a more rapid rate under oxygenic conditions. For more than 50 years now there has been an interest amongst microbiologists and biochemists to charaterise the microorganisms capable of degrading recalcitrant man-made compounds and elucidate the biochemical pathways involved in their dissimilation.

Triclosan is one such mass produced compound that is ubiquitous in the developed world and its degradation is of interest not merely because of its environmental significance but also because of its implication in the promotion of antibiotice resistance amongst the microbial community. In this review data has been summarized that deals with the concerns regarding this compounds presence in the environment and the efficacy of engineered and natural processes in dealing with its removal.

Microbial ecology is a growing field of interest due to our improved understanding of microbial processes and the critical role these play in maintaining a natural equilibrium. Something that microbiologists know little about is the influence that the presence of xenobiotics can have on the microbial biodiversity of an ecosystem. Whilst microbial removal of contaminants might be seen as a beneficial outcome, if the organisms that carry out these conversions prosper to the detriment or demise of microorganisms that play an important part in cycling processes then this could have potentially devastating effects on the equilibrium of a particular ecosystem. This is of particular relevance when considering what role this may play in further exacerbating the changes in natural processes being enacted through climate change.

So whilst microbial removal of contaminants might ostensibly appear to be a good thing, there could be unforeseen challenges created by unbalancing an established ecosystem. Although we have not considered this in detail here, excitingly, the molecular tools are now becoming available that will permit detailed study of natural microbial communities and begin to answer some of the questions surrounding these issues.

ACKNOWLEDGMENTS

This work was funded by Water Quality Research Australia (formerly the CRC for Water Quality and Treatment). The authors would like to acknowledge Dr. Alexandra Keegan (SA Water), Dr. Michael Heuzenroeder (IMVS) and Dr Zuliang Chen (UniSA) for their kind assistance and guidance on the laboratory work. We would also like to acknowledge the scholarships awarded to TQ by the IPRS program and the CRC for Water Quality and Treatment.

REFERENCES

Aiello, A. E., Marshall, B., Levy, S. B., Della-Latta, P. and Larson, E. 2004 'Relationship between triclosan and susceptibilities of bacteria isolated from hands in the community', *Antimicrobial Agents Chemotherapy, vol 48, no* 8, *pp.* 2973-2979.

Akiyama, T., Asfahl, K. L. and Savin, M. C. 2010 'Broad-host-range plasmids in treated wastewater effluent and receiving streams', *Journal of Environmental Quality, vol 39, no* 6, *pp.* 2211-2215.

Altenschmidt, U. and Fuchs, G. 1992 'Anaerobic toluene oxidation to benzyl alcohol and benzaldehyde in a denitrifying *Pseudomonas* strain', *Journal of Bacteriology, vol 174, no* 14, *pp.* 4860-4862.

Aranami, K. and Readman, J. W. 2007 'Photolytic degradation of triclosan in freshwater and seawater', *Chemosphere, vol 66, no* 6, *pp.* 1052-1056.

Arensdorf, J. J. and Focht, D. D. 1994 'Formation of chlorocatechol *meta* cleavage products by a pseudomonad during metabolism of monochlorobiphenyls', *Applied and Environmental Microbiology, vol 60, no* 8, *pp.* 2884-2889.

Baquero, F., Martínez, J. L. and Cantón, R. 2008 'Antibiotics and antibiotic resistance in water environments', *Current Opinion in Biotechnology, vol 19, no* 3, *pp.* 260-265.

Barry, M. A., Craven, D. E., Goularte, T. A. and Lichtenberg, D. A. 1984 '*Serratia marcescens* contamination of antiseptic soap containing triclosan: implications for nosocomial infection', *Infection Control, vol 5, no* 9, *pp.* 427-430.

Bartilson, M. and Shingler, V. 1989 'Nucleotide sequence and expression of the catechol 2,3-dioxygenase-encoding gene of phenol-catabolizing *Pseudomonas* CF600', *Gene, vol 85, no* 1, *pp.* 233-238.

Bernhard, M., Eubeler, J. P., Zok, S. and Knepper, T. P. 2008 'Aerobic biodegradation of polyethylene glycols of different molecular weights in wastewater and seawater', *Water Research, vol 42, no* 19, *pp.* 4791-4801.

Bester, K. 2003 'Triclosan in a sewage treatment process--balances and monitoring data', *Water Research, vol 37, no* 16, *pp.* 3891-3896.

Biegert, T. and Fuchs, G. 1995 'Anaerobic oxidation of toluene (analogues) to benzoate (analogues) by whole cells and by cell extracts of a denitrifying *Thauera* sp', *Archives of Microbiology, vol 163, no* 6, *pp.* 407-417.

Boyd, G. R., Reemtsma, H., Grimm, D. A. and Mitra, S. 2003 'Pharmaceuticals and personal care products (PPCPs) in surface and treated waters of Louisiana, USA and Ontario, Canada', *The Science of The Total Environment, vol 311, no* 1-3, *pp.* 135-149.

Braoudaki, M. and Hilton, A. C. 2004a 'Adaptive resistance to biocides in *Salmonella enterica* and *Escherichia coli* O157 and cross-resistance to antimicrobial agents', *Journal of Clinical Microbiology, vol 42, no* 1, *pp.* 73-78.

Braoudaki, M. and Hilton, A. C. 2004b 'Low level of cross-resistance between triclosan and antibiotics in *Escherichia coli* K-12 and *E. coli* O55 compared to *E. coli* O157', *FEMS Microbiology Letters, vol 235, no* 2, *pp.* 305-309.

Caldwell, M. E., Tanner, R. S. and Suflita, J. M. 1999 'Microbial metabolism of benzene and the oxidation of ferrous iron under anaerobic conditions: Implications for bioremediation', *Anaerobe, vol 5, no* 6, *pp.* 595-603.

Canosa, P., Morales, S., Rodríguez, I., Rubí, E., Cela, R. and Gómez, M. 2005 'Aquatic degradation of triclosan and formation of toxic chlorophenols in presence of low concentrations of free chlorine', *Analytical and Bioanalytical Chemistry, vol 383, no* 7, *pp.* 1119-1126.

Chau, W. C., Wu, J.-l. and Cai, Z. 2008 'Investigation of levels and fate of triclosan in environmental waters from the analysis of gas chromatography coupled with ion trap mass spectrometry', *Chemosphere, vol 73, no* 1, Supplement 1, *pp.* S13-17.

Christensen, E. G., Gram, L. and Kastbjerg, V. G. 2011 'Sublethal triclosan exposure decreases susceptibility to gentamicin and other aminoglycosides in *Listeria monocytogenes, Antimicrobial Agents and Chemotherapy, vol 55, no* 9, *pp.* 4064-4071.

Chuanchuen, R., Beinlich, K., Hoang, T. T., Becher, A., Karkhoff-Schweizer, R. R. and Schweizer, H. P. 2001 'Cross-resistance between triclosan and antibiotics in *Pseudomonas aeruginosa* is mediated by multidrug efflux pumps: exposure of a susceptible mutant strain to triclosan selects *nfx*B mutants overexpressing MexCD-OprJ', *Antimicrobial Agents and Chemotherapy, vol 45, no* 2, *pp.* 428-432.

Chuanchuen, R., Narasaki, C. T. and Schweizer, H. P. 2002 'The MexJK efflux pump of *Pseudomonas aeruginosa* requires OprM for antibiotic efflux but not for efflux of Triclosan', *Journal of Bacteriology., vol 184, no* 18, *pp.* 5036-5044.

Chung, W. W. P., Rafqah, S., Voyard, G. and Sarakha, M. 2007 'Photochemical behaviour of triclosan in aqueous solutions: kinetic and analytical studies', *Journal of Photochemistry and Photobiology A: Chemistry, vol 191, no* 2-3, *pp.* 201-208.

Coogan, M. A., Edziyie, R. E., La Point, T. W. and Venables, B. J. 2007 'Algal bioaccumulation of triclocarban, triclosan, and methyl-triclosan in a North Texas wastewater treatment plant receiving stream', *Chemosphere, vol 67, no* 10, *pp.* 1911-1918.

Cookson, B. D., Farrelly, H., Stapleton, P., Garvey, R. P. J. and Price, M. R. 1991 'Transferable resistance to triclosan in MRSA', *The Lancet, vol 337, no* 8756, *pp.* 1548-1549.

Copitch, J. L., Whitehead, R. N. and Webber, M. A. 2010 'Prevalence of decreased susceptibility to triclosan in *Salmonella enterica* isolates from animals and humans and association with multiple drug resistance', *International Journal of Antimicrobial Agents, vol 36, no* 3, *pp.* 247-251.

Cottell, A., Denyer, S. P., Hanlon, G. W., Ochs, D. and Maillard, J. Y. 2009 'Triclosan-tolerant bacteria: changes in susceptibility to antibiotics', *Journal of Hospital Infection, vol 72, no* 1, *pp.* 71-76.

D'Arezzo S, Lanini S, Puro V, Ippolito G and P, V. 2012 High-level tolerance to triclosan may play a role in *Pseudomonas aeruginosa* antibiotic resistance in immunocompromised hosts: evidence from an outbreak investigation. *BMC Research Notes.*

Dayan, A. D. 2007 'Risk assessment of triclosan [Irgasan] in human breast milk', *Food and Chemical Toxicology, vol 45, no* 1, *pp.* 125-129.

Di Gioia, D., Fava, F., Baldoni, F. and Marchetti, L. 1998 'Characterization of catechol- and chlorocatechol-degrading activity in the ortho-chlorinated benzoic acid-degrading *Pseudomonas* sp. CPE2 strain', *Research in Microbiology, vol 149, no* 5, *pp.* 339-348.

Downing, R. G. and Broda, P. 1980 'A cleavage map of the TOL plasmid of *Pseudomonas putida* mt-2', *Molecular and General Genetics, vol 177, no pp.* 189-191.

Ellison, M. L., Roberts, A. L. and Champlin, F. R. 2007 'Susceptibility of compound 48/80-sensitized *Pseudomonas aeruginosa* to the hydrophobic biocide triclosan', *FEMS Microbiology Letters, vol 269, no* 2, *pp.* 295-300.

Faoagali, J. L., George, N., Fong, J., Davy, J. and Dowser, M. 1999 'Comparison of the antibacterial efficacy of 4% chlorhexidine gluconate and 1% triclosan handwash products in an acute clinical ward', *American Journal of Infection Control, vol 27, no* 4, *pp.* 320-326.

Feakin, S. J., Blackburn, E. and Burns, R. G. 1994 'Biodegradation of s-triazine herbicides at low concentrations in surface waters', *Water Research, vol 28, no* 11, *pp.* 2289-2296.

Federle, T. W., Kaiser, S. K. and Nuck, B. A. 2002 'Fate and effects of triclosan in activated sludge', *Environmental Toxicology and Chemistry, vol 21, no* 7, *pp.* 1330-1337.

Ferreira Da Silva, M., Tiago, I., Verrissimo, A., Boaventura, R. A., Nunes, O. C. and Manaia, C. M. 2006 'Antibiotic resistance of enterococci and related bacteria in an urban wastewater treatment plant', *FEMS Microbiology Ecology, vol 55, no* 2, *pp.* 322-329.

Ferreira Da Silva, M., Vaz-Moreira, I., Gonzalez-Pajuelo, M., Nunes, O. C. and Manaia, C. M. 2007 'Antimicrobial resistance patterns in *Enterobacteriaceae* isolated from an urban wastewater treatment plant', *FEMS Microbiology Ecology, vol 60, no* 1, *pp.* 166-176.

Fewson, C. A. 1988 'Biodegradation of xenobiotic and other persistent compounds: the causes of recalcitrance', *Trends in Biotechnology, vol 6, no* 7, *pp.* 148-153.

Figueira, V., Vaz-Moreira, I., Silva, M. and Manaia, C. M. 2011 'Diversity and antibiotic resistance of *Aeromonas* spp. in drinking and waste water treatment plants', *Water Research, vol 45, no* 17, *pp.* 5599-5611.

Fine, D. H., Furgang, D., Bonta, Y., DeVizio, W., Volpe, A. R., Reynolds, H., Zambon, J. J. and Dunford, R. G. 1998 'Efficacy of a triclosan/NaF dentifrice in the control of plaque and gingivitis and concurrent oral microflora monitoring', *American Journal of Dentistry, vol 11, no* 6, *pp.* 259-270.

Galli, E. 1996 Alternative pathways for biodegradation of alkyl and alkenyl benzenes. In Nakazawa, T., Furukawa, K., Haas, D. and Silver, S. (Eds.) *Molecular biology of Pseudomonads: Proceedings of the Fifth International Symposium on Pseudomonads: Molecular Biology and Biotechnology, in Tsukuba, Japan, August 1995.* Washington, DC, ASM Press.

Gomez Escalada, M., Harwood, J. L., Maillard, J.-Y. and Ochs, D. 2005a 'Triclosan inhibition of fatty acid synthesis and its effect on growth of *Escherichia coli* and *Pseudomonas aeruginosa*', *Journal of Antimicrobial Chemotherapy, vol 55, no* 6, *pp.* 879-882.

Gomez Escalada, M., Russell, A. D., Maillard, J.-Y. and Ochs, D. 2005b 'Triclosan-bacteria interactions: single or multiple target sites?' *Letters in Applied Microbiology, vol 41, no* 6, *pp.* 476-481.

Haro, M.-A. and de Lorenzo, V. 2001 'Metabolic engineering of bacteria for environmental applications: construction of *Pseudomonas* strains for biodegradation of 2-chlorotoluene', *Journal of Biotechnology, vol 85, no* 2, *pp.* 103-113.

Hay, A. G., Dees, P. M. and Sayler, G. S. 2001 'Growth of a bacterial consortium on triclosan', *FEMS Microbiology Ecology, vol 36, no* 2-3, *pp.* 105-112.

Heath, R. J., Li, J., Roland, G. E. and Rock, C. O. 2000 'Inhibition of the *Staphylococcus aureus* NADPH-dependent enoyl-acyl carrier protein reductase by triclosan and hexachlorophene', *Journal of Biological Chemistry, vol 275, no* 7, *pp.* 4654-4659.

Heath, R. J., Rubin, J. R., Holland, D. R., Zhang, E., Snow, M. E. and Rock, C. O. 1999 'Mechanism of triclosan inhibition of bacterial fatty acid synthesis', *Journal of Biological Chemistry, vol 274, no* 16, *pp.* 11110-11114.

Heidler, J. and Halden, R. U. 2007 'Mass balance assessment of triclosan removal during conventional sewage treatment', *Chemosphere, vol 66, no* 2, *pp.* 362-369.

Hernández, A., Ruiz, F. M., Romero, A. and Martínez, J. L. 2011 'The binding of triclosan to *Sme*T, the repressor of the multidrug efflux pump *Sme*DEF, induces antibiotic resistance in *Stenotrophomonas maltophilia*', *PLoS pathogens, vol 7, no* 6, *pp.* 1-12, e1002103.

Hundt, K., Martin, D., Hammer, E., Jonas, U., Kindermann, M. K. and Schauer, F. 2000 'Transformation of triclosan by *Trametes versicolor* and *Pycnoporus cinnabarinus*', *Applied and Environmental Microbiology, vol 66, no* 9, *pp.* 4157-4160.

Jackson, J. and Sutton, R. 2008 'Sources of endocrine-disrupting chemicals in urban wastewater, Oakland, CA', *Science of The Total Environment, vol 405, no* 1-3, *pp.* 153-160.

Jakobsen, L., Sandvang, D., Hansen, L. H., Bagger-Skjøt, L., Westh, H., Jørgensen, C., Hansen, D. S., Pedersen, B. M., Monnet, D. L., Frimodt-Møller, N., Sørensen, S. J. and Hammerum, A. M. 2008 'Characterisation, dissemination and persistence of gentamicin resistant *Escherichia coli* from a Danish university hospital to the waste water environment', *Environment International, vol 34, no* 1, *pp.* 108-115.

Janssen, D. B., Dinkla, I. J. T., Poelarends, G. J. and Terpstra, P. 2005 'Bacterial degradation of xenobiotic compounds: evolution and distribution of novel enzyme activities', *Environmental Microbiology, vol 7, no* 12, *pp.* 1868-1882.

Kapoor, M., Mukhi, P. L., Surolia, N., Suguna, K. and Surolia, A. 2004 'Kinetic and structural analysis of the increased affinity of enoyl-ACP (acyl-carrier protein) reductase for triclosan in the presence of NAD+', *The Biochemical Journal, vol 381, no* Pt 3,*pp.* 725-733.

Kitayama, A., Achioku, T., Yanagawa, T., Kanou, K., Kikuchi, M., Ueda, H., Suzuki, E., Nishimura, H., Nagamune, T. and Kawakami, Y. 1996 'Cloning and characterization of extradiol aromatic ring-cleavage dioxygenases of *Pseudomonas aeruginosa* JI104', *Journal of Fermentation and Bioengineering, vol 82, no* 3, *pp.* 217-223.

Kobayashi, T., Murai, Y., Tatsumi, K. and Iimura, Y. 2009 'Biodegradation of polycyclic aromatic hydrocarbons by *Sphingomonas* sp. enhanced by water-extractable organic matter from manure compost', *Science of The Total Environment, vol 407, no* 22, *pp.* 5805-5810.

Kuo, M. R. 2003 'Targeting tuberculosis and malaria through inhibition of enoyl reductase', *The Journal of Biological Chemistry, vol 278, no* 23, *pp.* 20851-20860.

Kuster, M., Liez de Alda, M. J., Hernando, M. D., Petrovic, M., Marti-Alonso, J. and Barcel, D. 2008 'Analysis and occurrence of pharmaceuticals, estrogens, progestogens and polar pesticides in sewage treatment plant effluents, river water and drinking water in the Llobregat river basin (Barcelona, Spain)', *Journal of Hydrology, vol 358, no* 1-2, *pp.* 112-123.

Laturnus, F., Fahimi, I., Gryndler, M., Hartmann, A., Heal, M., Matucha, M., Schöler, H. F., Schroll, R. and Svensson, T. 2005 'Natural formation and degradation of chloroacetic acids and volatile organochlorines in forest soil. Challenges to understanding', *Environmental Science and Pollution Research, vol 12, no* 4, *pp.* 233-244.

Ledder, R. G., Gilbert, P., Willis, C. and McBain, A. J. 2006 'Effects of chronic triclosan exposure upon the antimicrobial susceptibility of 40 *ex-situ* environmental and human isolates', *Journal of Applied Microbiology, vol 100, no* 5, *pp.* 1132-1140.

Lefkowitz, J. R. and Duran, M. 2009 'Changes in antibiotic resistance patterns of *Escherichia coli* during domestic wastewater treatment', *Water Environment Research, vol 81, no* 9, *pp.* 878-885.

Leutwein, C. and Heider, J. 1999 'Anaerobic toluene-catabolic pathway in denitrifying *Thauera aromatica* : activation and β-oxidation of the first intermediate, (R)-(+)-benzylsuccinate', *Microbiology, vol 145, no* 11, *pp.* 3265-3271.

Levy, C. W., Roujeinikova, A., Sedelnikova, S., Baker, P. J., Stuitje, A. R., Slabas, A. R., Rice, D. W. and Rafferty, J. B. 1999 'Molecular basis of triclosan activity', *Nature, vol 398, no* 6726, *pp.* 383-384.

Levy, S. B. 2002 'Active efflux, a common mechanism for biocide and antibiotic resistance'. *Symposium series (Society for Applied Microbiology), 2002; (31): 65S-71S*

Li, C. H., Wong, Y. S. and Tam, N. F. Y. 2010 'Anaerobic biodegradation of polycyclic aromatic hydrocarbons with amendment of iron(III) in mangrove sediment slurry', *Bioresource Technology, vol 101, no* 21, *pp.* 8083-8092.

Lishman, L., Smyth, S. A., Sarafin, K., Kleywegt, S., Toito, J., Peart, T., Lee, B., Servos, M., Beland, M. and Seto, P. 2006 'Occurrence and reductions of pharmaceuticals and personal care products and estrogens by municipal wastewater treatment plants in Ontario, Canada', *Science of The Total Environment, vol 367, no* 2-3, *pp.* 544-558.

Lloyd-Jones, G., Laurie, A. D. and Tizzard, A. C. 2005 'Quantification of the *Pseudomonas* population in New Zealand soils by fluorogenic PCR assay and culturing techniques', *Journal of Microbiological Methods, vol 60, no* 2, *pp.* 217-224.

Lores, M., Llompart, M., Sanchez-Prado, L., Garcia-Jares, C. and Cela, R. 2005 'Confirmation of the formation of dichlorodibenzo-p-dioxin in the photodegradation of triclosan by photo-SPME', *Analytical and Bioanalytical Chemistry, vol 381, no* 6, *pp.* 1294-1298.

Louie, T. M., Ni, S., Xun, L. and Mohn, W. W. 1997 'Purification, characterization and gene sequence analysis of a novel cytochrome c co-induced with reductive dechlorination activity in *Desulfomonile tiedjei* DCB-1', *Archives of Microbiology, vol 168, no* 6, *pp.* 520-527.

Lovley, D. R. and Lonergan, D. J. 1990 'Anaerobic oxidation of toluene, phenol, and p-cresol by the dissimilatory iron-reducing organism, GS-15', *Applied and Environmental Microbiology, vol 56, no* 6, *pp.* 1858-1864.

McAvoy, D. C., Schatowitz, B., Jacob, M., Hauk, A. and Eckhoff, W. S. 2002 'Measurement of triclosan in wastewater treatment systems', *Environmental Toxicology and Chemistry, vol 21, no* 7, *pp.* 1323-1329.

McBain, A. J., Bartolo, R. G., Catrenich, C. E., Charbonneau, D., Ledder, R. G. and Gilbert, P. 2003 'Effects of triclosan-containing rinse on the dynamics and antimicrobial susceptibility of *in vitro* plaque ecosystems', *Antimicrobial Agents Chemotherapy, vol 47, no* 11, *pp.* 3531-3538.

McBain, A. J., Ledder, R. G., Sreenivasan, P. and Gilbert, P. 2004 'Selection for high-level resistance by chronic triclosan exposure is not universal', *Journal of Antimicrobial Chemotherapy, vol 53, no* 5, *pp.* 772-777.

McMurry, L. M., McDermott, P. F. and Levy, S. B. 1999 'Genetic evidence that InhA of *Mycobacterium smegmatis* is a target for triclosan', *Antimicrobial Agents and Chemotherapy, vol 43, no* 3, *pp.* 711-713.

McMurry, L. M., Oethinger, M. and Levy, S. B. 1998 'Triclosan targets lipid synthesis', *Nature, vol 394, no* 6693, *pp.* 531-532.

Meade, M. J., Waddell, R. L. and Callahan, T. M. 2001 'Soil bacteria *Pseudomonas putida* and *Alcaligenes xylosoxidans subsp. denitrificans* inactivate triclosan in liquid and solid substrates', *FEMS Microbiology Letters, vol 204, no* 1, *pp.* 45-48.

Meßmer, M., Reinhardt, S., Wohlfarth, G. and Diekert, G. 1996 'Studies on methyl chloride dehalogenase and O-demethylase in cell extracts of the homoacetogen strain MC based on a newly developed coupled enzyme assay', *Archives of Microbiology, vol 165, no* 1, *pp.* 18-25.

Miege, C., Choubert, J. M., Ribeiro, L., Eusebe, M. and Coquery, M. 2008 'Removal efficiency of pharmaceuticals and personal care products with varying wastewater treatment processes and operating conditions - conception of a database and first results', *Water Science and Technology, vol 57, no* 1, *pp.* 49-56.

Moon, J. H., Chang, H., Min, K. R. and Kim, Y. S. 1995 'Cloning and sequencing of the catechol 2,3-dioxygenase gene of *Alcaligenes* sp. KF711', *Biochemical and Biophysical Research Communications, vol 208, no* 3, *pp.* 943-949.

Nagulapally, S. R., Ahmad, A., Henry, A., Marchin, G. L., Zurek, L. and Bhandari, A. 2009 'Occurrence of ciprofloxacin-, trimethoprim-sulfamethoxazole-, and vancomycin-resistant bacteria in a municipal wastewater treatment plant', *Water Environment Research, vol 81, no* 1, *pp.* 82-90.

Neumann, A., Wohlfarth, G. and Diekert, G. 1995 'Properties of tetrachloroethene and trichloroethene dehalogenase of *Dehalospirillum multivorans*', *Archives of Microbiology, vol 163, no* 4, *pp.* 276-281.

Nojiri, H., Shintani, M. and Omori, T. 2004 'Divergence of mobile genetic elements involved in the distribution of xenobiotic-catabolic capacity', *Applied Microbiology and Biotechnology, vol 64, no* 2, *pp.* 154-174.

Okpokwasili, G. C. and Olisa, A. O. 1991 'River-water biodegradation of surfactants in liquid detergents and shampoos', *Water Research, vol 25, no* 11, *pp.* 1425-1429.

Okumura, T. and Nishikawa, Y. 1996 'Gas chromatography--mass spectrometry determination of triclosan in water, sediment and fish samples via methylation with diazomethane', *Analytica Chimica Acta, vol 325, no* 3, *pp.* 175-184.

Parikh, S. L., Xiao, G. and Tonge, P. J. 2000 'Inhibition of InhA, the enoyl reductase from *Mycobacterium tuberculosis*, by triclosan and isoniazid', *Biochemistry, vol 39, no* 26, *pp.* 7645-7650.

Pellegrini, C., Celenza, G., Segatore, B., Bellio, P., Setacci, D., Amicosante, G. and Perilli, M. 2011 'Occurrence of class 1 and 2 integrons in resistant enterobacteriaceae collected from a urban wastewater treatment plant: First report from central Italy', *Microbial Drug Resistance, vol 17, no* 2, *pp.* 229-234.

Pignato, S., Coniglio, M. A., Faro, G., Weill, F. X. and Giammanco, G. 2009 'Plasmid-mediated multiple antibiotic resistance of *Escherichia coli* in crude and treated wastewater used in agriculture', *Journal of Water and Health, vol 7, no* 2, *pp.* 251-258.

Poole, K. 2005 'Efflux-mediated antimicrobial resistance', *Journal of Antimicrobial Chemotherapy, vol 56, no* 1, *pp.* 20-51.

Prado, T., Pereira, W. C., Silva, D. M., Seki, L. M., Carvalho, A. P. D. A. and Asensi, M. D. 2008 'Detection of extended-spectrum β-lactamase-producing *Klebsiella pneumoniae* in effluents and sludge of a hospital sewage treatment plant', *Letters in Applied Microbiology, vol 46, no* 1, *pp.* 136-141.

Pries, F., van der Ploeg, J. R., Dolfing, J. and Janssen, D. B. 1994 'Degradation of halogenated aliphatic compounds: The role of adaptation', *FEMS Microbiology Reviews, vol 15, no* 2-3, *pp.* 279-295.

Qiu, T. 2012 *Biodegradation of triclosan as a representative of pharmaceuticals and personal care products (PPCPs) in the wastewater environment, PhD thesis,* Adelaide, University of South Australia.

Randall, L. P., Cooles, S. W., Piddock, L. J. V. and Woodward, M. J. 2004 'Effect of triclosan or a phenolic farm disinfectant on the selection of antibiotic-resistant *Salmonella enterica*', *Journal of Antimicrobial Chemotherapy, vol 54, no* 3, *pp.* 621-627.

Regos, J., Zak, O. and Solf, R. 1979 'Antimicrobial spectrum of triclosan, a broad-spectrum antimicrobial agent for topical application. II. Comparison with some other antimicrobial agents', *Dermatologica, vol 158, no* 1, *pp.* 72-79.

Sabaliunas, D., Webb, S. F., Hauk, A., Jacob, M. and Eckhoff, W. S. 2003 'Environmental fate of triclosan in the River Aire Basin, UK', *Water Research, vol 37, no* 13, *pp.* 3145-3154.

Sahlström, L., Rehbinder, V., Albihn, A., Aspan, A. and Bengtsson, B. 2009 'Vancomycin resistant enterococci (VRE) in Swedish sewage sludge', *Acta Veterinaria Scandinavica, vol 51, no* 24, *pp.* 1-9.

Saleh, S., Haddadin, R. N. S., Baillie, S. and Collier, P. J. 2011 'Triclosan - an update', *Letters in Applied Microbiology, vol 52, no* 2, *pp.* 87-95.

Samsøe-Petersen, L., Winther-Nielsen, M. and Madsen, T. 2003 Chemicals: Fate and Effects of Triclosan. *Environmental Project No. 861 2003.* Hørsholm, Danish Environmental Potection Agency.

Sanchez-Prado, L., Barro, R., Garcia-Jares, C., Llompart, M., Lores, M., Petrakis, C., Kalogerakis, N., Mantzavinos, D. and Psillakis, E. 2008 'Sonochemical degradation of triclosan in water and wastewater', *Ultrasonics Sonochemistry, vol 15, no* 5, *pp.* 689-694.

Schwartz, T., Volkmann, H., Kirchen, S., Kohnen, W., Schön-Hölz, K., Jansen, B. and Obst, U. 2006 'Real-time PCR detection of *Pseudomonas aeruginosa* in clinical and municipal wastewater and genotyping of the ciprofloxacin-resistant isolates', *FEMS Microbiology Ecology, vol 57, no* 1, *pp.* 158-167.

Schweizer, H. P. 2001 'Triclosan: a widely used biocide and its link to antibiotics', *FEMS Microbiology Letters, vol 202, no* 1, *pp.* 1-7.

Singer, H., Muller, S., Tixier, C. and Pillonel, L. 2002 'Triclosan: occurrence and fate of a widely used biocide in the aquatic environment: Field measurements in wastewater treatment plants, surface waters, and lake sediments', *Environmental Science and Technology, vol 36, no* 23, *pp.* 4998-5004.

Sinton, G. L., Fan, L. T., Erickson, L. E. and Lee, S. M. 1986 'Biodegradation of 2,4-D and related xenobiotic compounds', *Enzyme and Microbial Technology, vol 8, no* 7, *pp.* 395-403.

Slayden, R. A., Lee, R. E. and Barry, C. E. 2000 'Isoniazid affects multiple components of the type II fatty acid synthase system of *Mycobacterium tuberculosis*', *Molecular Microbiology, vol 38, no* 3, *pp.* 514-525.

Son, H. S., Choi, S. B., Zoh, K. D. and Khan, E. 2007 'Effects of ultraviolet intensity and wavelength on the photolysis of triclosan', *Water Science and Technology, vol 55, no* 1-2, *pp.* 209-216.

Stephen, K. W., Saxton, C. A., Jones, C. L., Ritchie, J. A. and Morrison, T. 1990 'Control of gingivitis and calculus by a dentifrice containing a zinc salt and triclosan', *Journal of Periodontology, vol 61, no* 11, *pp.* 674-679.

Stewart, M. J., Parikh, S., Xiao, G., Tonge, P. J. and Kisker, C. 1999 'Structural basis and mechanism of enoyl reductase inhibition by triclosan', *Journal of Molecular Biology, vol 290, no* 4, *pp.* 859-865.

Suller, M. T. and Russell, A. D. 2000 'Triclosan and antibiotic resistance in *Staphylococcus aureus*', *Journal of Antimicrobial Chemotherapy, vol 46, no* 1, *pp.* 11-18.

Tabak, M., Scher, K., Hartog, E., Romling, U., Matthews, K. R., Chikindas, M. L. and Yaron, S. 2007 'Effect of triclosan on *Salmonella typhimurium* at different growth stages and in biofilms', *FEMS Microbiology Letters, vol 267, no* 2, *pp.* 200-206.

Tatarazako, N., Ishibashi, H., Teshima, K., Kishi, K. and Arizono, K. 2004 'Effects of triclosan on various aquatic organisms', *Environmental Science, vol 11, no* 2, *pp.* 133-140.

Thompson, A., Griffin, P., Stuetz, R. and Cartmell, E. 2005 'The fate and removal of triclosan during wastewater treatment', *Water Environment Research, vol 77, no* 1, *pp.* 63-67.

Tong, A. Y., Peake, B. and Braund, R. 2011 'Disposal practices for unused medications around the world.' *Environment International, vol 37, no* 1, *pp.* 292-298.

Uyaguari, M. I., Fichot, E. B., Scott, G. I. and Norman, R. S. 2011 'Characterization and quantitation of a novel β-lactamase gene found in a wastewater treatment facility and the surrounding coastal ecosystem', *Applied and Environmental Microbiology, vol 77, no* 23, *pp.* 8226-8233.

van der Waarde, J. J., Kok, R. and Janssen, D. B. 1994 'Cometabolic degradation of chloroallyl alcohols in batch and continuous cultures', *Applied Microbiology and Biotechnology, vol 42, no* 1, *pp.* 158-166.

van Hylckama Vlieg, J. E. T. 2000 'Detoxification of reactive intermediates during microbial metabolism of halogenated compounds', *Current opinion in microbiology, vol 3, no* 3, *pp.* 257-262.

Vischer, W. A. and Regos, J. 1974 'Antimicrobial spectrum of triclosan, a broad spectrum antimicrobial agent for topical application', *Antibacterielles Wirkungsspektrum Von Triclosan, Einem Breitband Antimicrobicum Zur Lokalen Anwendung, vol 226, no* 3, *pp.* 376-389.

Vörösmarty, C. J., McIntyre, P. B., Gessner, M. O., Dudgeon, D., Prusevich, A., Green, P., Glidden, S., Bunn, S. E., Sullivan, C. A., Reidy Liermann, C. and Davies, P. M. 2010 'Global threats to human water security and river biodiversity', *Nature, vol 467, no* 7315, *pp.* 555-556.

Wackett, L. P. 2003 '*Pseudomonas putida* - a versatile biocatalyst', *National Biotechnology, vol 21, no* 2, *pp.* 136-138.

Walker, C., Borden, L. C., Zambon, J. J., Bonta, C. Y., DeVizio, W. and Volpe, A. R. 1994 'The effects of a 0.3% triclosan-containing dentifrice on the microbial composition of supragingival plaque', *Journal of Clinical Periodontology, vol 21, no* 5, *pp.* 334-341.

Webster, J. 1992 'Handwashing in a neonatal intensive care nursery: product acceptability and effectiveness of chlorhexidine gluconate 4% and triclosan 1%', *Journal of Hospital Infection, vol 21, no* 2, *pp.* 137-141.

Webster, J., Faoagali, J. L. and Cartwright, D. 1994 'Elimination of methicillin-resistant *Staphylococcus aureus* from a neonatal intensive care unit after hand washing with triclosan', *Journal of Paediatrics and Child Health, vol 30, no* 1, *pp.* 59-64.

Wiesmann, U., Choi, I. S. and Dombrowski, E.-M. 2007 *Fundamentals of Biological Wastewater Treatment: Fundamentals, Microbiology, Industrial Process Integration*, Wiley-VCH Verlag GmbH.

Xia, K., Bhandari, A., Das, K. and Pillar, G. 2005 'Occurrence and fate of pharmaceuticals and personal care products (PPCPs) in biosolids', *Journal of Environmental Quality, vol 34, no* 1, *pp.* 91-104.

Yazdankhah, S. P., Scheie, A. A., Høiby, E. A., Lunestad, B. T., Heir, E., Fotland, T.-Ø., Naterstad, K. and Kruse, H. 2006 'Triclosan and antimicrobial resistance in bacteria: an overview', *Microbial Drug Resistance, vol 12, no* 2, *pp.* 83-90.

Ying, G.-G., Yu, X.-Y. and Kookana, R. S. 2007 'Biological degradation of triclocarban and triclosan in a soil under aerobic and anaerobic conditions and comparison with environmental fate modelling', *Environmental Pollution, vol 150, no* 3, *pp.* 300-305.

Yu, J. C., Kwong, T. Y., Luo, Q. and Cai, Z. 2006 'Photocatalytic oxidation of triclosan', *Chemosphere, vol 65, no* 3, *pp.* 390-399.

Yu, P. and Welander, T. 1995 'Growth of an aerobic bacterium with trichloroacetic acid as the sole source of energy and carbon', *Applied Microbiology and Biotechnology, vol 42, no* 5, *pp.* 769-774.

Zafar, A. B., Butler, R. C., Reese, D. J., Gaydos, L. A. and Mennonna, P. A. 1995 'Use of 0.3% triclosan (Bacti-Stat) to eradicate an outbreak of methicillin-resistant *Staphylococcus aureus* in a neonatal nursery', *American Journal of Infection Control, vol 23, no* 3, *pp.* 200-208.

Zhang, Y., Marrs, C. F., Simon, C. and Xi, C. 2009 'Wastewater treatment contributes to selective increase of antibiotic resistance among Acinetobacter spp', *Science of the Total Environment, vol 407, no* 12, *pp.* 3702-3706.

In: Water Quality
Editor: You-Gan Wang

ISBN: 978-1-62417-111-6

Chapter 9

WATER QUALITY ASSESSMENT METHODS: THE COMPARATIVE ANALYSIS

Tatyana I. Moiseenko[1,*], Alexandr G. Selukov[2] and Dmitry N. Kyrov[2]
[1]Institute of Geochemistry and Analytical Chemistry, Russian Academy of Sciences, Russia
[2]Tyumen State University, Russia

ABSTRACT

A comparative analysis of approaches and methods of biological assessment of water quality is presented here. The method of ecotoxicological diagnostic of aquatic ecosystem "health" and water quality evaluation based on the physiological state of fish is substantiated as well in this chapter, based on our findings. Characteristics of the main symptoms of diseases in fish inhabiting freshwater bodies of Russia and pathologic disturbances in their organs and tissues, caused by water bodies' contamination with toxic substances, are also presented. In this chapter, the method of ecotoxicological assessment of water quality is shown to be both highly informative and easy-to-use in practical monitoring. For example, the dose-effect dependencies and critical levels of water pollution were determined for arctic lake Imandra on the basis of an ecotoxicological approach.

1. INTRODUCTION

The problem of qualitative depletion of water resources caused by their contamination has become especially acute for the last decades. The human factor affecting the formation of the chemical composition of water is becoming as important as the natural geochemical and biological processes. Transformation of catchment areas, transboundary flows, discharge of untreated industrial and domestic effluents, as well as non-sewage effluents lead to changes in

* Corresponding author.

the geochemical cycles of elements in the catchment area – water body system and the occurrence of toxic substances in the aquatic environment, which entails water quality deterioration.

The system of limitations, based on the concept of maximum permissible concentration (MPC) of pollutants (or Guideline Concentration) in the water, cannot fully protect aquatic ecosystems against degradation. Water quality criteria applied by water users differ depending on the aim of water use: industrial use, drinking water supply, natural or aquacultural fish reproduction.

From the viewpoint of the ecological paradigm, water is a vital resource for all live organisms. At the same time, water is a habitat for aquatic organisms. In the process of their vital activity, living water organisms, using water as both a resource and habitat, actively affect its properties. The relative natural stability of properties of water and their seasonal cycles in individual water bodies is supported due to the dynamic balance of physical, chemical, and biological processes — both in the water body itself and within its catchment area.

The habitat is characterized by certain conditions. According to E. Odum (1981), a condition is an ambient environmental factor, changing in time and in space, to which an organism responds depending on its strength. The conditions, necessary for living and reproduction of organisms inhabiting various water bodies, can differ largely. For instance, if typically northern species adapted to low-saline oligotrophic water are placed in "background" southern lakes, filled with water typical of such lakes it becomes clear that such water properties as low salinity and oligotrophic character will not be acceptable for them (even if thermal conditions in the southern lakes could be made similar to those in the northern ones); and vice versa. Specific ecosystems have formed in water bodies with unique properties of water (e.g., geothermal or brackish). For such communities, the properties under consideration are optimum. Therefore, while determining ecologically permissible human impact on aquatic ecosystems, it is necessary to take into consideration both the regional conditions of water formation and the sensibility of organisms and ecosystems as a whole to pollution.

Disengaging ourselves from subjective requirements of individual water users to water quality, we can say that the water quality definition from the viewpoint of the ecological paradigm appears to be more universal: "*Water quality is the totality of properties of water formed in the process of chemical, physical, and biological processes, occurring both in the water body itself and within its catchment area. Water quality in a water body can be regarded as good where it meets the requirements of preservation of the most vulnerable organism's health of species adapted to the existence in the given aquatic environment*".

At present, the necessity of developing unbiased ecological criteria of determining aquatic ecosystems state and assessment of water quality causes no doubt. Such criteria should be substantiated based on the response of individual organisms, their populations, and communities to the effect of pollutants. If properties of the water under consideration meet the requirements of the life and reproduction of the most sensitive aquatic organisms, then water quality (with the exception of certain cases) can be regarded as meeting the respective requirements for human health preservation, too.

This work is aimed at comparing biological methods of water quality assessment, substantiating the technique of ecotoxicological studies for water quality evaluation and regulation of human-induced contamination of water bodies.

1.1. Methods of Biological Evaluation of Water Quality

Biological methods of water quality assessment are based on studying different impacts on ecosystems and their structural elements (individual organisms, populations, and communities). The main methods of an estimation of the quality of water is presented on the block-diagram (Figure 1).

Biotesting methods understood as the evaluation of the potential hazard of pollutants (or concrete effluents or contaminated waters) entering the water body, based on ex-situ experimental laboratory studies. This method allows us to determine the lethal and sublethal concentrations of potential pollutants, industrial effluents, or contaminated waters for living organisms (test objects) under the laboratory conditions (Bioassey…, 1985; Canadian water quality.., 1994; Methodological recommendation, 1998). This approach was used to evaluate maximum permission concentration (MPCfish) for toxic substances in fish water body (Bespamyatnov, Krotov, 1985; List of Fishery-related…, 1999). This method is based on experimental determination of pollutant concentrations, causing most significant and easy-to-detect disturbances in aquatic organisms, such as morbidity, survivorship rate, physiological or pathological disturbances. The "threshold" value, causing visible abnormalities in the most sensitive group of organisms, is adopted as the MPC_{fish} for the given hazardous substance. Organisms belonging to different systematic groups (bacteria, algae, invertebrates, fish) are used as test objects (Bioassey…, 1985; Methodological recommendation…, 1998). The greater part of toxicological studies is carried out at the level of individual organisms. These experiments allow us to study the effect of most frequently occurring toxic substances on hydrobionts belonging to different systematic groups.

The main advantage of biontesting is the possibility to quickly obtain information on the toxicity of individual contaminants or industrial wastewater. However, it is not clear how rightful it will be to extrapolate the experimental results, obtained under the laboratory conditions, to natural objects.

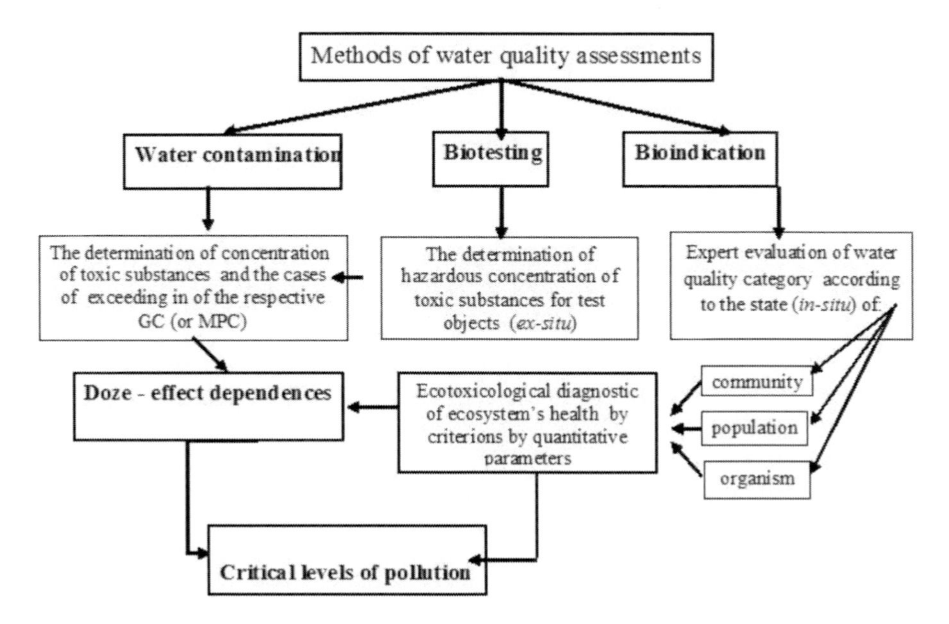

Figure 1. Block-diagram of the methods of the water quality assessment.

The behavior of pollutants in natural water bodies and their toxicological properties can largely differ from those demonstrated in aquaria; combined effects (both synergetic and antagonistic) can manifest themselves. Under the laboratory conditions, it is difficult to determine the transfer of toxins along food chains and their cumulative effects. In addition, individual organisms, used in the experiment, have very little in common with the respective natural populations and communities. Therefore, the MPC system, based on biotesting studies of individual elements and their compounds, does not present a scientific basis for the ecological standardization of pollutants entering natural water bodies.

Bioindication is aimed at evaluating water quality in natural water bodies or in the zone of contamination according to the state of indicator species or communities (*in-situ*). This method is widely used in monitoring of aquatic environment. The hydrobiological monitoring of freshwater ecosystems involves observations over the state of the major subsystems: microflora, periphyton, phytoplankton, macrophytes, zooplankton, and zoobenthos (Manual for hydrobiological..., 1992; Wang, Dixon, 1995; Environment quality.., 2001).

Detailed description of the techniques and advantages of using each group of organisms in the bioindication is presented elsewhere (Manual for hydrobiological..., 1992). The expert characteristic of the ecological state of a water body is based on the totality of all the features, including structural ones (species composition, quantity, biological diversity, the ratio of species of different ecological valency, and the characteristics of their saprobility) and functional characteristics of aquatic communities (production and destruction parameters, etc.).

Change in the structural and functional organization of communities is the most significant parameter describing variations in freshwater ecosystems under the impact of human factors. However, quantitative methods of ecosystem state evaluation were not paid due attention and did not become widely used in hydrobiological analyses. The indices and parameters deduced basing on the account of the species composition of biocenosis are very often subjective and depend on the biotope homogeneity and the season of the year. In addition, populations of different species differ, as concerns the degree of their polyfunctionality. The use of such indices becomes hampered where water bodies are eutrophic and contaminated with organic substances: the number of some communities begins to increase, while that of some others decreases. That is why, as a result, comparative estimates, expressed as grades, rating points, marks, and indices are obtained. These estimates occupy an intermediary position between the qualitative and quantitative parameters and depend on the qualification of experts.

Ecotoxicological diagnostic is aimed at obtaining an integrated assessment of water quality, based on symptoms of disturbance in the ecosystem "health" (*in situ*). Such diagnostic is based on theoretical principles of toxicology and ecology, synthesis of their methods, allowing the determination of environmental effects and mechanisms of the impact of hazardous substances on aquatic organisms. Aquatic ecosystems are stressed in all levels, ranging from individual and up to the population and community levels. For ecosystem health assessment the following four definitions have been used: i) cellular health, which describes the structural integrity of cellular organels and the maintenance of biochemical processes; ii) individual health, which presents structural and morphological health and functioning in terms of physiology of the entire organism; iii) population health, which measures the sustainability and maintenance of a population of a particular species; iv) community health, which describes a group of organisms and the relationships between species in that group.

Each method has its limitations and advantages, and the type of method used defines how we interpret the effect of a stressor on ecosystem health (Adams, Ryon, 1994; Cash, 1995; Arttril, Depledge, 1997; Moiseenko et al., 2006, 2008; Yeom, Adams, 2007).

In general, indicators at the biochemical and physiological levels provide information on the functional status of individual organisms, while intermediate-level responses, such as histopathological condition, are indicative of the structural integrity of tissues and organs. Symptoms of physiological changes and pathologic state of organisms, functional and structural disturbances in the state of populations and communities in natural water bodies reflect poor "health" of the ecosystem and unsatisfactory water quality. Community and population level measurements integrate the responses to a variety of environmental conditions, and therefore may be less reflective of toxic contaminant-induced stress in comparison to the level of organisms (Yeom, Adams, 2007). Unlike traditional methods of bioindication, ecotoxicological approach is aimed at revealing the dose-effect dependences between water contamination parameters and disturbances in biological ecosystems, serving as the basis for the determination of ecologically permissible concentrations of toxic substances in the water.

1.2. Fish Health as Criteria of Water Quality

Many groups of organisms can be used as indicators of environmental and ecological change. But numerous publications attest that fish (*in situ*) is a good indicator of environmental change and ecosystem health, especially in case of toxic water pollution (Whitfield, Elliott, 2002; Eliott et al., 2003; Moiseenko, 2009). Fish occupy the top level in the trophic system of aquatic ecosystems. Pathological changes in fish organ enable us to determine the toxicity of water and the potential danger of man-entering substances in water. Fish, in comparison with invertebrate, are more sensitive to many toxicants and are the convenient test-object for indication of ecosystem health. Physiological and biochemical parameters of the state of fish in water bodies, allowing us to reveal both short-term and long-term effects of the sublethal dose of pollutants, are also determined.

In the 1970s, methods of pathologic–physiologic research of fish were widely used in connection with more and more frequent cases of large-scale fish poisoning caused by natural-water contamination. Methods of clinical and postmortem examination of organisms applied in veterinary and medicine were used to study fish organisms in order to assess the consequences of water pollution with toxic substances.

Diagnostics of fish health requires system studies, which combine extensive data with disease diagnoses. A diagnostic system for fish state, suitable for practical monitoring, is presented elsewhere (Moiseenko et al., 2006; Moiseenko et al., 2008). The *macro-level* examination of individuals involves diagnosing of diseases on the basis of a visual examination of numerous organisms. Method of clinic and postmortem examination of organisms is used for large-scale examination of fish, carried out not later than within one hour after the fish has been caught. In the process of visual examination, special attention should be paid to the following: the intensity of color (the state of pigment cells–melanophores); the integrity of the fin edge and somactids; the total amount of mucus on the fish body; the state of squama, opercula, oral cavity, anus; the cases of hyperemia, subcutaneous hemorrhages, sores, or hydremia of the body; deformation of skull and skeleton bones; the state of eye crystalline lens and cornea. When the opercula are opened, gills are examined, in particular, their color, the presence and the amount of mucus, the state of gills petals (accretion, adhesion, dilatation, or thinning down). After the abdominal cavity is dissected, the state of fish muscles is studied (color, consistence, hemorrhages, attachment to bones), as well as the presence of exudate in the abdominal cavity, the amount of cavitary fat, its color and density. The topographic location of viscera (liver, kidneys, gonads, spleen, heart, stomach, intestines), their dimensions, color, density, edges, hemorrhages, zones of necrosis, etc. are studied. Mucous membranes of dissected stomach and intestines are examined, in addition to cerebrum, paying special attention to filling of vessels, their color and density. At this level, a preliminary diagnosis is made based on clinical and postmortem symptoms of intoxication.

The *micro-level* diagnostics include hematologic, histological, biochemical, instrumental, and other methods. These are labor-intensive and cannot be widely used, however they are essential to refine the diagnosis and estimate the consequences of pathologic changes in fish organisms.

The degree of disturbance in individual organisms is very important to diagnose the damage to fish organisms in the contaminated zone. For instance, up to 70% of individuals in the zones of contamination may be in a state close to the lethal threshold. If the concentration of toxic substances is not high, the percentage of affected individuals can be the same, but disturbances in fish organisms can be insignificant and will not be life threatening. In order to estimate the state of fish organism using the data of clinical and postmortem examination, experts suggest using different rating systems. In the process of macro-diagnostics, three stages of the disease can be singled out (0 denotes healthy individuals):

(1) Low disturbances, not threatening the life of the fish;
(2) Medium-level disturbances, causing a critical state of the organism;
(3) Distinct intoxication symptoms leading to inevitable death of the organism.

The overall index of morbidity in fish in the given zone of contamination can be presented as:

$$Z = (N_1 + 2N_2 + 3N_3) / N_{tot}.$$

Here N_1, N_2, and N_3 are the number of fishes in the first, second, and third stages of the disease, respectively; N_{tot} is the total number of the examined fishes, including healthy individuals. Note that $O \leq Z \leq 3$. If all the fish in the given water body does not demonstrate any intoxication symptoms, then $Z = 0$. The value of Z will increase with an increase in both the number of sick fishes and the extent of their diseases.

The table 1 presents typical health pathologies detected in fish inhabiting different polluted aquatic environment of lakes and rivers of Russia. These disturbances were diagnosed using the method of clinic and postmortem examination.

When toxic substances are discharged from non-point sources, zones of contamination, having different degree of hazard, are formed in the water bodies. Fish stock in the contaminated zones is non-homogeneous, as concerns morbidity, which is related to the concentration of pollutants in the zone under consideration and the duration of their effect, on the one hand, and the tolerance of individual fish organisms, on the other hand.

Table 1. Pathology disturbance in organ and tissues f fish inhabiting in polluted aquatic environment of lakes and rivers of Russian. This table is based on an analytical review of published studies (Chinareva, 1988; Savaitova et al., 1995; Lukin, Sharova, 2004; Moiseenko, Kudryavtseva, 2002; Moiseenko et al., 2006; Moiseenko et al, 2008, Moiseenko 2009)

Organ and tissues	Visual symptoms of intoxication	Changes in the cell structure and diagnosis	Fish species, source	*Disturbing factors*
Eyes	Opacity, hemorrhages in the cornea	Cataract or chemical scald	Bream in the Volga and Kama reservoirs; whitefish in lakes in the North Kola Peninsula; loach in water bodies of the Noril'sk Pyasinskaya system	Contamination with a set of toxic substances; wastewater of the copper and nickel metallurgical plant, with a high concentration of metals
Tectorial tissues	Change in the natural color of the body; specific tints of skin (yellowish, greenish); bristling of squama	–	Bream in the Volga and Kama reservoirs; whitefish in lakes in the North Kola Peninsula	Contamination with a set of toxic substances; effluents of ore-mining and ore-concen-trating plants
Heart	Increased dimensions of the heart, flabby structure of the heart; symptoms of obesity (adipose cover)	Decomposition of fibers and muscles of the heart, dystrophy of muscular fibers, lipoid dystrophy of myocardium	Bream in the Volga and Kama reservoirs; whitefish in lakes in the North Kola Peninsula; whitefish in the Pechora River; bream and whitefish in the Ladoga Lake	Contamination with a set of toxic substances; effluents of mining; metals, organic xenobiotic compounds, oil products; wastewaters of a pulp-and-paper plant
Gills	Anemic ring around the gill's arc; cyanotic tint of gills, filled with blood; conglomerated petals	Hyperplasia of epithelium of petals and stamens; exfoliation of epithelium; overfilling and stagnation in blood vessels; degeneration of cartilage tissue	Bream in the Volga and Kama reservoirs; whitefish in lakes in the North Kola Peninsula and the Pechora River; whitefish and bream in the Ladoga Lake and the Volkhov River	Contamination with a set of toxic substances; metals, organic xenobiotic compounds, oil products; wastewaters from a pulp-and-paper plant
Skull bones and spine bones	Insufficient ossification of cranium; curvature of the spine; accretion of vertebrae	Osteoporosis, scoliosis	Whitefish in Imandra Lake; whitefish in the Neva and Volkhov rivers in their lower sections; loach in water bodies of the Noril'sk Pyasninskaya system	Effluents of the ore-concen-trating plant, with a high concentration of Sr; wastewaters of a pulp-and-paper plant
Muscles	Flabby tissues, crumbled into myocepts	Myopathy	Whitefish in the Imandra Lake; sturgeon in the Caspian Sea	Contamination with a set of toxic substances: metals; POP's compounds, oil.

Table 1. (Continued)

Organ and tissues	Visual symptoms of intoxication	Changes in the cell structure and diagnosis	Fish species, source	*Disturbing factors*
Gonads	Constriction and twisting of gonads; insufficient saturation of egg-bearing plates with fish eggs, till their consistence becomes friable and transparent; hermaphroditism	Asynchronous development of fish eggs; lipoid and connective tissue degeneration of ovaries and testicles; resorption and degeneration of oocytes; lysis of their shells; laying of oocytes without nuclei; formation of cysts	Whitefish in the Imandra Lake; whitefish in the Pechora River; bream and whitefish in the Ladoga Lake; sturgeon in the Caspian Sea	The same
Liver	Appearance of bright spots or change in the color from brown (normal) to light yellow or sandy; increase in the dimension; intumescence or decrease in the dimensions and atrophy, change in the configuration	Cirrhosis (zones of necrosis inside the parenchyma, excessive development of the connective tissue, decomposition of hepatocyte nuclei, excessive proliferation of connective tissue, hemorrhages); lipoid degeneration (vacuolar lipoid dystrophy of the parenchyma; neoplasms	Whitefish in the Imandra Lake; whitefish in the Pechora River; loach in water bodies of the Noril'sk–Pyasinskaya system; bream and whitefish in Ladoga and Karelian lakes; bream in the Volga River; sturgeon in the Caspian Sea	The same
Kidneys	Increase in the dimension, considerable swelling of urine-excreting canaliculi in the distal section, with calculi of the diameter reaching 5 mm precipitated inside. The kidneys are of a bigger size as compared to the norm, have flabby structure, the uniform color changes to visible granulated structure	Nephrocalcinosis (disturbances in the structure of urine-excreting canaliculi, hyperplasia or disquamation of epithelium, with extraneous occlusions inside canaliculi); nephropathy (zones of necrosis); growth of connective tissue inside the parenchyma; neoplasms	Whitefish in Imandra Lake; whitefish in the Pechora River; loach in water bodies of the Noril'sk–Pya-sinskaya system; Ladoga and Karelian lakes; sturgeon in the Caspian Sea	The same

In case of acute intoxication, fishes quickly die; if fishes are exposed to a permanent impact of toxic substances, physiological disturbances manifest themselves in degenerative processes (sometimes, irreversible) occurring in fish organs and tissues. In the long run, uninterrupted effect of toxic substances results in delayed death of fishes.

Method of hematologic analysis is used to diagnose toxicoses in fish at early stages, because the circulatory system of fish is a sensitive indicator of the impact of unfavorable ambient conditions on organisms and responds to toxic agents by the appearance of disintegrating cells and their pathologic forms (Ziteneva, 1989). Changes in blood parameters manifest themselves earlier as compared to visual symptoms of pathologic diseases. In the blood samples thus taken, hemoglobin concentration, erythrocyte sedimentation rate (ESR), erythrocyte and leukocyte concentration. Blood smear examination allows the analysis of red blood composition, differential blood count, and the detection of occurrence of pathologic blood cells addition, other characteristics known in hematology can be used.

The exact interpretation of the results obtained needs taking into account the regularities of the development of toxicosis in fish, in which four stages can be arbitrarily singled out (Moiseenko, 1998):

i) *mobilization,*
ii) *destabilization,*
iii) *degradation.*

i) Investigation of the dynamics of hemathologic parameters (case study of whitefish inhabiting contaminated northern water bodies) have shown that at the first stage of toxicosis, young and reserve cells find their way to the blood channel, thus making the blood thicker. The amplitude of changes in the hemoglobin concentration in the blood grows: whereas the normal hemoglobin concentration in the fish blood varies from 80 to 130 g/l, certain individuals can be found, whose hemoglobin concentration reaches 180 g/l. Polychromasia and anacytosis can be diagnosed in fish, basing on blood smear analyses; abnormal cell division appears; the concentration of leukocytes grows, and the leukocyte formula begins to change, such changes manifesting themselves in an increase in the share of young lymphocytes, neutrophiles, and monocytes. All this testifies to the fact that, from the first stages of toxic substances impact, the system of hemogenesis in fish responds as a protective function by means of mobilizing young and reserve sells.

iii) At the stage of destabilization, the process of blood cells destruction is partly compensated for by reserve young cells finding their way in the blood channel, that is why hemoglobin concentration in the blood can be close to the normal value, but numerous pathologic and destroyed cell forms (schistocytes) begin to appear in the blood. Pycnosis of cell nuclei begins to manifest itself; anacytosis, polychromasia, and some other blood diseases become more pronounced. This stage can conventionally be adopted as a critical threshold of irreversible changes. For instance, for whitefish inhabiting the northern Kola Peninsula water bodies, this "threshold" state occurs, when hemoglobin concentration in the blood decreases to the value less than 80 g/l, and destructive and pathologic cells appear. After that, irreversible changes begin to occur in the system of hemogenesis.

iii) At the stage of degradation, when numerous cells are already destroyed, anemia, accompanied by destructive changes in other physiological systems of the organism (kidneys, liver, gills), begins to manifest itself. The above changes lead to the death of the organism. The scheme of toxicosis development described above cannot always be observed in fish inhabiting natural water bodies, because the adaptation threshold for different individuals and their resistance to the effect of sublethal doses of pollutants can vary.

Regularities of intoxication process development allow the conclusion that the use of such parameter as the average concentration of hemoglobin in the blood (without investigation of blood smears) for a group of individuals may fail to reflect the initial stages of intoxication. Variations in the feature and the number of fishes with hemoglobin concentration lower than the critical threshold are more informative.

Method of histological analysis is important for revealing disturbances in the morphologic and functional structure of organs and tissues, degenerative processes, and making the diagnoses of fish diseases according to the disturbances revealed. Pieces of organs and tissues with visible changes are sampled and fixed using traditional methods adopted in histology. In order to establish the normal physiological state, it is necessary to take organ and tissue samples from healthy individuals as well. The table 1 presents histopathological changes in tissues and organs, typical for fish diseases. Destructive changes in the cell structure of organs and tissues that develop most frequently in fish inhabiting natural water bodies contaminated by toxicants are presented below (Fish pathology…, 1994; Chinareva, 1988; Lukin, Sharova, 2004; Savaitova et al., 1995; Moiseenko, Kudryavtseva, 2002; Moiseenko et al., 2006; Moiseenko et al, 2008).

In liver, its cells change and its tissue becomes destroyed. Changes in liver histologic structure testify to the development of cirrhosis or lipoid degeneration of liver. In kidneys, pollutants cause hyperemia, dystrophic changes in the epithelium of ductules and capsules, frequently complicated by necrobiosis. A similar histologic pattern of disease is diagnosed as interstitial nephritis (fibroelastosis). Considerable pathologic disturbances can be observed in the structure of epithelium of gyrose kidney ductules: in some cases, epithelium can flatten, while in other cases it may increase even to become prismatic (the normal configuration of epithelium is cubic). In case of nephrocalcinosis, extraneous occlusions can be detected inside urine-excreting canaliculi. For gills, swelling of respiratory epithelium, hyperemia of petals and small petals, hemorrhages, desquamation, and necrosis of epithelium cells are typical. As for the heart, its expansion and overfilling with blood is observed, in addition to small hemorrhages in myocardium, dystrophic changes in muscular fibers and loss of transverse stripes, protein dystrophy and small-cell fatty degeneration of heart are also observed.

Pathologic changes in fish skeleton (osteoporosis or scoliosis) are most frequently related to the disturbance in mineral substances metabolism, and insufficient or low anabolism of calcium. In muscular tissue, myopathy often manifests itself in delamination of tissue.

Disturbances in the reproduction system manifest themselves in lipoid and connective-tissue degeneration of ovaries and testicles, resorption (observed in breams inhabiting the Volga River), degeneration of oocytes, the formation of cysts, and the occurrence of hermaphrodite individuals. Neoplasms, such as skin sarcoma in pike perch, as well as zones of tumor tissues in liver and kidneys, were also revealed.

Biochemical methods are very important when it is relevant to study the mechanisms responsible for the development of certain abnormalities in the organism. It is shown that, under the conditions of low concentration of pollutants, fishes inhabiting natural water bodies demonstrate activation of physiological systems responsible for detoxification. Intoxication is characterized by an increased concentration of enzymes, dispersion of fatty acids, cholic acids, triaglycerins (triacetins), lysosomatic ferments, acid phosphotasa forms, isoferments of lactadehydrogenasa, free aminoacids as a "quick response" group (Bitton and Dutku, 1985 Sidorov, and Yurovitskii, 1991). The general regularity of metabolism re-organization in fish under the impact of toxic substances corresponds to hemogenesis response. At the stage of contact, the primary response of the endocrine system is characterized by a stimulation of andrenoenergetic systems (leading to an increase in the concentration of catechilamines, adrenalines, and noradrenalines) and hypothalamus-hypophysis-epinephros centers (entailing an increase in the concentration of ATP and corticosterole), which testifies to an increase in the energy exchange rate as a response to toxic substance penetration in the organism.

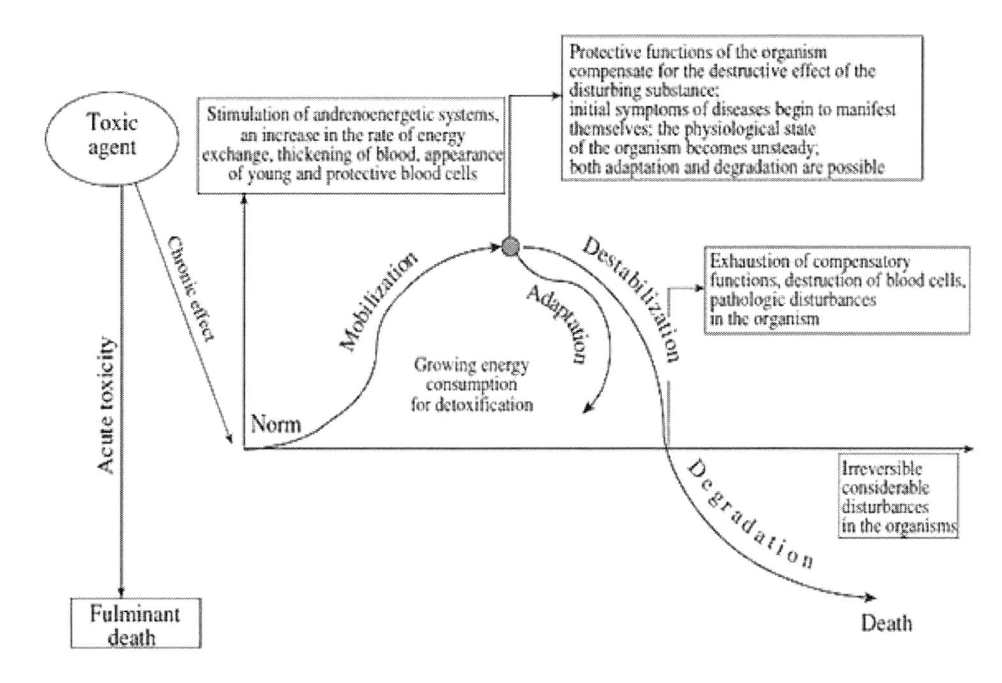

Figure 2. Scheme. of fish response on toxic stress

The secondary response is related to metabolic adaptation, ensuring steady elevated energy consumption for detoxification: increased concentration of catechilamines (the supply of energy to the organism by means of glycogenolysis) and cortisol (inhibiting protein synthesis and energy supply to the organism by means of catabolism of glycogen, lipids and proteins (Niimi, 1990; Sidorov, Yurovitsky, 1991; Werd, Komen, 1998; Nemova, Visotskaya 2004).

When the biochemical functions of detoxification are exhausted (stage of destabilization), these functions become de-regulated, which leads to the destruction and death of the organism. A generalized mechanism of the development of toxicosis in fish is presented in scheme (figure 2).

It is evident that methods of clinic and postmortem examination of fish inhabiting contaminated water bodies are most convenient. They are readily-available and easy-to-apply and can be widely used for express-diagnostic of the ecosystem's "health," because disturbances in physiologic systems of fish testify to unsatisfactory water quality, and these manifestations of poor water quality appear earlier than changes in the structural and functional organization of the community.

Criteria of water quality assessment are substantiated based on the general regularity of toxicosis development and parameters of fish stock heterogeneity in respect of morbidity. Specific diseases (e.g., scoliosis or nephrocalcinosis) can give an idea of their etiology - water contamination with heavy metals. While using hemathologic parameters, it is not the average concentration of hemoglobin in fish blood, but the number of fishes in the concrete habitat with hemoglobin concentration lower than the critical threshold value, delineating the third and the fourth stages of toxicosis (for whitefish, it is the number of fishes whose blood hemoglobin concentration is lower than 80%), in addition to the occurrence pathological forms, that are the most informative criteria.

1.3. Dose-Effect Dependencies and Critical Levels of Water Pollution: Case Study of Arctic Lake Imandra Served As Example

Among Arctic regions, the Russian Kola North (Murmansk region) is the most densely populated and industrially developed. Lake Imandra being one of the most contaminated lakes in Murmansk region served as an example. Development of copper-nickel industrial in the catchment of Imandra began in the 1940s. Large amounts of pollutants entered the lakes between 1940 and 1990. Transformation of economic conditions in Russia at the beginning of 1990s brought to a standstill of many industrial activities and, accordingly, slowed down surface water pollution. Some revival of the economy in last decade is spurred by technology modernization and more restrictions on pollution of the lake and the atmosphere. As a result, pollution levels entering the lake have decreased over the past 20 years, but nowadays it still remains high (Moiseenko et al., 2006).

The main pollutants were heavy metals (predominantly, nickel and copper), sulphates, chlorides and nutrients. The catchment areas were also polluted by airborne contaminants, metals and acid deposition. The highest contamination level of lakes in that region was observed in 1970-1980's when the copper and nickel production was the highest. The metal content in lake Imandra reflects the historical dynamics of water contamination. The maximum nickel contamination reached 290 μg/l (table 2).

Table 2. The metal concentration in the water of Lake Imandra in various years of researches (numerator – average values, denominator – minimum and maximum values)

Metal, μg/l	Years of investigation					
	1981	1986	1991	1996	2003	2007
Ni	42 16-84	61 13-290	25 13-49	15 13-29	10 7-27	11 4-16
Cu	3 0-5	11 1-130	15 4-38	6 5-18	5.5 4-10	4.4 3-9.9
Cd	n/d	n/d	n/d	0.27 0.12-0.60	<0.05	0.10 0.05-0.29
Pb	n/d	n/d	n/d	n/d	<0.3	<0.3 <0.3-0.5

Table 3. The characteristic of whitefish diseases and metals accumulations in condition of metal water pollution of Lake Imandra in various years of researches

Years of investigation	1981	1986	1991	1996	2003	2007
The main symptoms of fish diseases, % from number of the surveyed individuals						
Nephrocalcitoses	52	47	45	14	-	-
Fibroelastos	48	53	55	48	39	16
Lipoid degeneration of a liver and a cirrhosis	100	89	78	48	39	16
Anomalies of a structure gonads	34	27	8	-	-	3
Number investigated fish	n = 788	n=721	n=453	n=462	n=235	n=196

Unfortunately, there is no historical data on the content of cadmium and lead in water for that time period. The reasonable assumption is that the content of all concomitant metals was much higher.

Starting from 1980 there are studies aimed at the investigation of fish morbidity. As bioindicator white fish was used (*Coregonus lavaretus* L.). Table 3 summarizes the data about and fish morbidity in Imandra lakes and content of metals in fish in last 26 years.

The main symptoms of fish intoxication in the Kola lakes polluted by metals are as follows: change of the integument colour (depigmentation), tousling of scales, oedema gills and appearance of anemia rim, destructive changes of liver (increase of size, change colour and friability) and kidneys (colour, granulation, thickening of renal and presence of nephritic calculi), anomalies in gonad texture, etc. In the areas polluted by Kola copper-nickel smelters the fish endemical pathology most frequently met is nephrocalcitosis and fibroelastosis of kidney.

The kidney having nephrocalcitosis signs show clear changes in kidney cytomorphological structure. There have been visible fibrous granulomas formations and necrotic parts in the parenchyma. Connective tissue growing was found around the Boumen capsules, blood vessels and excretory duct. Significant pathologies were observed in epithelium texture of proximal tubule. There has also been found epithelial hyperplasia (up to the prismatic epithelial cells, in norm - cubic); at other part - epithelial hypoplasia (up to desquamation); inside it - inclusions of stones (Figure 3).

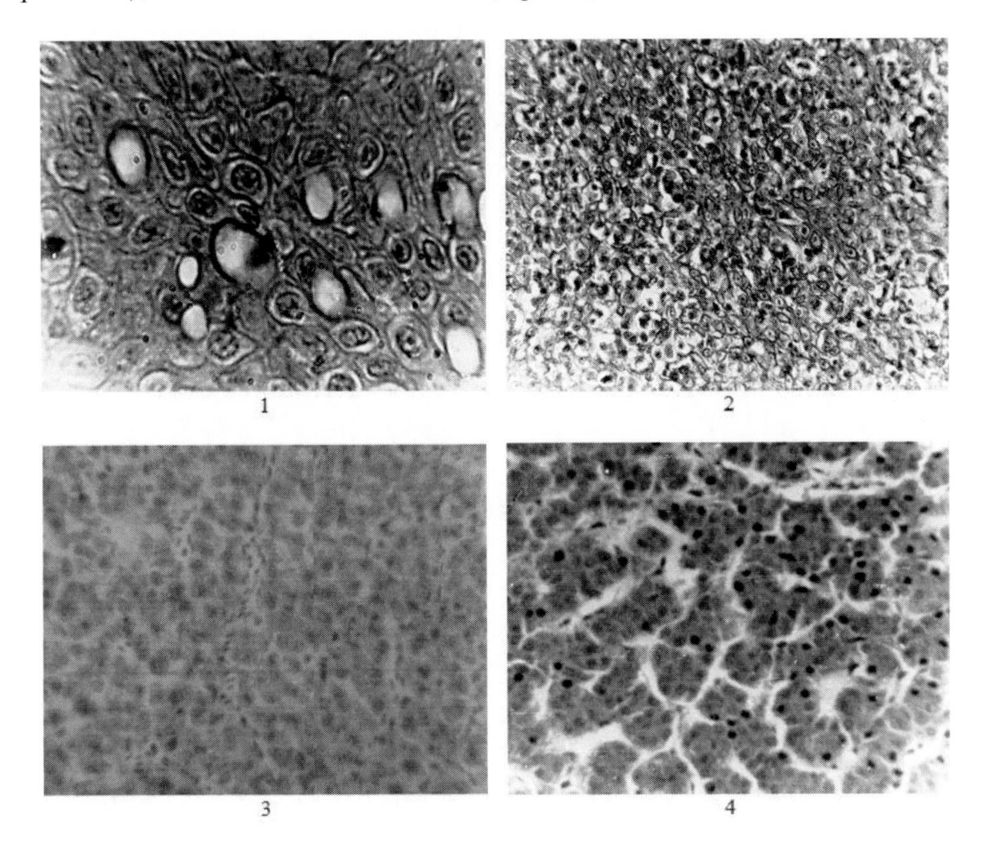

Figure 3. Liver histopathologies: 1 – drops of fat in the hepatocyte; 2 – lipoid degeneration of the liver; 3 – liver parenchyma hemorrhage; 4 – widening of inter-hepatocyte spaces.

Fish kidneys are capable of accumulating high levels of nickel. Copper being one of the essential elements does not exhibit the same trend; moreover, the copper content in liver during the period of highest contamination was low. The work of Moiseenko, Kudryavtseva (2002) reflects this phenomenon, which is explained by the degenerative alteration of the functional tissue of this organ.

Hystological analysis of hepatic diseases shows lipoid degeneration of its cells, in some places - signs of necrosis and degenerative damage caused by destruction of hepatocytes and overgrow of connecting tissue (cirrhosis). Different abnormalities of liver (from initial to debilitative) are also frequently detected in fish living in water polluted by metals (Figure 4).

Gill histopathology is presented on figure 5. In nature, aquatic organisms are exposed to a combined dose of all contaminates.

It is important to find a general numeric parameter describing the total impact of contaminates on biota. In case of Imandra lake the main pollution is heavy metals. An integrated impact dose of metals is determined by the number of metals, their concentration and toxic properties for each of them.

The values of Guideline Concentrations (GC) or Maximum Permeation Concentrations (MPC) largely differ by country, in spite of the fact that experimental research techniques to establish the MPCs are universal. For example, in Russia, the MPC values for Cu, V, Mn and some other elements are unreasonably underestimated, whereas the MPCs for Cd, As, Pb, and Al are overestimated.

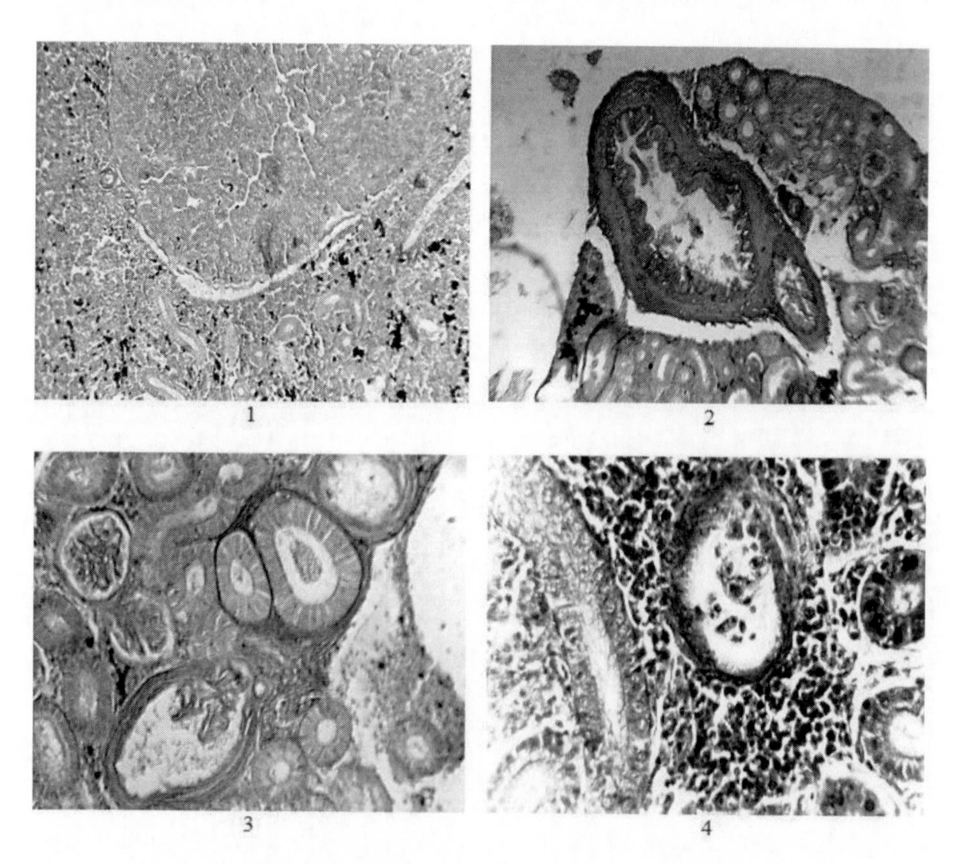

Figure 4. Kidney histopatologies: 1 – focal renal parenchyma necrosis; 2 – conjunctive tissue overgrowth; 3 – necrosis of glomerulus renalis vascular loops; 4 – renal channel necrosis.

Although accepted in Russia, as well as in other countries, water quality standards for metals in water do not take into account the integrated impact dose. Using data about toxicological properties of each metal based on GC, we can define the integrated impact dose by summing the excess of real concentration for each of metals to their GC or known threshold of impact as follows:

$$I_{tox} = \Sigma(C_i / GC_{i\text{-fish.}}).$$

I_{tox} – is the integrated toxicity index, C_i – are concentrations registered in water; $GC_{i\text{-fish}}$ – are GC for metals accepted in Russia for aquatic biota, which are more stringent than those for drinking water. Water quality may be considered good if $I_{tox\text{-}1}$ is no more than one ($0 < I_{tox} \leq 1$).

Despite our critique of the GC, since Imandra is in Russia, we used the following, legally binding in Russia, concentrations, (μg/l): Cd = 5, Ni = 10, Cu = 1, Pb = 10, Zn = 10, As = 50; Cd = 5 (List of Fishery-Related Standards…, 1999).

The rate of fish disease (measured in terms of all the three indicators considered above: percent of fish with physiological deviances from norm, Z - the morbidity index of health, and the percent of fish with anemia - decrease of hemoglobin in blood below 80 g/l) came out as a very sensitive measure of toxic pollution (Figure 6).

Analysis of these dose-effect dependencies shows that when the integrated toxicity index of water is more then 1 pathologies and dysfunctions are likely to appear in fish organisms.

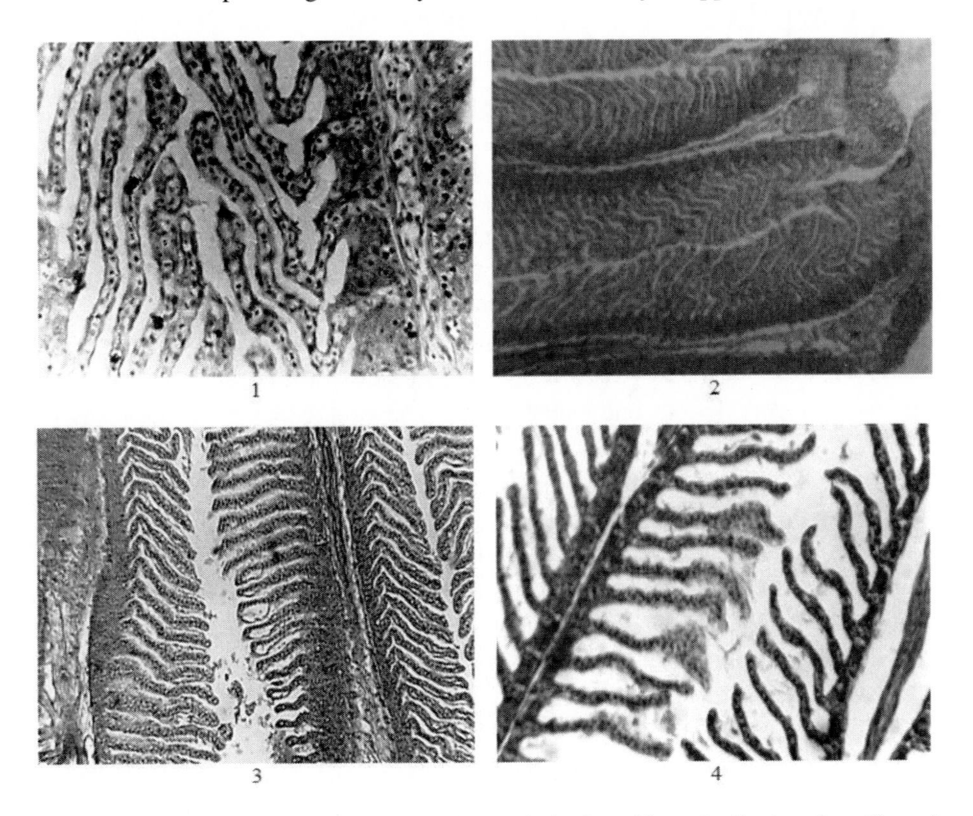

Figure 5. Gill histopathology: 1 – gill epithelium hyperplasia; 2 – gill petal adhesion; 3 – gill petal epithelium exfoliation; 4 – swelling of apical ends of gill petals.

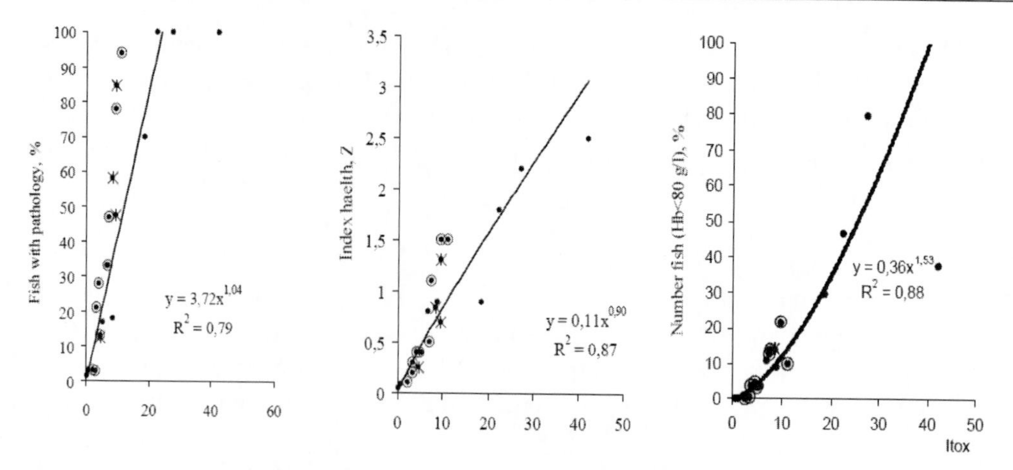

Figure 6. Measuring ecosystem health: dependence between an integrated parameter of toxic loading ($I_{tox} = \Sigma(C_i / GC_i)$ and indicators of fish diseases (% of fish with deviances, % of fish with anemia (decrease of haemoglobin in blood lower than critical 80% level).

Our results showed that for ecosystem health assessment the fish disease index is sensitive, and also demonstrated that the standards for water quality assessment adopted in Russia are inadequate for Arctic regions. The dose-effect dependencies (in case study of arctic lake Imandra) clearly show that heavy metal pollution must be significantly decreased. At the same time, for healthy ecosystem revitalization water pollution (integral parameter $I_{tox-fishl}$) must be decreased at least 5 times, first of all for nickel, which determines many water properties.

Conclusion

The necessity of determining ecological standards for water quality requires an improvement of the existing methods of its biological assessment and the development of new ones, which would be effective and readily-available for large-scale application. At present, two methods are used. The first one is the bioassay technique, implying water quality evaluation on the basis of the response of organisms to the impact of toxic substances during the experiment (*ex-situ*). The second method is bioindication, which is based on studying the state of the ecosystem's structural elements (individu-als, populations, communities *in situ*). Both these methods allow a qualitative assessment of water to be used for certain purposes. At the same time, these methods do not allow the determination of critical water pollution and standardization of toxic substances input in the water bodies.

At present, ecotoxicological approach to estimating the maximum permissible level of water contamination, which allows a comprehensive characteristic of the ecosystem "health," gradually wins the recognition of researchers. Fish, regarded as the upper trophic level of the aquatic ecosystem, is an indicator of toxic contamination of water bodies. Detection of changes in fish physiologic parameters ensures forecasting the consequences of toxic substances occurrence in the water for human beings as well. The state of fish can be determined using all the available methods, such as clinic, postmortem, hemathologic, or biochemical studies. As compared to parameters of structural and functional state of aquatic communities, these methods allow us to characterize water quality in a shorter period and to assess the effect of multicomponent long-term water contamination. Scientific substantiation of criteria of estimating the biological effects of water pollution (the establishment of ecological standards, the critical levels or hazard of water contamination) is the most important element in the system of human impact management. Basing on dose–effect dependences (between numerical indices of pathologic–physiologic states of fish and the chemical parameters of water quality, in particular, the total concentration of toxic substances in the water standardized to MPC), the critical water contamination can be determined.

In nature, aquatic organisms are exposed to a combined dose of all contaminates. It is important to find a general numeric parameter describing the total impact of contaminates on biota. In case of Imandra lake the main pollution is heavy metals. Parameters of fish physiological conditions are even more directly related to water quality. During the period of severe pollution there were mass fish diseases (nephrocalcinosis, lipid liver degeneration, cirrhosis, anemia, scoliosis and others). Based on fish intoxication symptoms, we can conclude that water quality is getting better in last period, but it is still far from recovery. In case study of arctic lake Imandra used ecotoxicological approach clearly show that heavy metal pollution must be significantly decreased.

Acknowledgments

The work was supported by the Russian Foundation for Basic Research (Projects no 10-05-00854) and grant of Russian Governments (№ 11G34.31.0036).

References

Adams, S.M, Ryon, M.G.A. (1994). Comparison of health assessment approaches for evaluating the effects of contaminant-related stress on fish populations. *J. Aquat. Ecosyst. Health* 3, 15-25.

Attrill, M.J, Depledge, M.H. (1997). Community and population indicators of ecosystem health: targeting links between levels of biological organization. *Aquat. Toxicol.* 38, 183 - 197.

Bespamyatnov GP, Krotov Yu.A. (1985). Maximum Permissible Concentrations of Chemical Compounds in the Environment, Leningrad: *Khimiya*, 163 p. (in Russian).

Bioassay Methods for Aquatic Organisms. (1985). In: *Standart Methods for the Examination of Water and Wastewater*, Washington, DC: Amer. Public Health Assoc., pp. 45–52.

Bitton, G. and Dutku, B.J. (1985). Introduction and Review of Microbial and Biochemical Toxicity Screening Procedures. In: *Standart Methods for the Examination of Water and Wastewater*, Washington, DC: Amer. Public Health Assoc., pp. 31–40.

Canadian Water Quality Guidelines. (1994). Published by Canadian Council of Ministry of Environment.

Cash, K.J. (1995). Assessing and monitoring aquatic ecosystem health - approaches using individual, population, and community/ecosystem measurements. In: *Northern River Basins Study Project*, Report No 45,68 p.

Chinareva, I.D. (1988). Pathohistological changes in fish of the Ladoga basin. In: *Pollution Impact on the Ladoga Ecosystem*. Leningrad: GosNIORKh, pp. 24–32 (in Russian).

Elliott M., Hemingway K.l., Krueger D., Thiel R., Hylland K., Arukwe A., Forlin l., Sayer M. (2003). From the individual to the Population an Community Responses to Pollution. In: *Effects of Pollution on Fish* (eds. Lawrence A.J., Hemingway) K.L. Blackwell Science Ltd. pp. 221 -255.

Environmental Quality Objectives for Hazardous Substances in Aquatic Enviroment. (2001). Berlin: UMWELTBUNDESAMT. 186p.

Fish Pathology, third edition. (1994). Ed./ Roberts, R.J. London.WB SAUNDERS.

Liney, K. E., Hagger, J. A., Tyler, Ch. R., Depledge, M. H., Galloway, T. S. and Jobling, S. (2006). Health Effects in Fish of Long-Term Exposure to Effluents from Wastewater Treatment Works. *Environ. Health Perspect.* 114, 81–89.

List of Fishery-Related Standards on Maximum Permissible Concentrations (MPC) and Safe Reference Levels of Impact (SRLI) of Hazardous Substances for Water Bodies. (1999). Used for Fishery and Aquatic Life). Moscow: VNIRO; 304 pp.

Lukin, A.A. and Sharova, Yu.N. (2004). Water Quality Estimation Based on Histological Investigations of Fish: Case Study of Kenozero Lake. *Water resour.* 4, 481–489.

Manual on Hydrobiological Monitoring of Freshwater Ecosystems (Ed. Abakumov, V.A.) (1992). St. Petersburg: *Gidrometeoizdat*, 308 p. (In Russian).

Methodological recommendations for establishing environmental and fishery standards for pollutants in water for aquatic life and Fishery. (1998) (Ed. Filenko OF), Moscow: VNIRO. (In Russian).

Moiseenko, T.I., and Kudryavtseva, L.P. (2002). Trace Metals Accumulation and Fish Pathologies in Areas Affected by Mining and Metallurgical Enterprises, *Environ. Poll*, 114 (2), 285 - 297.

Moiseenko, T.I. (1998). Hematologic Characteristics of Fish in the Assessment of Their Toxicoses, *Russian Journal of Ichthyology*, 2, 371–380.

Moiseenko, T.I., Voinov, A.A., Megorsky, V.V. (2006). Ecosystem and human health assessment to define environmental management strategies: the case of long-term human impacts on an Arctic lake. *Sc. Tot. Environ*. 369, 1-20.

Moiseenko T. I. Gashkina N., Sharova Yu., L.P. Kudryavtseva L. (2008). Ecotoxicological assessment of water quality and ecosystem health: A case study of the Volga River, *Ecotoxicology and Environment Safety*, 71, 837 -870.

Моисеенко T.I. (2009). Aquatic Ecotoxicology: Fundamental and Applied Aspects. M.: *Science*. 367p. (in Russian).

Nemova, N.N. and Visotskaya, R.U. (2004). Biochemical indication of fish health M.: *Nauka*. 230 p (In Russian).

Niimi, A. J. (1990). Review of biochemical methods and other indicators to assess fish health in aquatic ecosystems containing toxic chemicals. *J. Great Lakes Res.* 16, 529-541.

Odum, P. (1981). *Fundamentals Ecology*. EUGENE. Philadelphia-London-Toronto. 618 pp.

Savvaitova, K.A., Chebotarev, Yu.V., Pichugina, M.Yu., and Maksimov, S.V. (1995). Anomalies in Fish Organisms as Indications to the Environmental Conditions. *Russian journal of Ichthyology*, 2, 182–188.

Sidorov, V.S. and Yurovitskii, Yu.G. (1991). Perspectives of the Use of Biochemical Methods for Recording Ecological Modulations. In: Ecological Modifications and

Environmental Standardization Criteria, Leningrad: *Gidrometeoizdat*, pp. 264–278. (In Russian).

Werd J.H., Komen J. (1998). The effects of chronic stress on growth in fish: critical appraisal // *Comparative Biochemistry and Physiol.* 120, 107-112.

Whitfield, AK, Elliott, M. (2002). Fish as indicator of environmental and ecological changes within estuaries: a review of progress and suggestions for the future. *J. Fish Biol.* 61, 229-250.

Wong PTS, Dixon DG. (1995). Bioassessment of water quality. *Environ. Toxicol. Water Qual.*10, 9–7.

Yeom D-H. and Adams S.M. (2007). Assessment effects of across level of biological organization using an aquatic ecosystem health index. *Ecotoxicol. Environ. Saf.* 67, 286–295.

Zhiteneva, L.D., Poltavtseva, T.G., and Rudnetskaya, O.A. (1989). *Atlas of Normal and Pathologic Modified Blood Cells of Fish*, Rostov-on-Don: Rostovskoe Kn. Izd. 110 p (In Russian).

In: Water Quality
Editor: You-Gan Wang

ISBN: 978-1-62417-111-6

Chapter 10

Water Quality Impacts on Human Population Health in Mining-and-Metallurgical Industry Regions, Russia

***T. I. Moiseenko*[1,*], *N. A. Gashkina*[1], *V. V. Megorskii*[2], *L. P. Kudryavtseva*[2], *D. N. Kyrov*[3] *and S. V. Sokolkova*[3]**

[1]Institute of Geochemistry and Analytical Chemistry, Russian Academy of Sciences, Russia;
[2]Institute of Problems of Industrial Ecology of the North, Kola Research Center, Russia
[3]Tyumen State University, Russia

Abstract

The pollution of water sources and drinking water in some towns and settlements of the Kola North by metals, wastewaters, and airborne pollution effluents of mining-and-metallurgical industry is characterized in this chapter. Our statistical data on population morbidity are provided, as well. Over the course of our studies, relationships were found to exist between water quality indices and heavy metal accumulation in kidney and liver of postmortem-examined patients, and the results of their histological, clinical, and postmortem examination are given. The results of comprehensive studies are used to assess the effect of drinking water pollution on the population health in the region.

1. Introduction

The Arctic region is a part of the Planet where the territory is covered by a very great number of lakes and rivers. The high provision of the Arctic regions with water till recently has not caused any trouble about the state of the latter. At the same time, intensive development of the rich deposits of mineral recourses and trans-boundary pollutions lead to a

---

* Corresponding author.

rapid disturbance in the fragile environmental equilibrium already in many urbanized and industrial Arctic regions, which leads to qualitative depletion of the water resources.

The Kola peninsula of Russia is most densely populated and industrially developed region of Arctic. The spectrum of anthropogenic impacts on the surface water is wide: mining, metallurgy, refineries and chemical industries, nuclear power plants, etc. Industrial development of copper- nickel, rich apatite-nephelinite and iron deposits in the Kola Peninsula began in the 1930s. Large amounts of pollutants entered the lakes between 1940 and 1990. The main pollutants were heavy metals (predominantly nickel and copper), sulphates, chlorides and nutrients. The catchment areas were also polluted by airborne contaminants, metals and acid deposition. Since 1990, as a result of the economic crisis in Russia, the anthropogenic pressure on the lakes has decreased.

For more than 70 years the lakes have been used as a source of the technical and drinking water supply, for recreation, tourism and fishery. Paradoxically though it may seem, the water used for drinking water supply to major towns is often being taken from surface water bodies, receiving effluents from industrial plants or located in the zone of airborne pollution. The recent recovery (2000 -2010) of the economy goes on simultaneously with technological modernization and stricter controls of pollutant emissions into the lakes and the atmosphere. The map of the region and location of basic industries in it is shown in figure 1.

The aquatic environments are final collectors of all kinds of pollution. Contents of elements in water reflect airborne metal contamination and pathways of metals. Life in water bodies, as opposed to terrestrial conditions, is characterized by stronger relations between aquatic organisms and factors of the environment due to the high role of ecological metabolism in water ecosystems and high mobility of polluting substances in water. Four basic processes are distinguished that lead to high contents of metals in the surface water of region: (i) in waste waters of metallurgic manufactures; (ii) distribution with smoke emissions; (iii) acid leaching from surrounding rocks, especially from natural geochemical formations; iv) transboundary pollution to Arctic. Ni, Cu, As, Cd, Pb, Hg, Co, Zn, Mn and rare-earth metals will enter the environment as a result of industrial activities.

Many metals, derived from the rocks and enriched in technological treatment, will become toxic when they enter the environment. It is known that a number of human diseases are connected with increased metal concentrations. The surplus of trace elements in the human organism results in specific diseases: Hg causes a neurological effect, Cd and Pb have cancerogenic properties, Sr leads to pathologies of bone tissues, Cu - to anemia, etc. (Handbook of Metals…, 1994; Handbook of Ecotoxicology, 2005; Spry, Wiener, 1992). Difficulties in determination of dangerous metal levels for vitality are stipulated by the following factors: (i) many elements (Cu, Zn, Co, Sr, Se, Ni etc.) are significant, i.e. inherent to organisms and are present in organisms in microquantities; (ii) poisoning influence of metals is formed both due to direct effects and ability to be accumulated in organisms, causing remote consequences - mutagenic, embriotoxic, gonadotoxic, cancerogenic, etc.; (iii) toxicological properties depend on metal speciation, combinations of elements (phenomena of sinergetism and antagonism) and concomitant factors. The aquatic environments are final collectors of all kinds of pollution. Contents of elements in water reflect airborne metal contamination and pathways of metals. New geochemical provinces are shown to form under current conditions (Moiseenko, Kudryatseva, 2002), and the environmental consequences of this process require close attention, and especially—studying the effect of surface water pollution on human health.

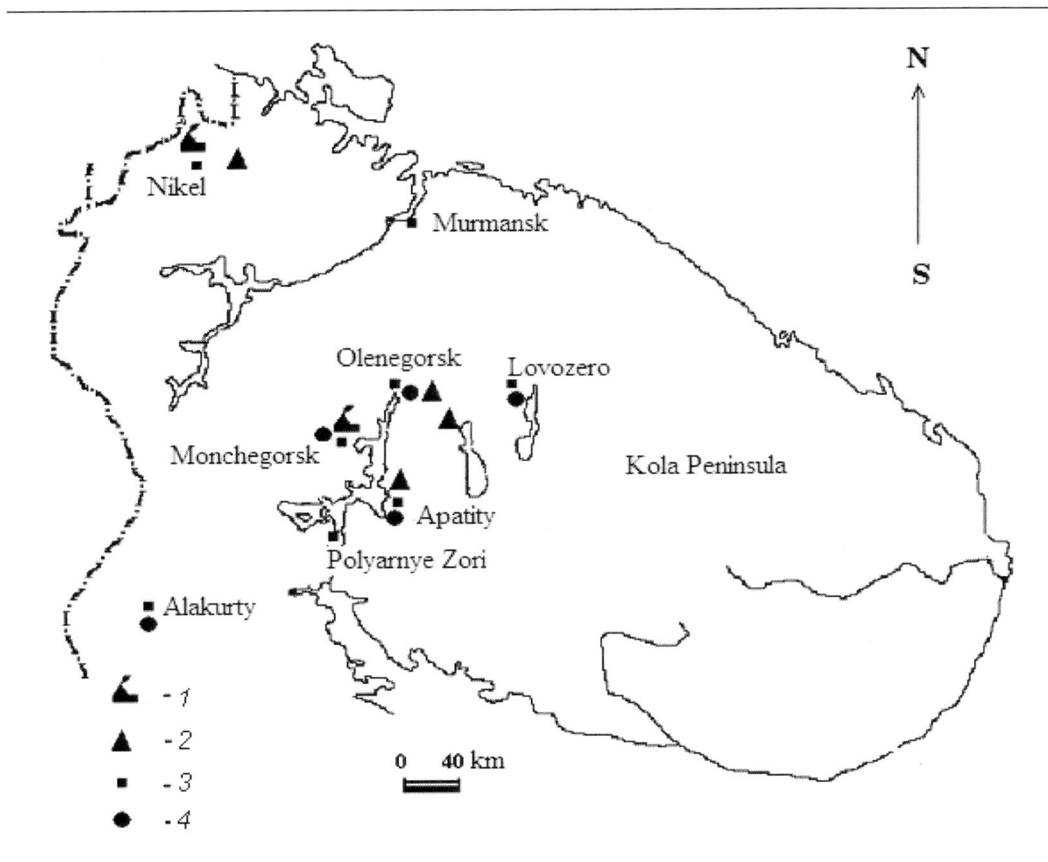

Figure 1. Layout of (*1*) copper_nickel works, (*2*) mines and mining and concentration plants, (*3*) population localities, and (*4*) drinking water intakes.

It is necessary to note, that hazardous substances enter the human organism with food, water, and air; therefore, it is often difficult to establish correlation between drinking water quality and population health. However, it should be taken into account that people in transpolar regions largely eat products delivered from southern regions; therefore, the input of pollutants with food made from local agricultural products has not significant effect on population morbidity.This region can serve as a model region for understanding key anthropogenically-inducted processes in lakes and its impacts for water quality and population health.

The objectives of this study were as follows:

- the identification of pollution of surface water used for drinking water supply by toxic heavy metals with six towns and settlements in the industrial Kola region of Arctic used as an example;
- efficiency assessment of water purification of metals in the process of water treatment;
- assessment of population morbidity in towns and settlements taking water for drinking water supply from surface water bodies;
- assessment of the effect of drinking water pollution and heavy metals accumulation in human organs on the population health, including the development of metal-induced pathologies in humans.

2. Materials and Methods

Multidisciplinary studies were carried out in industrially developed towns (Monchegorsk, Apatity, Polyarnye Zori, and Olenegorsk) and in more remote settlements (Alakurti and Lovozero), which take water for drinking water supply from surface sources. The layout of major industrial facilities and water intakes for drinking water supply is schematically shown in Figure 1. Water for drinking water supply to Monchegorsk population is taken from Monche Lake, which lies in the zone of aerotechnogenic pollution by emissions of Severonikel industrial complex; the towns of Apatity and Polyarnye Zori consume water originating from Lake Imandra.

The population morbidity in towns and settlements was assessed by using data from the Committee of Public Health, Murmansk region; Murmansk Region Cancer Detection Centre; and Moscow Research Oncological Institute.

Water intakes of Apatity and Polyarnye Zori are located in the zones of transit mixed flow of wastewaters from various industrial facilities (copper-nickel smelter, mining-and-processing of apatite and iron ores) and municipal wastes of Apatity and Monchegorsk towns. The distances between the various discharge sites and the water intakes are 50–100 km for Apatity and 100– 150 km for Polyarnye Zori. Water supply to Lovozero and Alakurti settlements also relies on surface water bodies; however, they are far enough from industrial centers (>200 km from the smoke emission plumes).

Drinking water samples for the analysis were taken from the source of water supply and from networks for water supply to the population, allowing changes in water chemistry after treatment and during its passage through pipes to be assessed.

Water samples were taken into Nalgen® polyethylene bottles, whose material has no sorption capacity. The bottles were thoroughly cleaned in the laboratory and twice rinsed by the lake's water before sampling. Once taken, the samples were placed into dark containers and cooled to about +4°C, at which temperature they were transported into the laboratory. Water for trace element analysis was filtered in the field with the use of Milipore plant; both filtered and unfiltered waters were acidified by nitric acid and in the prepared form send to the laboratory for further analysis.

To study heavy metals accumulation in the tissues of residents of the examined populated localities, samples were taken from the liver and kidneys of peoples who have lived in the area for no less than 10 years and did not work immediately at plants with high health hazards (110 postmortem-examined patients). In the case of chronic alcoholism or viral hepatitis, the appropriate samples were rejected. The reference group or a "norm" of the trace element composition of organs for the assessment of the extent of metal bioaccumulation was taken to be appropriate tissues of fetuses (dead-born and immature births). It was assumed that the transplacental barrier prevents metals from penetration into the organisms of developing fetus. Simultaneously, postmortem samples were taken for histological examination, which was further carried out by conventional methods. The consolidated conclusion regarding the population morbidity was based on the results of analysis of clinical, histological, and laboratory characteristics of the examined patients. The examined clinical features included anamnesis data, the results of clinical examinations while alive and the pathologist's report.

Determine of water chemistry are executed by uniform techniques according to the recommendations (Standart method…, 1992) - Ca^{2+}, Mg^{2+}, K^{+}, Na^{+}, Alk, SO_4^{2-}, Cl^{-}, color,

NO_3^-, NH_4, Ntot (TN), PO_4, Ptot (TP), Si; pH – by a Metrohm® pH-meter; conductivity (20°C) - by Metrohm®-conductivity; alkalinity - be using the Gran titration method, organic matter content – by the Mn oxidation method. The applied analytical techniques and results of the determination of the chemical composition of waters were verified using a common system of standard solutions under permanent strict intralaboratory control.

Samples for determining trace element concentrations in biological sample were prepared by wet decomposition in aquafortis with addition of hydrogen peroxide.

Concentration of elements (Sr, Al, Fe, Mn, Cr, Cu, Ni, Zn, Cd, Co, Pb, As) in water and in sample the tissues of residents were determined twice: atomic absorption spectrophotometry with graphite atomization Analyst-800 with Zeeman background corrector and by plasma atomic emission spectroscopy «Plasma Quad 3» of firm «Fisons Instruments Elemental Analisis» (Great Britain). Mercury concentration was determined on mercury analyzer FIMS-100 (Perkin-Elmer).

3. RESULTS

3.1. Water Quality

Surface water of Kola ecoregion in reference condition was once characterized by very clean water and oligotrophic conditions. Historical sampling indicated low concentrations of suspended material (0.7 – 1.0 mg/l), microelements (<1 μg/l), and nutrients (e.g total phosphorus < 2 μg/l). Biological uptake of any available nutrients was rapid, so phosphates during the vegetation period were almost gone. Water transparency was about 8m (Moiseenko, 1999).

Accordingly during these periods, the industries spawned the development of numerous towns and cities to support their operations, including Monchegorsk, Olenegorsk, Apatity, and Polyarnye Zori. Pollution from these areas, as well as from the industrial facilities, began to affect the condition of the lake. Yet still, Lake Imandra was (and still is) used as a source of drinking water for several cities, while serving as a sink of industrial and domestic wastewater. The source of water supply Dumps of waste rock and processed ores, mine and quarry water, wastewaters and dumped wastes of concentration plants are sources of anthropogenic migration of some elements. Acid precipitation, which form under the effect of atmospheric emissions of acid-forming agents by smelter and transboundary pollution, accelerate the chemical leaching of elements and their migration.

3.2. Characteristic of Water Pollution Near Water Intakes and When Supplied to the Population

Metal concentrations in water bodies involved in drinking water supply and in drinking water supplied to the population of towns and settlements are given in Table 1.

Lake Monche lies in the zone of aerotechno-genic pollution from the source of smoke emission of Severonikel mining-and-processing integrated works (<30 km). The lake's water is high in Ni, Cu, Cd, and other metals relative the background values.

Table 1. Metal concentrations in natural water bodies at water intake sites for water supply (top number) and in a distribution pipeline during the delivery to the population (bottom number) in towns and settlements in Murmansk province

Water supply source/populated locality	Ni	Cu	Cd	Pb	Sr	Cr	Co	$\Sigma C_i/MPC_i$
Lake Monche	11.0	15.5	0.30	<0.5	20	0.6	0.4	0.345
Monchegorsk	15.8	15.8	0.19	<0.5	17	0.2	0.5	0.233
Lake Imandra	6.0	3.0	0.10	1.5	73	0.1	0.3	0.174
Apatity	4.9	3.2	0.15	0.6	90	0.2	0.2	0.196
Lake Kuna	1.4	2.3	0.10	<0.5	35	0.2	<0.2	0.113
Olenegorsk	0.9	1.6	0.09	<0.5	34	0.3	<0.2	0.103
River Virma	0.5	0.7	0.10	<0.5	43	0.3	0.3	0.116
Lovozero Settl.	0.8	1.4	0.13	<0.5	49	0.3	<0.2	0.144
River Tuntassaioki	0.5	0.2	0.11	<0.5	160	0.1	<0.2	0.136
Alakurti Settl.	0.4	0.6	0.14	<0.5	138	0.1	<0.2	0.161

The water intake of Olenegorsk Town is also situated within the propagation zone of smoke emission, though at a larger distance (>50 km); accordingly, metal concentrations in the lake's water are lower. Lovozero and Alakurti settlements are supplied by water from lakes where metal content of water is near the regional background level. The high concentration of Sr in the water of Alakurti Settlment is accounted for by the geochemical features of rocks in the area. Water to Apatity Town is supplied from Lake Imandra, which is polluted by wastewater from Severonikel mining-and-processing integrated works. The concentrations of metals in the water intake site in Lake Imandra is lower than in Lake Monche because of the larger volume of the former lake, though they are far in excess of background values. The concentrations of Ni and Cu at the wastewater discharge sites is 5–10 times greater than those in the water intake zone (Moiseenko et al, 2006).

The water processed in the water treatment cycles is only slightly clearer than water in natural water bodies. In population localities, Fe concentration in water is found to increase during its passage through pipelines, because of Fe leaching from steel pipes. Higher Ni concentrations were established for water supplied to the population in Monchegorsk Town. Neither of metals was found to exceed the sanitary–hygienic standards for drinking water in any population locality. However, it should be taken into account that water is polluted by a complex of metals, which act in soft water against the background of very low Ca content, the feature that enhances their penetrability and the adverse impact on the health of living organisms, including humans.

3.3. Characteristics of Public Population Morbidity

According to statistical data, the most significant disease classes in the region are diseases of blood circulatory system; neoplasms; diseases of respiratory organs, urogenital system, and digestion organs, including hepatic cirrhosis. The highest morbidity rates are typical of the population of towns consuming water from lakes Imandra and Monche, where the highest metal concentrations were recorded in drinking water (Figure 2).

According to data of the Committee of Public Health, the mortality rate in Monchegorsk region increased by 14% over the past five years. The most unfavorable is the situation with the growth of malignant neoplasms. Notwithstanding the relatively small percentage of oncopathology among other diseases, its significance in the analysis of the population health is high, since neoplasms have the highest death and invalidization rates as their consequences. The identified classes of diseases, which are more inert to the effect of the etiologic factor, can be largely dependent on the cumulative impact of chronic doses of heavy metals.

Clearly the effect of drinking water on humans is to one extent or another combined with all environmental factors, as well as the social and economic conditions. Humans in the Extreme North live under stress climate conditions.

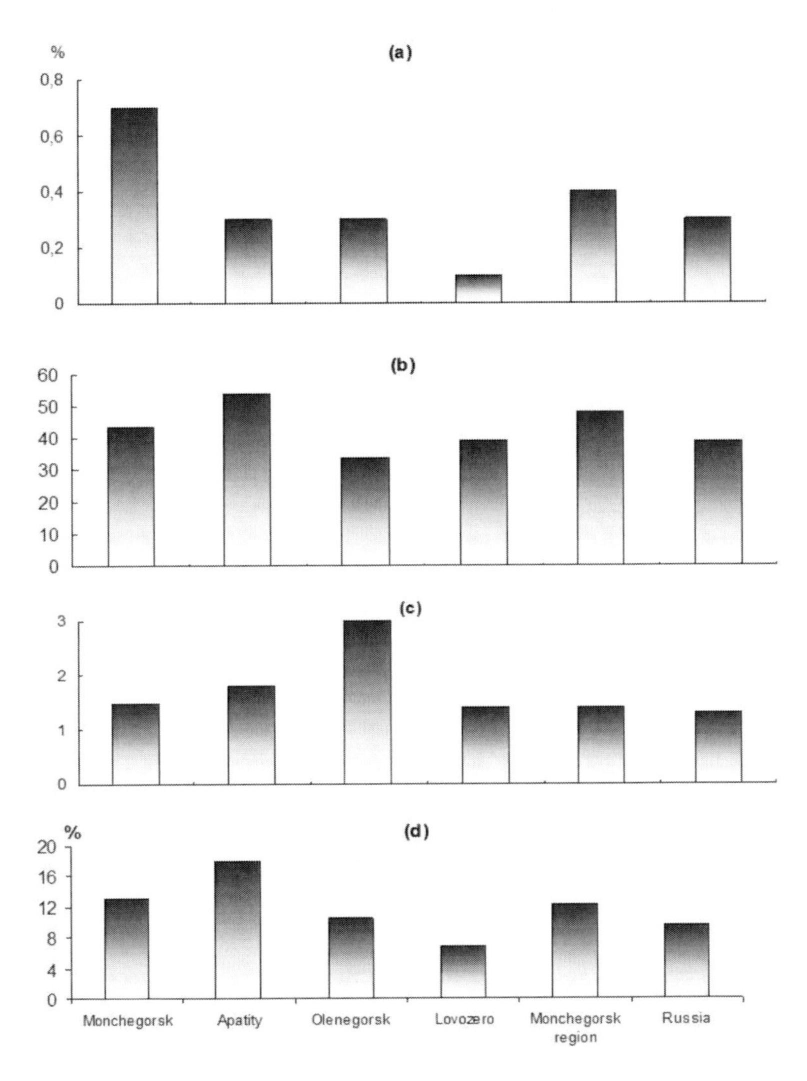

Figure 2. Average statistical data on occurrence of major diseases in adult population in cities and Murmansk region as a whole (a – disease of organs of digestion, b – diseases of genitourinary system, c – urolithiasis, d – neoplasms). Occurrence measured as number of first diagnosed per 1000 population.

It is possible that the complex of factors, including the effect of metals in low-mineralization water, results in higher population morbidity in the towns under consideration as compared with statistical data on Russia (Figure 2). The highest morbidity of population in the towns consuming water from Imandra and Monche lakes suggests that human organisms accumulate technogenically introduced metals, which can increase the population morbidity.

3.4. Accumulation of Heavy Metals and Accompanying Pathologies of Organism Systems

More detailed studies in the system metals in water–their bioaccumulation–histopathology–the diagnostics of pathologies with toxic etiology were carried out with the aim to prove the reliability of the effect of drinking water pollution on the population health. The concentrations of metals in the liver and kidneys of the examined patients are given in Figure 3.

The highest indices of heavy metal accumulation in the liver were recorded in Monchegorsk inhabitants, where the concentrations of many metals, especially, Ni, Cu, Cr, Cd, and Pb, are 2–10 times larger than the norm. The highest concentration in the kidney tissue were recorded for Cr and Cd – 10–50 times the norm.

Human organisms (liver and kidneys) mostly accumulate Cu, Cr, and Cd in Apatity region; Cu, Cd, and Pb in Olenegorsk region; and Cu and Cd in Lovozero region and Alakurti Settlement.

Despite the fact that Ni and Cd are the major drinking water pollutants in the region, Cd features the largest ability to accumulate in human kidneys. It should be taken into account that the Kola Region is subject to acid precipitation, which actively leach this element into water. Lake Monche, where a municipal water intake is operated, lies in the zone subject to a strong impact of acid precipitation. High concentrations of Cd were detected in people living in Olenegorsk and Apatity.

By the complex accumulation of heavy metals in the liver and kidneys of inhabitants, the towns can be ranked in the following order: Monchegorsk > Apatity > Olene-gorsk > Lovozero > Alakurti. This series is in agreement with the order of decreasing HM concentrations in the drinking water of the towns.

To assess the consequences of the chronic impact of subtoxic doses of metals and their accumulation on human health, histologic specimens taken from extracted organs of postmortem examined patients were analyzed for heavy metal content. Parallel to that, clinical records and autopsy protocols were examined. The main objective was to identify the latent forms (which were not revealed by life-time diagnostics) of diseases associated with compromised liver and kidney function; it is therefore most likely that their etiology and pathogenesis are associated with chronic metal intoxication.

Pathological processes in liver not diagnosed in the life-time were recorded in 24 cases out of 110. The his-tological pattern of liver pathogenesis is most often represented by fatty degeneration of hepatocytes, their sporadic necroses, autolytic decay in the periphery with the formation of fat–protein detritus.

One case of hemosiderosis, five cases of toxic destruction, four different types of dystrophy, and 10 cases of fatty degeneration were recorded. Most detected dystrophies are of toxic etiology.

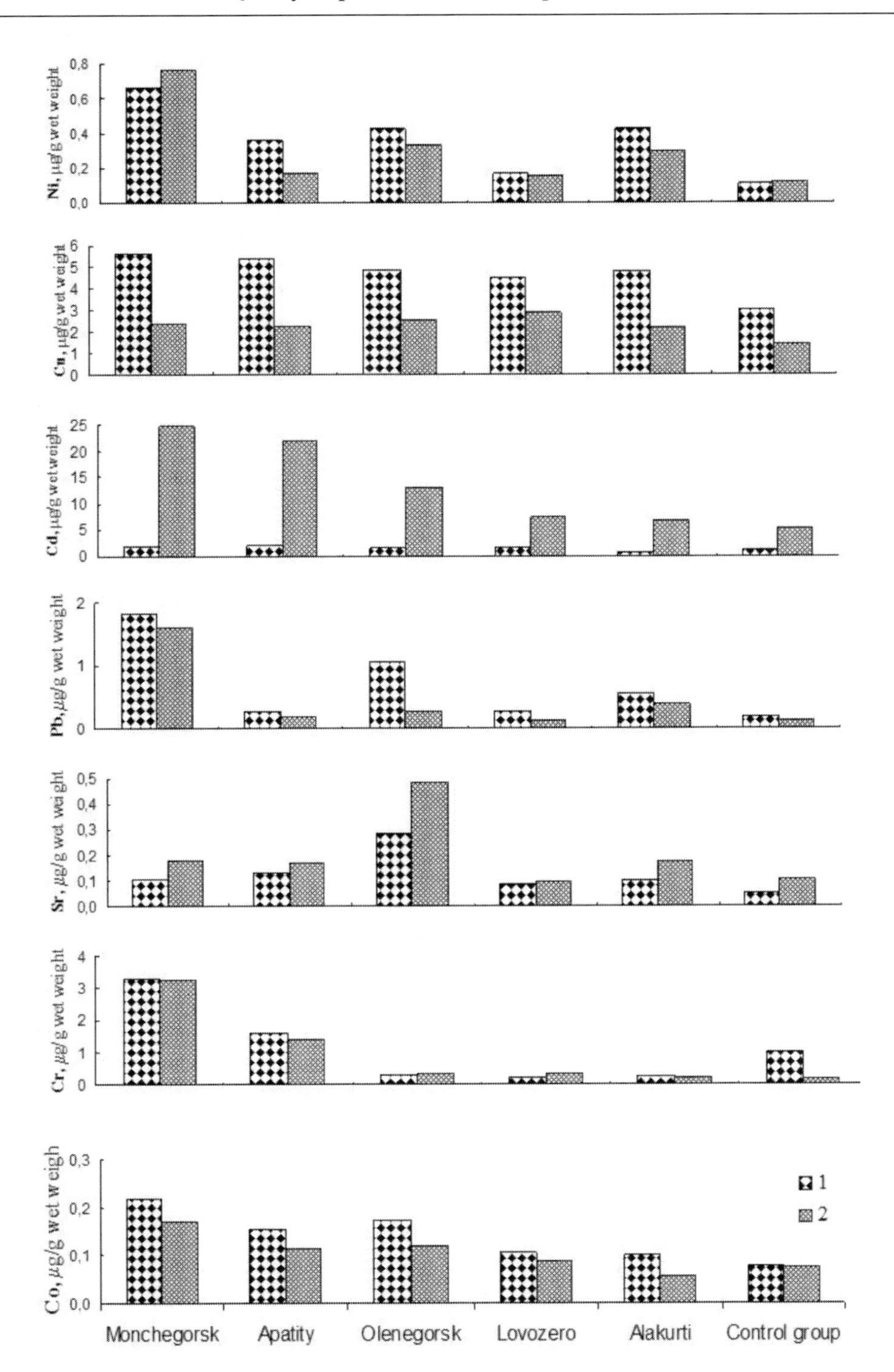

Figure 3. Metal content in liver (*1*) and kidneys (*2*) of the examined patients in cities and settlements of Murmans region.

The histologic picture of kidney pathogenesis is represented by nephrosclerosis at the initial stage of its development, as well as by nephrosclerosis and amy-loidosis. All these diseases are noninfectious and polyetiologic, which does not rule out their toxic patho-genesis under the effect of heavy metals. In addition to inflammatory changes in kidneys, specific focal dystrophic and necrobiotic changes in blood vessels (capillaries, precapillaries, capillary veins) were recorded. The morphological changes include small blood vessel diseases, an increase in vascular permeability, resulting in tissue edema and serous or serosanguineous inflammation in the tissues of the cortex and medulla of kidney. Moreover, 36 cases of urolithiasis not revealed by life-time diagnostics were identified, accounting for 12% of the specimens examined. These results are in agreement with statistical data on the incidence of urolithiasis in the region. Latent cases of pathologies with toxic etiology can weaken the organism and cause the development of accompanying diseases and an increase in population death rates.

CONCLUSION

Hazardous substances enter the human organism with food, water, and air; therefore, it is often difficult to establish correlation between drinking water quality and population health. However, it should be taken into account that people in transpolar regions largely eat products delivered from southern regions; therefore, the input of pollutants with food made from local agricultural products has not significant effect on population morbidity.

Water bodies are the terminal accumulators of all types of pollution, including fallouts from the polluted atmosphere. They determine, to one extent or another, the total dose of pollutants entering the environment. In the northern region under study with widely developed metallurgical and ore mining and processing industry, the leading environmental factor adversely affecting the population health is the pollution of water bodies used as sources for drinking water supply to the population. Pollutants can penetrate into the organism with drinking water and accumulate in it, provoking diseases. According to medical–demographic data, the Kola Region features higher incidence of urolithiasis and cholelithiasis in people.

The toxic impact of chemical elements on human organism is governed by their chemical nature, amount, and composition, as well as the individual features of the organism. The threshold concentrations of individual elements vary depending on other elements present in the environment and the organism. Surface continental waters in the extreme north regions are generally low in Ca and mineralization. The toxicity of metals increases in low-mineralization water, especially in ionic forms, which ensures their maximal penetrating capacity and toxicity for living organisms.

The water processing system used at treatment plants is ineffective with respect to metals. According to the authors' data, their concentrations in the water supplied to the population are near and sometimes even higher than those in natural waters near the water intake. Earlier detailed studies (Kudryavtseva, 1999) of drinking water and changes in metal concentrations at different stages of water treatment showed that metal concentrations (in particular, Fe, Zn and Mn) not only fail to drop, but even increase because of leaching from pipelines during water delivery to the population (Figure 4).

Correlation analysis of population morbidity indices in towns and settlements with averaged data on drinking water quality and the extent of metal accumulation confirm the leading effect of drinking water pollution by metals on population health. Table 2 gives correlation coefficients between Ni, Co, and Cu accumulation in human liver and their concentrations in drinking water.

Table 3 gives significant correlation coefficients between statistical data on human morbidity and drinking water quality in the appropriate towns and settlements.

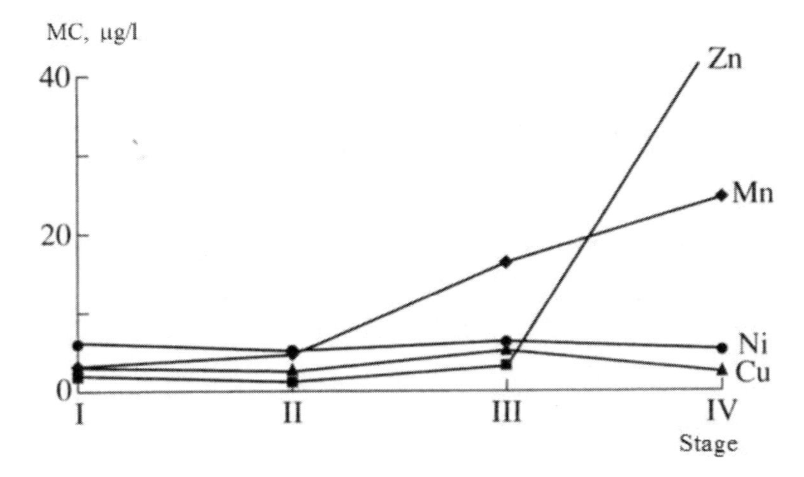

Figure 4. Dynamics of metal concentrations (MC) in drinking water in the process of treatment (Kudryavtseva, 1999). (I) Water before treatment at the station, (II) water after treatment at the station, (III) water from distribution hydrants, (IV) water from indoor water supply net-works.

Table 2. Correlation coefficients between metals accumulated in the liver and kidneys of people and their concentrations in drinking water in different towns and settlements in Murmansk region (here and in Table 3, significant correlation coefficients are shown in bold type)

Organ	Ni	Cu	Cd	Pb	Sr	Cr	Co
Liver	**0.78**	**0.78**	0.30	–0.45	–0.30	0.26	**0.81**
Kidney	**0.87**	–0.19	0.62	–0.29	–0.30	0.31	**0.83**

Table 3. Correlation coefficients between diseases and pathologies of systems and organs in the population of Kola Peninsula towns and microelement concentrations in drinking water

	Organs		Systems		
	kidney	liver	gastrointestinal	hematopoietic	cardiovascular
Ni	0.40	**0.63**	**0.89**	0.64	**0.73**
Cu	0.43	**0.64**	**0.91**	**0.67**	**0.77**
Co	**0.93**	**0.75**	**0.85**	**0.90**	**0.91**
Cr	0.60	0.53	0.51	**0.76**	**0.70**
Sr	0.56	0.18	-0.03	0.08	0.01
Cd	**0.87**	**0.76**	**0.87**	**0.77**	**0.83**
Pb	**0.97**	0.58	0.58	**0.69**	**0.70**
Zn	0.49	0.59	**0.81**	**0.61**	**0.68**
$\sum$Ci/MPCi	**0.76**	**0.77**	**0.95**	**0.79**	**0.86**

Interestingly, the incidence of kidney pathology has a high correlation coefficient with the concentration in water of elements such as Cd, Co, and Pb, as well as with the index of total exceedance of MPC_{dw}. Similarly, the incidence of gastrointestinal, liver, and vascular diseases correlates well with metal concentrations in drinking water.

Priority metal pollutants in surface continental waters of Kola region are Ni and Cu. Co is an accompanying element of copper–nickel ores. Its excessive concentrations cause diseases of heart, liver, and organs of vision and keratitis. Ni has a carcinogenic and gonadotoxic effect (Sidorenko, Itskova, 1980). Epidemiologic examination of workers involved in Ni production has shown that it can provoke cancerous diseases of nose, throat, and lungs. Malignant tumors formed in warm-blood animals after Ni had been introduced in their organisms. Higher incidence of nephrolithiasis and cholelithiasis was recorded in Kola residents who use for drinking water taken from Ni-polluted water bodies. Cu, as well as Co is not an acutely toxic element for humans.

When in very high concentration in water, Cu compounds can be toxic for warm-blood animals and humans: liver diseases, anemia, and yellow sickness occur. MPC_{dw} of Cu for drinking water is 1 mg/l, while its concentration in drinking water in the Kola Region is two orders of magnitude less. However, the correlations given above suggest that higher Cu concentrations in drinking water have their effect on gastrointestinal diseases in people in combination with other metals. Special attention should be paid to the high accumulation of Cd in human kidneys, where its concentration exceeds those of other heavy metals.

This element is inessential and does not occur in the organism. The authors' studies have shown its small amounts to occur in babies, which does not exclude its penetration through the transplacental barrier. The body of data suggesting the high toxicity of Cd for humans has increased in the recent years. Cd can cause pathological changes in organs and tissues; it affects the progress of diseases, such as diabetes, hypertension, osteoporosis, leukemia, and the development of neoplasms (Nishijo et al, 2000). Small doses of Cd result in its accumulation in lungs, kidneys, and adrenal glands of humans, provoking kidney pathology (Nishijo, 1999). The increase in cardiovascular diseases, including hypertension, in the group of people, is associated with kidney pathology provoked by Cd. Measurements of Cd concentration in urine covering a group of 22000 people in USA showed about 2.3% of them to have Cd concentration in the organism exceeding the admissible level (Satarug et al., 2003).

Thus, there is a good reason to suppose that the inadequate quality of drinking water in industrial regions of Extreme North provokes diseases in the population, a decrease in the immunity, and perhaps, higher death rate.

Six towns and settlements in the Kola North are used to study the effect of drinking water pollution on population health. Relationships are established in the system metal content of drinking water → metal accumulation in human liver and kidneys → metal-induced pathologies in organisms.

It was established that waters in the region are polluted by heavy metals because of the ore mining and smelting works functioning here, the major polluting element there being Ni. The higher concentrations of Cd (against the background) in waters of the region are associated with this element leaching into surface water sources. The existing water processing system fails to remove technogenic elements from the withdrawn water. Water in Extreme North region has low mineralization and a high capacity for leaching elements from pipelines (the presence of free carbon dioxide in the water), thus facilitating the high migration activity of metals.

The fact of accumulation of metals in human liver and kidneys has been established. The high level of metal concentrations in the liver and kidneys of people (relative to a reference group) is typical of residents of Monchegorsk Town, whose water intake is located in the propagation zone of smoke emissions from copper–nickel smelting works. Notwithstanding the fact that Ni concentrations are the highest in drinking water, the most abundant element in human kidney tissue is Cd (50 times the norm). This element is highly toxic and provokes the development of some pathologies in the organism. The highest morbidity rate in the population (neoplasms, urolithiasis, cholelithiasis, glouronephritis) was shown to be characteristic of Monchegorsk, where drinking water are higher in Ni, Cd, Pb, and other elements. The population morbidity rate in other towns decreases with decreasing concentrations of elements in drinking water.

Though the concentrations of toxic metals in water supplied to the population does not exceed the standards accepted in Russia for drinking water, their long action and accumulation in the systems of human organisms can cause pathologies and increase the population morbidity.

The most significant classes of diseases in that region are blood circulatory system diseases, neoplasms, and diseases of respiratory organs, urogenital. Analysis of statistical data on drinking water quality, population morbidity rate, metal accumulation in kidneys and liver in postmortem examined patient, as well as on their pathogenesis gives good grounds to suppose that a direct relationship exists between the increase in the morbidity rate and drinking water quality indices, and the etiology of the diseases depends on a prolonged impact of metals on human organisms. These conclusions suggest the need to carry out more thorough studies with the aim to correct drinking water quality standards with allowance made for the specific regional conditions and factors of environmental pollution and to develop more elaborate water treatment systems to reduce the risk of diseases in the population.

ACKNOWLEDGMENTS

The work was supported by the Russian Foundation for Basic Research (Projects no 10-05-00854), grant of Russian Governments (№ 11G34.31.0036) and Program 21 of Presidium RAS “Basic Sciences to the Medicine”.

REFERENCES

Handbook of Ecotoxicology. (2005). / eds. Hoffman D.J., Rattner B.A., Burton G.A, Cairnce J.Jr. N.Y.: Lewis Publishers, 501-556.

Handbook of Metals in Clinical and Analytical Chemistry. (1994). / eds. Seile, H.G.., Sigel, A.., and Sigel, N.Y.: Dekker. 471 p.

Kudryavtseva, L.P. (1999). Assessment of Drinking Water Quality in the City of Apatity, *Water Resour. (Engl. Transl.)*, 26(6), 659-665.

Moiseenko, T.I., (1999). A Fate of Metals in Arctic Surface Waters. Method for Defining Critical Levels. *The Science of the Total Environ.* 236, 19-39.

Moiseenko, T.I., and Kudryavtseva, L.P. (2002). Trace Metals Accumulation and Fish Pathologies in Areas Affected by Mining and Metallurgical Enterprises, *Environ. Poll*, 114 (2), 285-297.

Moiseenko, T.I., Kudryavtseva, L.P., and Gashkina, N.A. (2006). Trace Elements in Surface Continental Waters: Technophilia, Bioaccumulation, and Ecotoxicology. Moscow: *Nauka*, 261 p. (in Russian).

Nishijo M., Nakagawa H., Morikaw M. et al. (1999). Relationship between Urinary Cadmium and Mortality among Inhabitants Living *in a Cadmium Polluted Area in Japan, Toxicol. Lett.*, 108. 321-327.

Nishijo, M., Hakagawa, H., and Kid, T. (2000). Environmental Cadmium Exposure and Hypertension and Cardiovascular Risk, in *Metals in Biology and Medicine*, Paris: JA John Libbey Eurotext, 6, 365-367.

Satarug, S., Baker, J.R., Urbenjapol, S., et al. (2003). A Global Perspective on Cadmium Pollution and Toxicity in *Non-Occupationally Exposed Population, Toxicol. Lett.*, 137, 65-83.

Sidorenko, G.I. and Itskova, A.I. (1980). Nickel: Hygienic Aspects of Environmental Protection, Moscow: *Meditsina.* 176 p. (in Russian).

Spry D.J., Wiener J.G. (1991). Metal Bioavailability and toxicity to fish in low-alkalinity lakes: a critical review. *Environ. Pollution*, 71, 243-304.

Standard Methods for the Examination of Water and Wastewater, (1992). American Public Health Association, Washington. (D.C.), 1195 p.

In: Water Quality
Editor: You-Gan Wang

ISBN: 978-1-62417-111-6

Chapter 11

CHITOSAN BIOPOLYMER FOR WATER QUALITY IMPROVEMENT: APPLICATION AND MECHANISMS

Xinchao Wei*[*] *and F. Andrew Wolfe
Dept. of Engineering, Science and Mathematics, The State University of New York Institute of Technology, Utica, NY

ABSTRACT

Chitosan is a cationic biopolymer derived from chitin, the second most abundant natural fiber (next to cellulose), which is found in the shells of shrimp and crab. Chitosan has drawn great attention as an effective biosorbent for various dissolved contaminants mainly due to its high density amino groups ($-NH_2$) and hydroxyl groups (-OH). Chitosan has the highest adsorption capacity among the biopolymers. Chitosan can be a low-cost alternative to granular activated carbon (GAC) and has been used to remove heavy metals (Cu, Cd, Hg, Pb, Cr, As, and Se etc.) and anionic contaminants (phosphate, nitrate, fluoride, perchlorate). This chapter highlights the recent research and development in the application of chitosan for water quality improvement. It summarizes the advances and results in removal of various contaminants using chitosan as an adsorbent. The mechanisms involved in the adsorption of cationic and anionic species by chitosan are presented and discussed.

INTRODUCTION

To ensure safe drinking water, governmental agencies undertake important actions to assess and manage risks posed by various water quality contaminants, including man-made and naturally occurring chemicals in surface water, groundwater, and wastewater. Water and wastewater treatment plays a critical role in protecting public health and safeguarding the environment. To meet the water quality standards and accelerate the advancement of sustainable drinking water protection, innovative treatment technologies or materials are

[*] E-mail address: weix@sunyit.edu.

needed for public water systems to remove or mitigate groups of contaminants or contaminant precursors from the water source. There are a wide range of technologies available to remove contaminants from water, including physical, chemical, and biological processes. The specific type of the treatment process or process combination to be employed depends upon the nature of the contaminants, the scale of the plants, and the treatment costs. In general, the adsorption process with granular activated carbon (GAC) is considered the most versatile method, which can remove organics/inorganics, metals/nonmetals, and cations/anions. Moreover, the GAC adsorption process can be conveniently operated [1-4].

However, the wide use of GAC in water treatment systems is often restricted due to its relatively high cost [5] and the large carbon footprint during the production and regeneration of GAC [6]. Therefore, it is desired to develop effective and economical adsorption processes employing low-cost adsorbents. There are many nonconventional materials which have been tested to adsorb various contaminants with some success. In particular, some abundant and environmentally friendly biosorbents have demonstrated great promise in water and wastewater treatment. Chitosan is a cationic biopolymer derived from chitin, the second most abundant natural fiber (next to cellulose), which is found in the shells of shrimp and crab. Chitosan has drawn great attention as an effective biosorbent for various dissolved contaminants mainly due to its high density amino groups ($-NH_2$) and hydroxyl groups (-OH) [5, 7-10]. Chitosan has the highest adsorption capacity among the biopolymers [7, 11]. Chitosan can be a low-cost alternative to GAC and has been used to remove dissolved organic contaminants and dissolved inorganics, such as metals and anions. Considering the wide spectrum of contaminants which can be present in the water supply, there is a great advantage to using chitosan as an adsorbent water and wastewater treatment.

Chitosan and Its Properties

Chitosan is typically derived from chitin, which is the second most abundant natural polymer, next to cellulose, with a wide range of sources [5, 12]. However, chitosan is primarily manufactured from crustaceans (crab and shrimp), because of the easy availability of a large amount of the crustacean exoskeletons as a by-product of food processing [13, 14]. Chitosan is a partially deacetylated polymer of acetylglucosamine and its structure is shown in Figure 1. Chitosan is a poly(aminosaccharide) consisting mainly of poly($1 \rightarrow 4$)-2 amino-2-deoxy-D-glucose units [8]. The mole fraction of deacetylated units (glucosamine), defined as the degree of deacetylation, is usually 70–90%. The molecular weight of commercial chitosan varies significantly within the range 10,000–1,000,000 Da, depending on the processing conditions.

While chitin is insoluble in most solvents, chitosan is soluble in water at low pHs (<6). The water solubility of chitosan in acidic solutions is primarily due to the presence of amino groups with a pK_a value of ~6.3. At low pHs, the amino radicals undergo protonation and become a water soluble polymer with positive charges. At high pHs (>6), the amino radicals deprotonate and the cationic polymer becomes insoluble and subsequently precipitates as solids. Chitosan can be solubilized in inorganic acids such as hydrochloric acid and nitric acid. However, chitosan is insoluble in sulfuric and phosphoric acids [15]. Chitosan can also be solubilized in organic acids such as, acetic, formic, lactic, and oxalic acids.

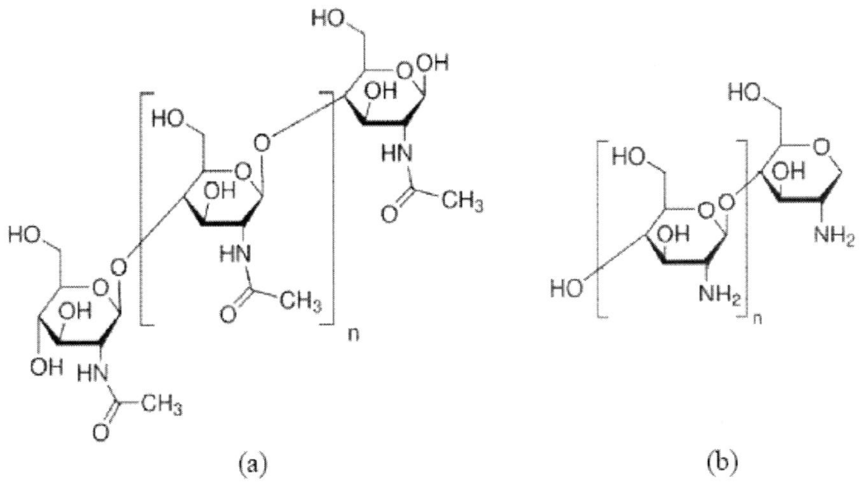

(a) (b)

Figure 2. Chemical structure of chitin (a) and chitosan (b).

The solubility change of chitosan might affect its stability as an adsorbent in treating water with low pHs. On the other hand, the solubility change with pH provides a convenient way to prepare small chitosan beads for adsorption by spraying solubilized chitosan into an alkaline solution media [16-18] or to coat other adsorbents with a layer of chitosan polymer [11, 19, 20].

The adsorptive properties of chitosan depend on its deacetylation degree, crystallinity, and molecular weight [10]. The free electron doublet of nitrogen on amino groups are the active sites for adsorption of contaminants by chitosan and the degree of deacetylation dictates the fraction of free amino groups available [21, 22]. The residual crystallinity in chitosan after derivation from chitin influences the accessibility of contaminants to the sorption sites and low crystallinity is preferred for adsorption.

One unique property of chitosan is that it can be readily modified physically or chemically to improve its mechanical or chemical properties. With respect to the use of chitosan as an adsorbent for contaminant removal, its disadvantage is due to its relatively lower specific surface area (2-30 m^2/g), as opposed to that of most commercial GAC based adsorbents (800-1,500 m^2/g)[23]. For example, chitosan flakes and powders with low surface area and no porosity are typically not effective in adsorbing contaminants from water. However, the adsorptive property can be improved when chitosan is prepared in such forms as beads, fibers (solid or hollow), sponges, membranes, and nanoparticles [10, 23]. Chitosan gel beads with controlled drying processes have enhanced volumetric adsorption capacity and improved kinetic performance for metal removal [24]. The adsorption selectivity can be improved by template formation or imprinting methods [25, 26].

Chemical modification is done to enhance the surface properties of chitosan as an adsorbent, while maintaining its fundamental skeleton. For example, chemical modification can prevent chitosan from dissolution into solutions, which is particularly important in removing metals from acidic waters. Chemical modification can also lead to increased adsorption capacity or enhanced selectivity for some specific contaminants. Chemical adsorption is typically achieved by grafting new functional groups in the chitosan backbone [10, 23]. The cross-linking agents may include bi-functional group (e.g. glutaraldehyde), mono-functional group (e.g. epichlorhydrin), tri-polyphosphate groups, carboxylic groups, amine groups and sulfonic groups [23]. The choice of cross-linking agents is determined by the specific use and application in dealing with water with different characteristics.

Removal of Heavy Metals

Heavy metals are the most ubiquitous and persistent contaminants in the aquatic environment. They are toxic to aquatic plants and animals and they pose a threat to human health either through direct consumption or bioaccumulation through the food-chain pyramid. Heavy metals occur naturally in surface water and groundwater, and increasing amounts of metals are released into the environment by anthropogenic sources such as the mining or chemical industry. Many processes can be used to remove heavy metals from water including precipitation, cationic exchange, reverse osmosis, and adsorption [27]. Chitosan is one of the most extensively studied adsorbents for the removal of heavy metals from water and wastewater.

Chitosan actually has the highest adsorption capacity for metals among the biopolymers [11], mainly due to its high density amino and hydroxyl functional groups with high adsorptive potential for a host of aquatic contaminants [5, 12]. The high adsorption capacity of chitosan for metals can be attributed to (1) the high hydrophilicity due to a large number of hydroxyl groups of glucose units, (2) the presence of a large number of functional groups (acetamido, primary amino and/or hydroxyl groups) (3) the high chemical reactivity of these groups and (4) the flexible structure of the polymer chain [12, 28]. Two mechanisms have been clearly established for the interpretation of metal adsorption on chitosan materials, i.e. chelation of metal cations to amine groups and ion exchange/electrostatic attraction [7].

Copper Removal

Copper (Cu) as one of heavy metals in the environment has been of great concern because of their increased discharge, toxic nature and other adverse effects on receiving waters [29]. Excessive Cu intake by humans can result in Cu accumulation in the livers and Cu is toxic to aquatic organisms even at very small concentrations in natural waters [30] The potential sources of copper pollution include metal cleaning and plating baths, pulp, paper and paper board mills, wood pulp production, and the fertilizer industry [29].

There are many studies on Cu removal using various chitosan products with different adsorption performances. Popuri et al. [31] coated polyvinyl chloride (PVC) beads with a layer of chitosan to develop a biosorbent and the chitosan modified beads were examined in batch and column experiments to remove Cu from solutions. It was found that the maximum monolayer adsorption capacity of the biosorbent was 87.9 mg/g for Cu(II). Kyzas et al. [32] prepared chitosan sorbents by cross-linking and grafting with amido or carboxyl groups. The calculated maximum sorption capacity of the carboxyl-grafted chitosan was 318 mg/g at pH 6. The desorption experiments demonstrated that the sorbent can be regenerated without significant loss in sorption capacity. Kannamba et al. [33] studied the removal of Cu(II) using chitosan which were first cross-linked with epichlorohydrin and then chemically modified as xanthate chitosan. The Cu(II) adsorption by chitosan was pH and temperature dependent The maximum adsorption capacity of modified chitosan was 43.47 mg/g at pH 5.0 and 50 °C. The desorption tests of loaded chitosan by sulfuric acid, hydrochloric acid and EDTA indicated the adsorbent can be used for many adsorption and desorption cycles. The adsorption mechanism involved the complexation of Cu(II) ions with the amine and thiol functional groups in the modified chitosan. Futalan et al. [34] immobilized chitosan onto the surfaces of bentonite and used the fixed-bed column to study the adsorption of Cu(II) from the aqueous solutions. Chitosan immobilized bentonite was effective in the removal of Cu(II) in the packed bed systems. The breakthrough curves and adsorption capacity was strongly dependent upon the bed height, flow rate and influent concentrations. At the optimum conditions, the adsorption capacity value is 14.92 mg/g with breakthrough time of 24 hr and exhaustion time of 35 h [34]. The fixed bed results could provide essential information in the practical design of permeable reactive barriers for the removal of heavy metals from groundwater. ***Lopes*** [35] chemically modified chitosan through reaction with ethylenediamine and diethylenetriamine. They found the effectiveness of such surface modification depended on a prior reaction to incorporate cyanuric chloride as an intermediate biomaterial. In general, chitosan derivatives were more effective than the original chitosan in interacting with the copper cation, demonstrating that the chemically modified chitosans could be successfully employed for copper removal from wastewater or industrial effluents [35]. In order to increase the Cu sorption capcity of raw chitosan beads, Gandhi et al. [36] chemically modified chitosan into protonated chitosan beads (PCB), carboxylated chitosan beads (CCB) and grafted chitosan beads (GCB). The sorption capacities for PCB, CCB, and GCB were 52, 86 and 126 mg/g, respectively, as opposed to that of raw chitosan beads (40 mg/g). Thermodynamic studies revealed that copper sorption onto chitosan bead products is spontaneous and endothermic.

The mechanism of copper sorption onto the modified chitosan beads is governed by adsorption, ion-exchange and chelation [36]. Chen and Wang [37] prepared magnetic chitosan nanoparticles by chemical co-precipitation of Fe^{2+} and Fe^{3+} ions by NaOH in the presence of chitosan, followed by hydrothermal treatment. The nanoparticles were super-paramagnetic and the size was in the range of 8–40 nm. The Langmuir maximum sorption capacity for Cu(II) was calculated to be 35.5 mg/g and more than 90% Cu(II) ions could be desorbed from magnetic chitosan nanoparticles using 0.02–0.1 M EDTA. FTIR analysis suggested that the removal mechanism of Cu(II) by chitosan involved $-NH_2$ and $-OH$ groups [36]. Al-Karawi et al. [37] investigated the graft copolymerization of acrylamide onto chitosan using potassium persulfate as initiator, and the grafted chitosan was used to remove Cu(II). It was found that the Cu(II) adsorption equilibrium from Cu(II) followed the Langmuir isotherm.

Cadmium Removal

Cadmium (Cd) is a heavy metal widely used in many industries including metallurgy, surface treatment, dye synthesis and battery production. Due to its toxicity, Cd in aqueous effluents has been a great concern for many decades. Cd in industrial wastewater can usually be removed through chemical precipitation, leading to the production of highly toxic sludges, which must be further treated before environmentally safe disposal. Other treatment options include ion exchange, adsorption, electrodeposition and membrane systems [38].

Erosa et al. [38] studied the kinetic and equilibrium of Cd adsorption by chitosan and found that chitosan was an effective adsorbent in removing Cd through chelation mechanisms involving amine groups of chitosan. The adsorption capacity exceeded 150 mg/g at optimum pH 7 and both intraparticle diffusion and external diffusion contributed to the kinetic control of Cd(II) adsorption by chitosan. Hasan et al. [39] prepared chitosan-coated perlite beads via the phase inversion of a liquid slurry of chitosan dissolved in oxalic acid and perlite to an alkaline bath. The perlite beads revealed a porous chitosan structure. The Cd(II) adsorption on chitosan was pH-dependent, and the maximum adsorption capacity of the chitosan-coated perlite beads was determined to be 178.6 mg/g of bead at the pH 6.0 (the capacity was 558 mg/g of chitosan) [39]. Desorption was achieved by dilute HCl or EDTA solution. A novel chitosan adsorbent was obtained via chemical modification by introducing a xanthate group onto the backbone of chitosan [40]. The chemically modified chitosan showed superior adsorption capacity (357.14 mg/g at the optimum pH 8) to the plain chitosan (85.46 mg/g). The regeneration experiments demonstrated the chemically modified chitosan could be used in Cd laden wastewater. The same adsorbent was also tested in fixed bed columns for Cd removal from electroplating wastewater [41]. The adsorption column treated 367 bed volumes of electroplating wastewater using the adsorbent, reducing the concentrations of Cd(II) from 10 to 0.1 mg/L [41]. Filho et al. [42] chemically modified chitosan with ethylene sulfide, under solvent free conditions, to obtain an adsorbent with a high content of thiol groups. The modified chitosan showed a good adsorption for Cd(II) with a capacity of 1.94 mmol/g. Copello et al. [43] prepared a layer-by-layer silicate–chitosan composite biosorbent. The films were evaluated on their stability regarding polymer leakage and their capability in the removal of Cd(II) from an aqueous solution. For Cd(II) adsorption, the greatest adsorption was obtained at pH 7. This non-covalent immobilization method allowed chitosan surface retention and did not affect its adsorption properties [43]. Chitosan was also used to modify the surfaces of zeolite [44], which functions as a porous support for chitosan. The modified zeolite was used to remove Cd(II) from a micro-polluted water source. It was demonstrated Cd (II) was more effectively removed by the modified zeolite powder than by natural zeolite powder [44].

Mercury Removal

Mercury (Hg) is one of the most toxic heavy metals which persists in the environment and creates long term contamination problems in the water, land and air [12]. The primary anthropogenic sources of Hg pollution are urban runoff, mining discharge, coal combustion, and industrial discharge. Once released into the environment, Hg typically undergoes complex physical, chemical and biological transformation. The Hg toxicity depends strongly on its redox states [45], while the most toxic form of Hg in the aquatic environment is the high reactive Hg(II)[12]. Although there are a variety of technologies which can be used to remove Hg(II) from water and wastewater, including precipitation, ion exchange, coagulation, membrane process and adsorption, adsorption with different adsorbents seems to be the most versatile and widely used technology for Hg removal.

Extensive research has been performed using chitosan as an adsorbent to remove Hg(II) from the aqueous phase, and the Hg removal performance depends on the chitosan source, the degree of deacetylation, and the experimental conditions such as pH, particle size, and ionic strength, etc [12]. Benavente [46] studied the adsorption kinetic and equilibrium using chitosan derived from shrimp shells and found that chitosan had the largest adsorption capacity for Hg(II) compared with those for Cu, Zn and As. The Hg adsorption on chitosan followed the pseudo-second-order equation and Langmuir isotherm model with a maximum adsorption capacity of 106.4 mg/g at pH 4. NaCl solutions of 1M were effective to desorb Hg for chitosan adsorption regeneration. Shafaei et al [47] evaluated Hg(II) removal using chitosan with >85% deacetylation of three different sizes: 0.177mm, 0.5mm, and 1.19mm. In general, higher pHs and smaller particle sizes favored Hg adsorption on chitosan. The chitosan of 0.177mm showed the Langmuir maximum capacity of 1,127.1 mg/g (or 5.62 mmol/g) at pH 6, which was significantly higher than those reported by other researchers [47]. Gamage and Shahidi [48] prepared three types of chitosan with different degrees of deacetylation from fresh crab processing discards: 91.3%, 89.3%, and 86.4%. The chitosan products were evaluated for their adsorption performance for several divalent metals at different pHs in batch and column tests. It was found pH 7 served best for Hg adsorption and higher degree of deacetylation resulted in better Hg(II) adsorption. As opposed to other metals (Mn(II), Co(II), Cd(II), Pb(II), Cu(II) and Zn(II)), chitosan demonstrated the best adsorption potential for Hg(II). Chitosan (molecular weight of 54,000 g/mol) of a low degree of deacetylation (20%) was prepared from red lobster shells by Polimeros et al [49] and used for Hg(II) removal. The chitosan product was in the form of flakes with an average length from 0.35 mm to 0.45 mm. The Langmuir maximum adsorption capacity of the chitosan for Hg(II)was 1.8 mmol/g and 2.3 mmol/g (or ~360 mg/g and ~460 mg/g), respectively. FTIR analyses revealed both amine ($-NH_2$) and –OH groups were involved in the retention of Hg(II) on chitosan, due to complex formation. Lopes et al. [50] used the chitosan of 93% deacetylation to prepare membrane pieces and investigated the Hg(II) adsorption on the chitosan membrane. The Hg(II) adsorption kinetic did not follow the traditional Lagergren model. The control step of Hg(II) adsorption by chitosan seemed to be the migration of Hg anions into the pores of the chitosan membrane and the interaction of the ions with the available adsorptive sites on the interior of the membrane. Peniche-Covas et al. [51] studied the removal of Hg(II) from aqueous solutions using chitosan derived from lobster shells. They found the intraparticle diffusion was the rate-limiting step in the adsorption of Hg(II) on chitosan. The adsorption followed the Langmuir isotherm with a maximum capacity of 429.3 mg/g. The column experiments demonstrated that chitosan could be used to remove Hg(II) from solutions in the absence of high levels of chlorides. In order to increase the adsorption capacity of chitosan for Hg(II), Jeon and Holl [52] evaluated several chemical modifications and prepared chitosan beads cross-linked with glutaraldehyde. They found the aminated chitosan beads prepared via chemical reaction with ethylenediamine had the best adsorption capacity of 2.3 mmol/g (~460 mg/g) at pH 7, which was attributed to the increased number of amine groups. The chitosan beads showed the characteristic of competitive sorption between Hg (II) and hydrogen ions [52].

Lead Removal

Lead (Pb) is one of the ubiquitous toxic metals which can be released into the aquatic environment from natural and anthropogenic sources. The effect of Pb on neurobehavioral development and brain cell function has been long recognized [53] and the bioaccumulation of Pb in bones, muscles, kidney and brain can interfere with the normal physical activities of humans. Due to its high toxicity, the U.S. Environmental Protection Agencies has set the maximum contaminant level (MCL) of Pb ions in drinking water as 0.015 mg/L. Different technologies has been developed for Pb removal from water and wastewater including chemical precipitation, ion exchange, and adsorption.

Ng et al. [53] evaluated the removal of Pb(II) from aqueous solutions using chitosan of three size ranges: 250–355 μm, 355–500 μm and 500–710μm. Chitosan with the smallest sizes showed the best adsorption of Pb(II), and at pH 4.5, the Langmuir maximum adsorption capacity was 0.558 mmol/g (~116 mg/g). Rangel-Martdez et al. [54] prepared chitosan from shrimp heads by homogenous deacetylation and evaluated its effectiveness in removing Pb(II). At pH 4, the chitosan adsorption for Pb(II) was better than that for Cd(II) and Cu(II). Asandei et al [55] studied the adsortpon of Pb(II) using chitosan of 20.8 % deacetylation at different conditions. The optimum pH for Pb(II) adsorption was found to be pH 6 with a Langmuir maximum capacity of 47.39 mg/g. To address the high cost of chitosan beads, Wan et al. [11] studied the immobilization of chitosan on a low-cost material (sand). The chitosan coated sand demonstrated high efficiency in removing Pb(II) and higher pH favored Pb(II) adsorption. The maximum adsorption capacity for Pb(II) was 13.32 mg/g, and the binding strength between metal ions and sorbents showed high stability under neutral pHs. The chitosan coated sand could potentially be used to build inexpensive large-scale filters as a permeable reactive barrier for Pb(II) removal in contaminated groundwater. In order to enhance adsorption performance of chitosan for Pb(II), chemical modification of cross-linked chitosan through the introduction of xanthate was performed to remove Pb(II) in acidic solutions [56]. Optimum adsorption was observed at pH 4 to 5 with a maximum capacity of 322 mg/g at pH4, which is higher than those from other reports. The enhanced adsorption was attributed to the involvement of xanthate groups beside amine groups in the adsorption process. The developed chitosan was also successfully applied to remove Pb(II) from actual battery wastewater [56]. Yan and Bai [57] examined the adsorption behavior of chitosan hydrogel beads for Pb(II) and the effect of the presence of humic acid (HA). Lead adsorption on chitosan hydrogel beads was found to be strongly pH-dependent in the pH range of 5-7.5. The amine group in chitosan was attributed to the adsorption of Pb(II) in water. Previously adsorbed HA on chitosan beads led to enhanced Pb(II) adsorption while previously adsorbed Pb(II) resulted in decreased HA adsorption. The formation of chitosan-HA-Pb structure had a more favorable structure than that of chitosan-Pb-HA [57]. In a mixed Pb(II)-HA solutions, the adsorption of chitosan for both species was significantly reduced, indicating the interference of HA for Pb(II) removal. Jin and Bai [58] also investigated the mechanisms of lead adsorption on chitosan/PVA (poly(vinyl alchohol)) beads in aqueous solutions. Lead adsorption on chitosan/PVA beads was strongly pH-dependent with a maximum uptake capacity at pH ~4. The chitosan/PVA beads exhibited positive ζ potentials at pH <6.3 and negative ζ potentials at pH > 6.3. The adsorption mechanisms involved complexation, ion exchange, and electrostatic interactions. Futalan et al. [34] studied the comparative and competitive adsorption of Pb(II) with Cu(II) and Ni(II) using chitosan immobilized on bentonite (CHB). It was observed that the adsorption capacity is highest for Pb(II) over Cu(II) and Ni(II) for single and binary systems. The competition between metal ions for the binding sites on CHB was strong as indicated from the low adsorption isotherm constants in binary systems. To enhance the adsorption of Pb(II) on chitosan granules, Li and Bai [59] successfully grafted poly(acrylic acid) (PAAc) on cross-linked chitosan granules (CTS) through a simple two-step reaction in a solution. The PAAc modified CTS demonstrated significantly greater adsorption capacity for Pb(II) (294.12 mg/g) and shorter equilibrium time (5 hr), in contrast to CTS (95.15 mg/g and 8 hr). The enhanced performance of Pb(II) adsorption of CTS-PAAc was attributed to the carboxyle groups grafted on CTS. In addition, the adsorbed Pb(II) could be effectively desorbed and the CTS-PAAc could be regenerated and reused without any loss of adsorption capacity [59].

Chromium Removal

Chromium (Cr) can be present in the aqueous environment in its trivalent (Cr(III)) and hexavalent (VI) states. While Cr(III) is an essential nutrient, Cr(VI) is highly toxic and considered carcinogenic. Unlike the Cr(III) cation, Cr(VI) typically exists in water as chromate anion (CrO_4^{2-}) or dichromate anion ($Cr_2O_7^{2-}$). Cr is a very important metal with a very broad application such as metal finishing, hide tanning, and the manufacturing of alloys, pigments and dyes. Typical Cr treatment processes include reduction, precipitation, ion exchange, reverse osmosis and adsorption with activated carbon.

Rojas et al. [60] evaluated the Cr(III) and Cr(VI) removal from aqueous solutions using chitosan powder (21-850 μm) and cross-linked chitosan flakes (850-2000 μm) with a molecular weight of 125,000 Da and 87% deacetylation. Cross-linked chitosan was efficient in removing Cr(VI) (adsorption capacity of 215 mg/g) and less efficient in eliminating Cr(III) (adsorption capacity 6 mg/g) due to the protonated active sites interacting mainly with anions at low pHs. de Castro Dantas et al. [61] examined the Cr(III) removal from aqueous solutions by chitosan impregnated with a microeulsion. A significant increase in Cr(III) adsorption capacity was achieved compared to the untreated chitosan. Dynamic column experiments demonstrated that Cr(III) adsorption is pH-dependent and the amount of Cr(III) sorbed increased with increasing initial concentrations. The presence of Cu(II) interfered with Cr(III) adsorption. Schmuhl et al. [62] studied the Cr(VI) adsorption by chitosan and chitosan cross-linked by epichlorohydrin. The adsorption behavior of Cr(VI) was described by the Langmuir isotherm. The maximum capacity was 78 mg/g for non-cross-linked chitosan and 50 mg/g for cross-linked chitosan, indicating cross-linking did not improve Cr(VI) adsorption. When chitosan was cross-linked, some of the adsorptive sites were used for cross-linking [62]. Baroni et al. [63] tested the Cr(VI) removal using chitosan and chitosan cross-linked by glutaraldehyde and epichlorohydrin. They found both types of cross-linked chitosan showed significantly higher adsorption capacity for Cr(VI) than the plain chitosan, contradictory with the findings of Schmuhl et al. [62]. The maximum Cr(VI) adsorbed was about 1,400 mg/g for epichlorohydrin cross-linked chitosan at pH 6.0, which is the highest number reported in the literature for Cr(VI). The desorption tests with NaCl also showed chitosan cross-linked by epichlorohydrin was more effective for Cr(VI) removal at pH 2 than pH 6. Spinelli et al. [64] chemically modified the surface of chitosan with glycidyl trimethyl ammonium chloride in order to produce a chitosan quaternary ammonium salt that works as a strongly basic exchanger. In order to study the adsorption of Cr(VI) ions in the form of an oxyanion, they cross-linked the chitosan salt with glutaraldehyde making it insoluble in aqueous solution. The maximum adsorption capacity was 68.3 mg/g (1.31 mmol/g) at pH 4.5 and the sorbed Cr(VI) can be desorbed by 1 M NaCl/NaOH solution [64]. Similarly, a novel chitosan adsorbent was developed by chemically modifying chitosan through grafting glutaraldehyde, xanthante group onto the backbone of chitosan [65]. Both modified chitosan beads and flakes showed improved adsorption for Cr(VI) with maximum capacities of 625 mg/g and 256 mg/g, respectively. The desorption test indicated that the modified chitosan products can be reused after at least 10 cycles without any significant change in adsorption capacity. Boddu et al. [66] developed a new composite chitosan biosorbent by coating chitosan onto ceramic alumina and studied the Cr(VI) adsorption with the adsorbent in both batch isothermal and continuous column adsorption experiments. The effect of competing anions such as sulfate and chloride was also investigated as well as that of pH. The ultimate capacity obtained from the Langmuir model is 153.85 mg/g, indicating great adsorption capacity for Cr(VI). Hasan et al. [67] prepared chitosan coated perlite beads by drop-wise addition of chitosan solution to an alkaline bath. The beads containing 23% chitosan showed a good affinity for Cr(VI) with an adsorption capacity of 104 mg/g based on the beads (452 mg/g based on chitosan itself), which was higher than that of the chitosan only adsorbents (11.3-78 mg/g).

Arsenic Removal

Arsenic (As) is a highly toxic and carcinogenic metalloid which can be released into the aquatic environment naturally or anthropogenically. The World Health Organization (WHO) recommends the maximum As level of 10 μg/L in drinking water. Long-term consumption of arsenic contaminated water can lead to severe and permanent impairment of human health. In the aquatic environment, the most common and mobile As species are As(III) and As(V), existing as arsenite (As(III) or AsO_3^{3-}) and arsenate (As(V), or AsO_4^{3-}) anions. Due to the stringent As regulations, there has been extensive research on arsenic removal technologies demonstrating varying levels of success. As removal can be achieved using precipitation, coagulation, oxidation, ion exchange and adsorption with iron-based adsorbents. In general, As(V) is relatively easy to remove from the aqueous phase while As(III) is more troublesome and requires oxidation as a pretreatment to achieve the satisfactory results.

Boddu et al. [68] prepared a composite chitosan biosorbent (CCB) by coating chitosan on ceramic alumina via a dip-coating process and evaluated the adsorption capacity of CCB for As(III) and As(V) under equilibrium and dynamic experimental conditions. CCB contained 21.1 wt % of chitosan with a specific surface area of 125.24 m^2/g. The adsorbent showed an impressive adsorption capacity for As(III) and As(V) at pH 4 (56.50 and 96.46 mg/g, respectively, based on Langmuir isotherm data).The better adsorption for As(V) was explained by the speciation of arsenic at pH 4. The adsorbent was also suitable for column application with breakthrough bed volumes of 40 and 120 for As(III) and As(V), respectively, and the column bed could be regenerated using 0.1M NaOH solutions. Gupta et al. [69] prepared iron coated chitosan flakes (ICF) and iron doped chitosan beads (ICB) and used them as adsorbents to remove As(III) and As(V) from actual groundwater in batch and column experiments. The monolayer adsorption capacities of ICF were 22.47 mg/g for As(V) and 16.15 mg/g for As(III), which were considerably higher than those obtained for ICB (2.24 mg/g for As(V) and 2.32 mg/g for As(III)). The competing anions such as sulfate, phosphate and silicate at the levels typical of groundwater did not cause serious interference in the adsorption behavior of As(III) and As(V). The column regeneration studies were carried out for two sorption–desorption cycles for both As(III) and As(V) using ICF and ICB as sorbents. The adsorbents were also successful in treating the actual arsenic contaminated groundwater below <10 μg/L (total As) and the adsorbent could be regenerated using 0.1M NaOH solution. Gerente et al [70] investigated the sorption As(V) onto chitosan. They proposed the fixation of As(V) onto chitosan included the following steps: (i) the increase of amine protonation and zeta potential at the surface of the chitosan by a decrease of the pH, (ii) the fixation by electrostatic attraction between a positive surface charge located on the amine function and the anionic form $H_2AsO_4^-$, (iii) thermodynamically spontaneous and exothermic biosorption reaction. dos Santos [71] evaluated the effectiveness of chitosan-Fe-crosslinked complex (Ch-Fe) in removing As(III). The maximum adsorption capacity of Ch-Fe was 13.4 mg/g at the optimum pH 9.0 and adsorption kinetics followed the pseudo-first-order kinetic equation. Recently, Miller et al. [72] investigated As(III) and As(V) removal using TiO_2-impregnated chitosan beads (TICB) to maximize the capacity and kinetics of arsenic adsorption. It was found that the adsorption capacity of TICB was influenced by pH and TiO_2 loading within the bead and As removal was enhanced with exposure to UV light. The effective pH for As(V) removal was $pH < 7.25$ while it was $pH < 9.2$ for As(III). The reduction in bead size and exposure to UV light improved the adsorption rate. The common background groundwater ions, except phosphate, did significantly interfere As(V) adsorption onto TICB. Gupta et al. [73] prepared chitosan iron nanoparticles (CIN) by reducing Fe(III) with $NaBH_4$ in the presence of chitosan as a stabilizer and evaluated the As adsorption on CIN. Remarkably high adsorption capacities of CIN was observed in a wide pH range for both As(III) and As(V), i.e. 94 mg/L and 119 mg/L, respectively. The interfering anions commonly found in groundwater such as sulfate, phosphate and silicate only marginally influenced the adsorption of both As species onto CIN. The CIN adsorbent was successfully regenerated for five cycles with 0.1M NaOH solution without a significant sacrifice in adsorption capacity, indicating CIN could be a potential candidate for arsenic filtering units for groundwater treatment or remediation. Gang et al. [74] developed an iron-impregnated chitosan granular adsorbent and evaluated its ability to remove As(III) from water through batch and column studies. The impregnation of iron into chitosan significantly increased the As(III) adsorption capacity. The maximum adsorption capacity increased from 1.97 to 6.48 mg/g at pH = 8 as the initial concentration of As(III) increased from 0.3 to 1 mg/L.

Selenium Removal

Selenium (Se) is an emerging contaminant for many regions worldwide. It is an essential micronutrient for humans and animals, but considered toxic when ingested in amounts higher than those needed for optimum nutrition. Se regulations vary from country to country. For drinking water, most countries adopt the 10 μg-Se/L limit of World Health Organization (WHO) guideline. In the US, 50 μg-Se/L is still the current Environmental Protection Agency (EPA) limit for both the maximum contaminant level (MCL) and the MCL goal (MCLG), although a new limit of 5 μg-Se/L is being proposed. For surface water, the US Clean Water Act (CWA) lists Se as a priority toxic pollutant and adopts freshwater acute and chronic criteria of 20 μg-Se/L and 5 μg-Se/L, respectively. Se levels exceeding the freshwater criteria can pose a serious risk to aquatic life and humans due to the likelihood of bioaccumulation through the food chain. In addition to its natural occurrence in the environment, Se can be released as a result of anthropogenic processes such as the mining of minerals, combustion of coal, metal smelting, oil refining and utilization, and agricultural irrigation In aquatic environments, Se can be present in four different oxidation states viz. selenide (Se^{2-}), elemental selenium (Se^0), selenite (SeO_3^{2-}), and selenate (SeO_4^{2-}), of which selenite or Se(IV) (SeO_3^{2-}) and selenate or Se (VI) (SeO_4^{2-}) are more soluble and mobile. Technologies in removing Se from water and wastewater include: 1) conventional water treatment practices such as lime neutralization, softening and ferric coagulation , 2) ion exchange and membrane processes such as reverse osmosis, nanofiltration, and emulsion liquid membranes, 3) Se reduction, 4) biological processes, and 5) adsorption processes using such adsorbents as alumina, activated carbon, manganese nodule leached residues sulphuric acid-treated peanut shell and various iron oxides/hydroxides [75].

Bleiman and Mishael [76] prepared chitosan-montmorillonite composite to remove Se(VI) from water. Compared with Al-oxide and Fe-oxide, the chitosan composite exhibited higher capacity for Se(VI) (18.4 mg/g). Furthermore, Se(VI) adsorption by the composite was not pH dependent while its adsorption by the Fe- or Al- oxides decreased at high pHs. The experiments with actual well water indicated that the chitosan composite was able to reduce Se level to below 10 μg/L and the adsorption of Se was relatively selective in the presence of 13 mg/L sulfate. Yang et al. [77] developed chitosan-coated quartz sand and examined its performance in removing Se(IV) from groundwater. The effects of pH, adsorption time, temperature, and initial Se concentration were studied. It was found, at a dose of 40 g/L, the chitosan coated sand reduced the Se(IV) to 8.66 μg/L and monolayer adsorption is the primary mechanism for Se(IV) adsorption. Sabarudin et al. [78] synthesized a chitosan resin functionalized with 3,4-diamino benzoic acid (CCTS-DBA resin) using a cross-linked chitosan (CCTS) as base material. It was found that Se(VI) was strongly adsorbed at pH 2 and pH 3 as an oxyanion of SeO_4^{2-}, while selenium(IV) as $HSeO_3^-$ was adsorbed on the resin at pH 3. The adsorption capacities were 64 and 88 mg/g for Se(IV)and Se(VI), respectively. The common anions such as chloride, sulfate, phosphate and nitrate did not affect the Se adsorption when their concentrations were less than 20 ppm.

Removal of Anionic Contaminants

Nitrate Removal

Nitrate (NO_3^-) is one of the most common contaminants in surface water and groundwater. Excessive amount of nitrate can cause eutrophication problems in surface waters and outbreaks of infectious disease. Nitrate concentration higher than the regulation allowed level (10 mg/L, U.S. EPA) in drinking water can lead to potential human health issues especially for babies and children. Nitrate removal methods include biological denitrification, chemical reduction, reverse osmosis, electrodialysis, and ion exchanges [79].

Chatterjee and Woo [79] prepared chitosan hydrobeads of 2.5 mm by a drop-wise addition of chitosan solution to an alkaline coagulating mixture and examined the nitrate adsorption on the chitosan hydrobeads. The adsorption process was pH and temperature dependent. Lower pHs favored nitrate adsorption because a pH decrease in solution resulted in more protons available for the chitosan amine group for a more positive charge at the chitosan surface. The enhanced adsorption of nitrate was due to electrostatic interactions between negatively charged nitrate groups and positively charge amine groups. When the pH was above 6.4, the physical forces were attributed to the appreciable amount of nitrate adsorbed by chitosan hydrobeads. The maximum adsorption capacity was 92.1 mg/g at 30ºC. The kinetic results corresponded well with the pseudo-second-order rate equation and intra-particle diffusion played an important part at the initial stage of the adsorption process. The loaded nitrate was desorbed by increasing pH to the alkaline range and 87% desorption was achieved at pH 12. To improve the nitrate adsorption capacity, chitosan beads were modified by cross-linking with epichlorohydrin (ECH) and surface conditioning with sodium bisulfate [80]. The maximum adsorption capacity was found at a cross-linking ratio of 0.4 and conditioning concentration of 0.1mM $NaHSO_4$. The maximum adsorption capacity was 104.0mg/g for the conditioned cross-linked chitosan beads at pH 5, as opposed to 90.7mg/g for normal chitosan beads. The high adsorption capacity values for all adsorption systems in acidic solutions (pH 3–5) were attributed to the strong electrostatic interactions between its adsorption sites and the nitrate. Arora et al. [81] modified the zeolite surfaces with a layer of chitosan to prepare chitosan coated zeolite (Ch-Z). The analyses found that chitosan did not provide a full coating of the zeolite particles. Ch-Z exhibited a comparable capacity to other weak anion exchangers with a nitrate ion exchange capacity 0.74 mmol/g (~46 mg/g). Athough the nitrate exchange capacity was reasonable, Ch-Z was shown to be more selective towards sulfate and chloride than nitrate. Jaafari et al. [82] prepared chitosan gel beads and cross-linked with glutaraldehyde and evaluated nitrate adsorption by the chitosan beads. It was found protonated cross-linked chitosan was able to remove nitrate from model water and contaminated surface and groundwater to meet the drinking water standard. The reactive process involved the total volume of the chitosan beads, not only the surfaces. Fluoride was the main competing anions while chloride and sulfate did not interfere with nitrate adsorption.

Phosphate Removal

Excess phosphate (PO_4^{3-}) is believed to be the cause of eutrophication problems in surface waters and can also lead to other water quality problems. The sources of phosphate pollution include domestic source, agricultural runoff, urban runoff, industrial effluent, municipal wastewater discharge and mining drainage etc. Typical phosphate removal technologies include biological process, chemical precipitation, and adsorption.

Dai et al [83] examined the feasibility of phosphate removal by using the spent chitosan beads after copper adsorption. It was found that the spent chitosan beads loaded with copper were stable and suitable for phosphate removal in a wide pH range and the maximum adsorption capacity (28.86 mg-P/g) was achieved at pH ~5. Among the species of phosphates, dihydrogen phosphate was the preferred species for Cu(II)-loaded chitosan beads to adsorb. The effect of competing anions on adsorption capacity indicated that chloride and sulfate interfered with phosphate adsorption on the chitosan beads. The adsorption equilibrium and kinetic study indicated that the adsorption behavior was mainly chemical monolayer adsorption for phosphate facile to bind with Cu(II). Fierro et al. [84] investigated the use of chitosan immobilized algae for phosphate removal from water. Dark green microalgal colonies were observed in the outer and inner portions of chitosan beads. The chitosan immobilized algae showed a very high efficiency of phosphate removal(>94%) at an initial phosphate concentration of 6 mg-P/L, which was better than plain chitosan bead (60% removal)[84].

Chung et al. [85] evaluated the feasibility of using chitosan of different molecular weights to simultaneously remove various pollutants from the discharge of an eel culture pond. Experimental results indicated chitosan of a low molecular weight excelled at removing phosphate as well as ammonium from the wastewater. The best performance of chitosan was 99.1% for phosphate removal. The best removal of chitosan for turbidity, suspended solids, BOD, COD, NH_3 and bacteria was 87.7%, 62.6%, 52.3%, 62.8%, 91.8%, and 99.998%, respectively.

These results indicated that it was feasible to use chitosan to treat the aquaculture effluents to minimize deterioration of receiving water quality. Shimizu et al. [86] studied the removal of phosphate with a chemically modified chitosan/metal-ion complex. It was found that Cu(II) and Fe(III) based chitosan complexes had the highest adsorption ability toward phosphate. Fagundes et al. [87] investigated the adsorption of phosphate on iron(III)-cross-linked chitosan in an experiment of solid-phase extraction. A batch study showed that phosphate adsorption reached equilibrium and pH 7.0 is the optimal pH for phosphate adsorption.

The maximum adsorption capacity determined by Langmuir equation was 131 mg/g. For the column study, an increase in the flow-rate resulted in reduced breakthrough volume (180 and 230 mL) and breakthrough sorption capacity (18 and 23 mg/g). Zheng and Wang [88] prepared a chitosan-*g*-poly(acrylic acid)/vermiculite composite by aqueous dispersion polymerization using chitosan as the stabilizer, acrylic acid as the monomer and vermiculite as the inorganic additive. The composite was then cross-linked with common divalent or trivalent cations to obtain adsorbents with a higher affinity for phosphate ions. The results demonstrated that the trivalent ion cross-linked hybrid exhibited a phosphate potential for the removal.

The adsorption of phosphate ions onto the developed adsorbent was pH-dependent and a lower pH led to a higher adsorption capacity. The maximum adsorption capacity was 22.64 mg/g, comparable with those reported for other adsorbents. Desorption studies indicated that the adsorbent was relatively difficult to regenerate for reuse.

Fluoride Removal

While fluoride (F^-) is an essential element for dental health, fluoride concentration above 1.5 mg/L in drinking water can be detrimental to human health, leading to dental or skeletal fluorosis [23]. Consequently, the World Health Organization (WHO) has set a desirable and permissible fluoride limit range of 0.5 – 1.0 mg/L for drinking water [23]. The main fluoride source for drinking water is naturally occurring fluoride minerals in geological formations. Anthropogenic release of fluoride into the environment is from various engineering processes including semiconductor manufacturing, coal power generation, electroplating, rubber and fertilizer production, etc. [23, 89]. Fluoride can be removed from water by various methods such as ion exchange, coagulation, membrane process, and adsorption with various adsorbents.

Miretzky and Cirelli [23] have provided an excellent review on the use of chitosan for fluoride removal. Various chitosan and chitosan derivatives has been used for drinking water treatment and they report maximum adsorption capacity ranges from 2.22 mg/g to 44 mg/L [23]. The high variability in the performance of chitosan is mainly due to the disuniformity in chitosan property, chitosan processing, and modification (physical and chemical). This review is focused on the recent research and development in fluoride removal using chitosan since 2010.

Thakre et al. [90] synthesized lanthanum incorporated chitosan beads (LCB) using a precipitation method under optimized conditions and examined their adsorption performance in fluoride removal from drinking water. It was found that parameters for the synthesis of LCB such as complexation time, precipitation time, ammonia strength and lanthanum loading had a significant effect on fluoride removal. LCB effectively reduced the fluoride concentration below the level of 1.5 mg/L. The fluoride adsorption capacity of LCB was 4.7 mg/g which was much greater than that of the commercially used activated alumina (1.7 mg/g).

LCB also possessed other advantages such as relatively fast kinetics, high chemical and mechanical stability, high resistance to attrition, negligible lanthanum release, and suitability for column applications. The same group of researchers also prepared lanthanum incorporated chitosan flakes by lanthanum impregnation to enhance the fluoride removal capacity of chitosan [91].

It was observed that the synthesis parameters have significant influence on development of LCF and in turn on fluoride removal capacity. The LCF prepared under the optimal condition showed a maximum adsorption capacity of 1.27 mg/g, which was lower than LCB. SEM of LCF showed the presence of spherical particles spread over the chitosan matrix [91]. Viswanathan and Meenakshi [92] prepared a hydrotalcite/ chitosan composite using a co-precipitation method to enhance fluoride removal. The composite showed an adsorption capacity of 1,255 mg/kg, which was better than plain hydrotalcite (1,030 mg/kg) and plain chitosan (52 mg/kg). Field trial studies indicated the hydrotalcite/chitosan composite was an effective defluoridation agent. The same researchers prepared an alumina/chitosan composite by incorporating alumina particles in the chitosan polymeric matrix [93]. The composite displayed a maximum adsorption capacity of 3,809 mg /kg than the alumina (1566 mg/kg) and chitosan (52 mg F^-/kg). The fluoride removal by the chitosan composite was mainly governed by electrostatic adsorption/ complexation mechanism.

Jagtap et al. [94] used chitosan as a template to prepared mesoporous alumina (MA450) with improved properties fluoride removal from water. MA450 showed highly porous structure of amorphous alumina with some partially converted chitosan residue. It was observed that MA450 was effective over a wide range of pH (3-9) and showed a maximum adsorption capacity 8.264 mg/g at an initial fluoride concentration of 5 mg/L, much better than the conventional alumina. The order of anions interfering fluoride adsorption was observed as HCO_3^- > SO_4^- > NO_3^- > Cl. MA450 also demonstrated significantly high fluoride removal in field water. Vijaya et al. [95] developed a novel biosorbent, chitosan coated calcium alginate (CCCA), by coating chitosan onto an anionic biopolymer calcium alginate for the fluoride removal from aqueous solutions under batch equilibrium and column flow experimental conditions. The adsorption process was optimized through the study of the effects of pH, contact time, concentration of fluoride, and biosorbent dosage. The maximum monolayer adsorption of fluoride on plain calcium alginate and CCCA were found to be 29.3 and 42.0 mg/g, indicating the enhancement of chitosan for fluoride adsorption. The breakthrough curves were obtained from column flow tests and the experimental results demonstrated that chitosan coated calcium alginate beads could be used for the defluoridation of drinking water.

Perchlorate Removal

Perchlorate (ClO_4^-) is an emerging contaminant that has been detected in soil, surface water and groundwater. As a toxic species, perchlorate can inhibit iodine uptake by the thyroid gland and disturb normal metabolism, leading to physical or mental retardation or other diseases such as neurological damage or anemia [96, 97]. Perchlorate is both a naturally occurring and man-made contaminant and the anthropogenic source is the main concern. Perchlorate has been used in products such as explosives, rocket fuels, fireworks, air bags, bleaches, and fertilizers. Perchlorate, once released into the environment, can persist for several decades due to its high solubility, non-reactivity, and poor adsorption to soil matrix [96, 98]. Currently, U.S. EPA does not have regulation limit for perchlorate. However, it is planning to develop national limits due to its toxicity. The technologies for perchlorate removal include anionic exchange, biological reduction, chemical reduction, membrane filtration, and adsorption with activated carbon.

Xie et al. [96] examined the perchlorate removal from aqueous solutions using chitosan cross-linked with glutaradehyde in batch and column tests. The maximum monolayer adsorption capacity was 45.5 mg/g and the optimum pH was determined to be pH 4. Column adsorption indicated that the proper contact time was 8.1 minutes, indicating a rapid adsorption. The effluent perchlorate concentration was kept below 24.5 μg/L for up to 95 bed volumes with the influent perchlorate concentration of 10 mg/L. However, the presence of competing anions, sulfate in particular, negatively influenced the perchlorate adsorption. The adsorbent could well be regenerated with NaOH solution at pH 12 and reused for 15 cycles. Electrostatic attraction as well as physical forces was believed to be the driving force for perchlorate adsorption on cross-linked chitosan.

REMOVAL OF OTHER CONTAMINANTS

Ammonium Removal

Ammonium (NH_4^+) is commonly present in various wastewaters and it can be transformed into nitrite and nitrate in the aquatic environment. Therefore, ammonium, as part of the nitrogen nutrient along with phosphorus, is an important contaminant responsible for eutrophication problems in surface waters. At high concentrations, ammonium itself is toxic for many aquatic plants, fishes, and animals. The main source of ammonium pollution is from wastewater effluents and agricultural and urban runoffs. Ammonium can be removed from water by biological nitrification/denitrification, supercritical water oxidation, ion exchange and adsorption.

Zheng and Wang [99] evaluated the removal of ammonium ions from aqueous solution using a hydrogel composite chitosan grafted poly (acrylic acid)/ rectorite prepared from *in-situ* copolymerization. The ammonium adsorption equilibrium can be reached within 3–5 min, indicating rapid ammonium adsorption kinetics. The hydrogel composite had a higher adsorption capacity for ammonium (from 62 mg-N/g to 109 mg-N/g) in a wide pH levels ranged from 4.0 to 9.0. No significant changes in the adsorption capacity were found over the temperature range studied. Multivalent cations coexisting with ammonium in the solution had some negative effects on the ammonium adsorption capacity. The regeneration condition for the composite adsorbent was mild and the regenerated adsorbent was suitable for reuse in ammonium removal. The electrostatic attraction between $-COO^-$ and NH_4^+ was believed to be the main adsorption mechanism. The incorporation of inorganic clay particles improved the hydrogel strength and enhanced the thermal stability. The same research group [100] also developed a composite adsorbent with three-dimensional cross-linked polymeric networks based on chitosan and attapulgite via *in situ* copolymerization in aqueous solution. The efficacy of the composite adsorbent was examined for removing ammonium from synthetic wastewater using batch adsorption experiments. At natural pH, the composite adsorbent had an adsorption capacity of 1.0 mg-N/g, far higher than the other adsorbents (such as clay and powdered activated carbon). Desorption of ammonium was achieved using 0.1 M NaOH within 10 min. The results demonstrated the as-prepared composite was a fast-responsive and high-capacity adsorbent for ammonium removal [100]. Chitosan was also used as the backbone to prepare halloysite hydrogel composite adsorbents with an adsorption capacity of 32.87 mg-N/g [101].

Humic Acid Removal

Humic acid (HA) is a subclass of humic substances commonly present in surface waters. It is macromolecular material possessing both hydrophobic and hydrophilic groups as well as other functional groups such as carboxyl, phenolic, carbonyl and hydroxyl groups [102]. Humic acid in the aquatic environment predominantly carries negative charges due to the existence of carboxylic and phenolic groups [103]. The presence of humic acid in surface water has been of great concern in the water supply community, mainly due to the consumption of disinfectant and the generation of disinfection byproducts. Therefore, it is of practical significance to minimize the humic acid level in drinking water.

Maghsoodloo [104] investigated the equilibrium and kinetic adsorption of HA onto chitosan treated granular activated carbon (GAC). The adsorption performance was compared with that of GAC. The HA adsorption onto chitosan treated GAC followed the Langmuir isotherm while Freundlich isotherm was better fitted for HA adsorption onto plain GAC, indicating the chitosan coating changed the predominant adsorption mechanisms. Monolayer HA capacities onto plain GAC and chitosan modified GAC were 55.8 mg/g and 71.4 mg/g, respectively. Film diffusion and intraparticle diffusion were simultaneously participating in the HA adsorption onto chitosan coated GAC. Ngah et al. [105] evaluated the HA adsorption onto chitosan-H_2SO_4 beads in a batch study.

Based on the Langmuir isotherm model, the maximum adsorption capacities attained by chitosan-H_2SO_4 beads were from 342 mg/g to 377 mg/g with low temperature favoring HA adsorption. The same research group also examined HA adsorption onto chitosan beads cross-linked by epichlorohydrin [106]. The optimum HA adsorption on cross-linked chitosan beads was obtained at pH 6.0. The maximum adsorption capacity determined from the Langmuir model was 44.84 mg/g. Zhang and Bai [107] coated the surfaces of polyethyleneterephthalate (PET) granules with a layer of chitosan through a dip and phase inversion process and used them as an adsorbent for HA removal. The uniform coverage of chitosan onto PET granules with numerous open pores on the surface was confirmed by scanning electron microscopic (SEM) images. The maximum adsorption capacity was 0.407 mg per gram of chitosan coated

PET granules. Adsorption of humic acid onto chitosan-coated granules was pH dependent and significant amounts of humic acid could be adsorbed under acidic and neutral pH conditions. Chitosan-coated granules were found to have positive zeta potentials at pH < 6.6, mainly due to the protonation of the amino groups in chitosan. The adsorption process involved protonation of the amino groups in chitosan followed by attachment of humic acid onto the protonated amino sites on the surface. Under low-pH conditions, the adsorption process is transport-controlled, but under high-pH conditions, both transport and attachment [107].

Conclusion

Improving water quality to protect human health and the environment is a challenging task in view of ever increasing contaminants (in type and quantity) from industrial, agricultural, and domestic sources. Chitosan, the second most abundant natural polymer, has been demonstrated to be an effective biosorbent for many contaminants including metals, anions, and other contaminants. The versatile adsorption properties of chitosan are mainly due to its high content of amine and hydroxyl functional groups exhibiting high affinity to various water contaminants.

Although chitosan has a very low specific surface area (2-30 m^2/g) compared to commercial activated carbon (800-1,500 m^2/g) [12], the outstanding adsorption capacity makes chitosan an attractive, low cost alternative to activated carbon because of its unique attributes such as physicochemical characteristics, high reactivity, excellent chelation ability, and high selectivity towards contaminants. Chitosan can be easily made into different shapes to satisfy different applications such as beads, flakes, microspheres, films, membranes, nanoparticles, and magnetic particles. Chitosan can also be modified by chemical or physical processes to improve the mechanical and chemical properties for specific applications. However, the disadvantages of chitosan as an adsorbent are its high solubility in acidic water and the lack of selectivity for some contaminants as indicated by the interferences of competing cations or anions.

References

[1] R.C. Bansal, M. Goyal, *Activated Carbon Adsorption*, CRC press, Bota Raton, FL (2005).

[2] EPA, *Technologies for Upgrading or Designing New Drinking Water Treatment Facilities*, EPA/625/4-89/023, EPA Office of Drinking Water, Cincinnati, OH (1990).

[3] EPA, *Small Community Water and Wastewater Treatment,* EPA/625/R-92/010, EPA Office of Research and Development, Washington, DC (1992).

[4] EPA, *Drinking Water Treatment for Small Communities*, EPA/640/K-94/003, EPA Office of Research and Development, Washington, DC (1994).

[5] Bhatnagar and M.Sillanpää, *Adv. Colloid Interf. Sci.* 152, 26 (2009).

[6] P. Bayer, E. Heuer, U. Karl and M. Finkel, *Water Res.* 39, 1719 (2005).

[7] G. Crini, and P.-M. Badot, *Prog. Polym. Sci.* 33, 399 (2008).

[8] C. Gerente, V.K.C. Lee, P.Le. Cloirec and G. McKay, *Cri. Rev. Environ. Sci. Technol.* 37, 41(2007).
[9] A.J. Varma, S.V. Deshpande and J.F. Kennedy, *Carbohydr. Polym.* 55, 77 (2004).
[10] E. Guibal, *Sep. Purif. Technol.* 38, 43 (2004).
[11] M.-W. Wan, C.-C. Kan, B.D. Rogel and M.L.P. Dalida, *Carbohyd. Polym.* 80, 891 (2010).
[12] P. Miretzky and A.F. Cirelli, *J. Hazard. Mater.* 167, 10 (2009).
[13] E. Guibal, M.Van. Vooren, B.A. Dempsey and J. Roussy, *Sep. Sci. Technol.* 41, 2487 (2006).
[14] M.N.V.R. Kumar, *React. Funct. Polym.* 46, 1 (2000).
[15] C.K.S. Pillai, W. Paul and C.P. Sharma, *Prog. Polym. Sci.* 34, 641 (2009).
[16] S. Chatterjee and S.H. Woo, *J. Hazard. Mater.* 164, 1012 (2009).
[17] X.-F. Sun, S.-G. Wang, X.-W. Liu, W.-X. Gong, N. Bao and Y. Ma, *Colloids. Surf. A Physicochem. Eng. Asp.* 324, 28 (2008).
[18] W.L. Yan and R. Bai, *Water Res.* 39, 688 (2005).
[19] M. Arora, N.K. Eddy, K.A. Mumford, Y. Baba, J.M. Perera and G.W. Stevens, *Cold Reg. Sci. Technol.*, 62, 92 (2010).
[20] V.M. Boddu, K. Abburi, J.L. Talbott, E.D. Smith and R. Haasch, *Water Res.* 42, 633 (2008).
[21] M.S.D. Erosa, T.I.S. Medina, R.N. Mendoza, M.A. Rodriguez, and E. Guibal, *Hydrometallurgy* 61, 157 (2001).
[22] B. Benguella and H. Benaissa, *Colloids Surf. A: Physicochem. Eng. Aspects* 201, 142 (2002).
[23] P. Miretzky and A.F. Cirelli, *J. Fluo. Chem.* 132, 231 (2011).
[24] M.A. Ruiz, A.M. Sastre and E. Guibal, *Sep. Sci. Technol.* 37, 2143 (2002).
[25] T.W. Tan, X.J. He and W.X. Du, *J. Chem. Tech. Biotechnol.* 76, 191 (2001).
[26] Y. Baba, K. Masaaki and Y. Kawano, *React. Funct. Polym.* 36, 167(1998).
[27] W.W. Eckenfelder, *Industrial Water Pollution Control*, 3rd ed., McGraw-Hill, New York, NY (2000).
[28] G. Crini, *Prog. Polym. Sci.* 30, 38 (2005).
[29] W.S. Wan Ngah, C.S. Endud and R. Mayanar, *React. Funct. Polym.* 50, 181 (2002).
[30] V.K Gupta. *Ind. Eng. Chem. Res.* 37, 192 (1998).
[31] S.R. Popuri, Y. Vijaya, V.M. Boddu and K. Abburi, *Bioresour. Technol.* 100, 194 (2009).
[32] G.Z. Kyzas, M. Kostoglou and N.K. Lazaridis, *Chem. Eng. J.* 152, 440 (2009).
[33] B. Kannamba, K. Laxma Reddy and B.V. AppaRao, *J. Hazard. Mater.* 175, 939 (2010)
[34] C.M. Futalan, C.-C. Kan, M.L. Dalida, C. Pascua, M.W. Wan, *Carbohydr. Polym.* 83, 697 (2011).
[35] E.C.N. Lopes, K.S. Sousa and C. Airoldi, *Thermochim. Acta* 483, 21 (2009).
[36] Y. Chen and J. Wang, *Chem. Eng. J.* 168, 286 (2011).
[37] J.M. Al-Karawi, Z.H.J. Al-Qaisi, H.I. Abdullah, A.M. A. Al-Mokaram and D.T.A. Al-Heetimi, *Carbohydr. Polym.* 83, 495 (2011).
[38] M.S.D. Erosa, T.I.S. Medina, R.N. Mendoza, M.A. Rodriguez and E. Guibal, *Hydrometallurgy* 61, 157 (2001).
[39] S. Hasan, A. Krishnaiah, T.K. Ghosh and D.S. Viswanath, *Ind. Eng. Chem. Res.*, 45, 5066 (2006).

[40] N. Sankararamakrishnan, A.K. Sharma and R. Sanghi, *J. Hazard. Mater.* 148, 353 (2007).
[41] N. Sankararamakrishnan, P. Kumar and V.S. Chauhan, *Sep. Purif. Technol. 63*, 213 (2008).
[42] E.C.D. S. Filho, P.D.R. Monteiro, K. S. Sousa and C. Airoldi, *J. Therm. Anal. Calorim.* 106, 369 (2011).
[43] G.J. Copello, F. Varela, R.M. Vivot and L.E. Diaz, *Bioresour. Technol.* 99, 6538 (2008).
[44] M. Li, X. Zhu, D. Lin, W. Chen and G. Ren, *Environ. Eng. Sci.* 28, 735 (2011).
[45] T.W. Clarkson, *Environ. Health Perspect.* 100, 31, (1992).
[46] M. Benavente, Adsorption of metallic ions onto chitosan: equilibrium and kinetic studies, *Licentiate Thesis*, Royal Institute of Technology, Department of Chemical Engineering and Technology, Stockholm, Sweden (2008).
[47] Shafaei, F.Z. Ashtiani and T. Kaghazchi, *Chem. Eng. J.* 133, 311 (2007).
[48] Gamage and F. Shahidi, *Food Chem.* 104, 989 (2007).
[49] E. Taboada, G. Cabrera and G. Cardenas, *J. Chile Chem. Soc.* 48, 7 (2003).
[50] E. Lopes, F. dos Anjos, E. Vieira and A. Cestari, *J. Colloid Interf. Sci.* 263, 542 (2003).
[51] Peniche-Covas, L.W. Alvarez and W. Arguelles-Monal, *J. Appl. Polym. Sci.* 46, 1147 (1992).
[52] Jeon and W.H.H. Holl, *Water Res.* 37, 4770 (2003).
[53] J.C.Y. Ng, W.H. Cheung and G. McKay, *Chemosphere* 52, 1021 (2003).
[54] J.R. Rangel-Mendez, R. Monroy-Zepeda, E. Leyva-Ramos, P.E. Diaz-Flores and K. Shirai, *J. Hazard. Mater.* 162, 503 (2009).
[55] Asandei, L. Bulgariu and E. Bobu, *Cellulose Chem. Technol.* 43, 211 (2009).
[56] Chauhan and N. Sankararamakrishnan, *Bioresour. Technol.* 99, 9021 (2008).
[57] W.L. Yan and R. Bai, *Water Res.* 39, 688 (2005).
[58] L. Jin and R. Bai, *Langmuir* 18, 9765 (2002).
[59] N. Li and R. Bai, *Ind. Eng. Chem. Res.* 45, 7897 (2006).
[60] Rojas, J. Silva, J.A. Flores, A. Rodriguez, M. Ly and H. Maldonado, *Sep. Purif. Technol.* 44, 31 (2005).
[61] T. N. de Castro Dantas, A. A. Dantas Neto, M. C. P. de A. Moura, E. L. Barros Neto, and E. de Paiva Telemaco, *Langmuir* 17, 4256 (2001).
[62] P. Baroni, R.S. Vieira, E. Meneghetti, M.G.C. da Silva and M.M. Beppu, *J. Hazard. Mater.* 152, 1155 (2008).
[63] R. Schmuhl, H.M. Krieg and K. Keizer, *Water SA* 27, 1 (2001).
[64] V. A. Spinelli, M.C.M. Laranjeira and V.T. Fávere, *React. Funct. Polym.* 61, 347 (2004).
[65] N.Sankararamakrishnan, A. Dixit, L. Iyengar and R. Sanghi, *Bioresour. Technol.* 97, 2377 (2006).
[66] V.M. Boddu, K. Abburi, J.L. Talbott and E.D. Smith, *Environ. Sci. Technol.* 37, 4449 (2003).
[67] S. Hasan, A. Krishnaiah, T.K. Ghosh, D.S. Viswanath, V.M. Boddu and E.D. Smith, *Sep. Sci. Technol.* 38, 3775 (2003).
[68] V.M. Boddu, K. Abburi, J.L. Talbott, E.D. Smith and R. Haasch, *Water Res.* 42, 633 (2008).
[69] Gupta, V.S. Chauhan and N. Sankararamakrishnan, *Water Res* 43, 3862 (2009).

[70] Gérente, Y. Andrès, G. McKay and P. Le Cloirec, *Chem. Eng. J.* 158, 593 (2010).
[71] H.H. dos Santos, C.A. Demarchi, C.A. Rodrigues, J.M. Greneche, N. Nedelko and A. Slawska-Waniewska, *Chemosphere* 82, 278 (2011).
[72] S.M. Miller, M.L. Spaulding and J.B. Zimmerman, *Water Res.* 45, 5754 (2011).
[73] Gupta, M. Yunus and N. Sankararamakrishnan, *Chemosphere* 86, 150 (2011)
[74] D.D. Gang, B. Deng and L.S. Lin, *J. Hazard. Mater.* 182, 156 (2011).
[75] X. Wei, S. Bhojappa, L.-S. Lin and R.C. Viadero, *Environ. Eng. Sci.* (2011) (in press).
[76] N. Bleiman and Y.G. Mishael, *J. Hazard. Mater.* 183, 590 (2010).
[77] W. Yang, H. Chi, B. Sun, H. Zhao and Z. Wei, *J. Shenyang Jianzhu Univ.* 26, 744 (2010).
[78] Sabarudin, K. Oshita, M. Oshima and S. Motomizu, *Anal. Chim. Acta* 542, 207 (2005).
[79] S. Chatterjee and S.H. Woo, *J. Hazard. Mater.* 164, 1012 (2009).
[80] S. Chatterjee, D.S. Lee, M.W. Lee , S.H. Woo, *J. Hazard. Mater.* 166, 508 (2009).
[81] M. Arora, N.K. Eddy, K.A. Mumford, Y. Baba, J.M. Perera and G.W. Stevens, *Cold Reg. Sci. Technol.* 62, 92 (2010).
[82] K. Jaafari, S. Elmaleh, J. Coma and K. Benkhouja, *Water SA* 27, 9 (2004).
[83] J. Dai, H. Yang, H. Yan, Y. Shangguan, Q. Zheng and R. Cheng, *Chem. Eng. J.* 166, 970 (2011).
[84] S. Fierro, M. del, Pilar Sánchez-Saavedra and C. Copalcúa, *Bioresour. Technol.* 99, 1274 (2008).
[85] Y.-C. Chung, Y.-H. Li and C.C. Chen, *J. Environ. Sci. Health A* 40, 1775 (2005).
[86] Y. Shimizu, S. Nakamura, Y. Saito and T. Nakamura, *J. Appl. Polym. Sci.* 107, 1578 (20080.
[87] T. Fagundes, E.L. Bernardi and C.A. Rodrigues, *J. Liq. Chromatogr. Related Technol.* 24, 1189 (2006).
[88] Y. Zheng and A. Wang, *Adsorpt. Sci. Technol.* 28, 89 (2010).
[89] Lee, C. Chen, S.T. Yang, W.S. Ahn, *Micropor. Mesopor. Mater.* 127, 152 (2010).
[90] D. Thakre, S. Jagtap, A. Bansiwal, N. Labhsetwar and S. Rayalu, *J. Fluo. Chem.* 131, 373 (2010).
[91] S. Jagtap, M.K.N. Yenkie, S. Das and S. Rayalu, *Desalination* 273, 267 (2011).
[92] N. Viswanathan and S. Meenakshi, *Appl. Clay Sci.* 48, 607 (2010).
[93] N. Viswanathan and S. Meenakshi, *J. Hazard. Mater.* 178, 226 (2010).
[94] S. Jagtap, M.K.N Yenkie, N. Labhsetwar and S. Rayalu, *Microporous Mesoporous Mater.* 142, 454 (2010).
[95] Y. Vijaya, S.R. Popuri, G.S. Reddy and A. Krishnaiah, *Desalin. Water Treat.* 25, 159 (2011).
[96] Y. Xie, S. Li, F. Wang and G. Liu, *Chem. Eng. J.* 156, 56 (2010).
[97] F.X. Li, L. Squartsoff and S.H. Lamm, *J. Occup. Environ. Med.* 43, 630 (2001)
[98] B.E. Logan, *Environ. Sci. Technol.* 35, 482A (2001).
[99] Y. Zheng and A. Wang, *J. Hazard. Mater.* 171, 671 (2009).
[100] Y. Zheng and A. Wang, *Chem. Eng. J.* 171, 1201 (2011).
[101] Y. Zheng and A. Wang, *Ind. Eng. Chem. Res.* 49, 6034 (2010).
[102] K. Ghosh and M. Schnitzer, *Soil Sci.* 129, 266 (1980).
[103] P.K. Cornel, R.S. Summers and P.V. Roberts, *J. Colloid Interface Sci.* 110, 149 (1986).
[104] S. Maghsoodloo, B. Noroozi, A.K. Haghi and G.A. Sorial, *J. Hazard. Mater.* 191, 380 (2011).

[105] W.S.W. Ngah, S. Fatinathan and N.A. Yosop, *Desalination* 272, 293 (2011).
[106] W.S.W. Ngah, M.A.K.M Hanafiah and S.S. Yong, *Colloids Surf. B* 65, 18 (2008).
[107] X. Zhang and R. Bai, *J. Colloid Interface Sci.* 264, 30 (2003).

INDEX

A

B

C

D

E

F

G

H

I

J

K

L

M

N

O

P

Q

R

S

T

U

V

W

Y

Z